D0445070

Bored, Lonely, Angry, Stupid

Bored, Lonely, Angry, Stupid

Changing Feelings about Technology, from the Telegraph to Twitter

Luke Fernandez
Susan J. Matt

Harvard University Press
Cambridge, Massachusetts, and London, England 2019

Printed in the United States of America

First printing

Library of Congress Cataloging-in-Publication Data

Names: Fernandez, Luke. | Matt, Susan J. (Susan Jipson), 1967- author.
Title: Bored, lonely, angry, stupid : changing feelings about technology, from the telegraph to Twitter / Luke Fernandez and Susan J. Matt.
Description: Cambridge, Massachusetts : Harvard University Press, 2019. | Includes bibliographical references and index.
Identifiers: LCCN 2018043580 | ISBN 9780674983700 (alk. paper)
Subjects: LCSH: Technology—Social aspects—United States. | Technology—United States—Psychological aspects. | Technological innovations—Social aspects—United States. | Technological innovations—United States—Psychological aspects.
Classification: LCC T14.5 .F385 2019 | DDC 303.48/30973—dc23
LC record available at https://lccn.loc.gov/2018043580

For Our Parents

James and Renate Fernandez
Joseph (1920–2007) and Barbara Matt

who taught us the joys and virtues
of collaboration

Contents

Introduction

On a bright November day, aboard a ferry crossing San Francisco Bay, Eli Gay found herself troubled by her thoughts. "We were on the boat to Alcatraz. It was a beautiful day. And I was like 'Oh, I should take a picture of my cool life. . . . I have to get a picture of this for Facebook!' And that was kind of disturbing." A tax preparer living in Oakland, Eli had not been on Facebook that long, and she noted, "I could already tell how it was changing my psychology." She recalled that Facebook was hard to leave aside: "Say I'm waiting for a ride. . . . I have ten minutes to wait here; I'll go on Facebook and just start scrolling through things. And sometimes right before bed . . . just one more page, just one more page. . . . I could see the time and I'd be like 'OK, bye. The top of the hour that's going to be the end.'" She was hooked.[1]

Over the next seven years, Eli struggled with her feelings about social media, deactivating and reactivating her Facebook account multiple times. She was trying to figure out what her true self was, and how to present it, protect it, shape it, in the midst of the digital revolution.

What was clear from our conversation with Eli in 2016, and with the dozens of other people we interviewed for this book, is that Americans are living through a time of digital upheaval and rapid technological change. These changes, which have happened very quickly, are paradoxical, in that they seem ordinary but at the same time extraordinary. They have become integrated into so many aspects of life and often so seamlessly that at times people stop bothering to reflect on their significance. Yet, every so often, they are prodded to consider these transformations when they recognize in themselves a new emotion, a new behavior, as Eli did aboard the boat to Alcatraz, or, more commonly, when they read articles with provocative titles like "Is Google Making Us Stupid?," "Is Facebook Making Us Lonely?," and "Is Social Media to Blame for the Rise in Narcissism?"[2] When those questions come to people's attention, they are prompted to remember a time before the internet, and to consider how their phones and iPads, their laptops and selfie sticks, are changing their lives—and themselves.

Bored, Lonely, Angry, Stupid grapples with that concern as it examines a new American emotional style that is taking shape today. It takes as its starting point six closely related questions, which have received a great deal of media attention and provoked significant debate: Is social media making people narcissistic? Is the internet causing loneliness? Have digital devices made individuals incapable of enduring boredom? Are people losing the ability to concentrate in an age of ceaseless distraction and multitasking? Are they so exposed to digital spectacles that they have become jaded and incapable of experiencing awe? Is social media fomenting anger? In essence, such questions point to a more fundamental query: have Americans' feelings and their sense of self been radically remade by digital technologies?

This book examines those questions, but unlike other discussions of them, it places them in a broader historical context, as it considers how earlier generations coped with the innovations of their day, and how contemporary Americans are faring today. It investigates how women and men from a range of races and classes have felt about their technologies, and how their technologies have made them feel, from the telegraph to Twitter.

Ultimately, Americans' emotional lives have changed dramatically in the last two hundred years—and they are still changing today. As they struggle with narcissism, loneliness, boredom, distraction, anger, and awe, many in contemporary society are developing a new sense of self, a new emotional style, and a new set of expectations and ideas about what it means to be human.

For most of US history, Americans saw limits and constraints in themselves and in the world around them, and they had a more circumscribed set of expectations about what they could do, feel, express. As sixteen-year-old Caroline Cowles Richards wrote in her diary in 1857, "what can't be cured, must be endured."[3] Such an outlook reflected a belief that some hardships, feelings, and conditions could not be resolved or eliminated. This sense of limitations was reinforced by preachers' admonitions, moralistic myths, and fables. In biblical tales like the fall of Adam, in classical myths such as the story of Icarus, and in sermons that reminded listeners of their own mortality, men and women learned of their fallibility and their finiteness, and heard again and again of the folly of transgressing certain imagined limits. This sense of limitedness was also present both in the vocabularies Americans used to describe themselves and in the emotional styles they used to express the self.

Consequently, when it came to their inner lives, earlier generations of Americans felt and experienced the world differently than people do today. Most expected that they sometimes would be lonely, for they believed this was part of the human condition. They knew they were mortal and would perish; clergy constantly reminded them that it was a vain, futile, and, ultimately, short life that humans led. Because of this, they should not be too absorbed with themselves, for the self was ephemeral. They also knew that as mortals they had finite physical and mental powers—in fact, nineteenth-century medical authorities and religious leaders believed it was risky and perhaps immoral to try to exceed the natural limits of the brain and the body. Likewise, Americans expected drudgery and monotony and were not surprised when they encountered tedium. They often expressed anger but worried they might provoke God's wrath if they displayed too much of their own. This widespread emotional style, which prevailed up through the early twentieth century, taught Americans that they were small, bounded, finite, limited, and it reminded them that in the universe there were forces larger and grander than themselves.

This sense of smallness has changed, however, as technology has grown ever more powerful and ubiquitous. New inventions have revised many Americans' sense of limits, making them seem more imagined than real, as an array of seemingly concrete biological and intellectual constraints have fallen by the wayside. An early exponent of this new sensibility was John Perry Barlow, who in 1996 declared cyberspace "the new home of the mind," describing it as a "global social space," in which anyone could express any thought or sentiment, for traditional societies and governments had no power or jurisdiction over it.[4] Like Barlow, many people have found dramatic new ways of expressing, amusing, and projecting themselves—ways that seem to give them new capacities and to suggest that old notions of humanness may no longer apply. They have come to believe that it is possible to have few constraints on their online identities, on their modes of expression, on their social connections. Because the use of new technologies like the smartphone is so absorbing and frequently involves a mental if not physical retreat from the outer world about them, many users feel, at least fleetingly, that they have escaped the boundaries of social life that formerly constrained them. As a result, Americans increasingly believe they can have easy and infinite social connections, endless diversion and stimulation, unceasing affirmation of their worth, and boundless intellectual capacity.

This view of unbounded opportunity and an unlimited self is also reinforced by technology companies. Modern technologies (as well as the media that surrounds them and the companies that market them) have inspired hopes of limitlessness. Google's name, for example, is a reference to googol, which is 1×10^{100}. The name suggests that the search engine is capable of sorting through almost infinitely large datasets. Underscoring the meaning of the brand name is Google's mission, which is "to organize the world's information and make it universally accessible and useful." Google's ambitions have only gotten larger as it ventures into self-driving cars and artificial intelligence. It has also hired and encouraged Ray Kurzweil, the chief evangelist of what he terms "the singularity," to pursue and promote his hopes for immortality and seemingly limitless life. Following closely at Google's heels are the artificial intelligence efforts of the other Silicon Valley "Big 5," which are all racing to create technologies that will ostensibly augment human intelligence to unprecedented degrees.[5] Facebook founder Mark Zuckerberg and his wife articulated such high hopes in a letter they wrote in 2015 to their daughter Max. There they declared their commitment to "pushing the boundaries on how great a human life can be," and expressed the hope that their daughter would be able to "learn and experience 100 times more than we do today."[6] Such pronouncements reveal a hope for a new type of human, unconstrained by past limits.

This new attitude about the self is not solely the result of technological innovation, however, for contemporary psychological theories have also promoted a hope that Americans today might enjoy more fulfilling lives than their ancestors did. During the mid-twentieth century, psychologists began to pathologize a number of emotions and behaviors, from loneliness to narcissism, began to measure and quantify them, and also began to offer putative cures for them. The rise of positive psychology at the end of the twentieth century further fostered the idea that it might be possible to optimize one's moods through simple changes in one's daily habits as well as through the use of technology. In a range of books, articles, and websites, psychologists suggested that emotional fulfillment might be within reach, that traditional limits might be overcome. As historian Daniel Horowitz notes, "Positive psychology promised tens of millions of ordinary people that they could rely on individual experiences to bypass, temporarily forget, or transcend social, political, and economic difficulties."[7] Many find to their frustration, however, that this limitless self, promised by technology, by psychology, and by popular culture, is harder to realize than they expected.

This psychological transformation from a small and limited sense of self to a large, unbounded one—and the emotional changes that undergird it—has significant social and political implications, although they are often overlooked.[8] As numerous pundits of technology have excitedly observed, a digital revolution is occurring.[9] Yet, amid the excitement, there are also anxieties that the revolution is concentrating wealth, eroding privacy, and facilitating authoritarianism.[10] In contrast, worries about emotions and mental states seem like "first world problems" that are the afflictions of a coddled elite. They may seem precious and not as obviously political as worries about income distribution or the rise of a surveillance state. In actuality, however, they often are crucibles through which Americans express abiding values and work out their commitments to themselves and to others. To ask how a technology makes us feel, how it changes our sense of self, often leads to larger questions about who we are as a society, where we are going, and who or what is leading us there.

To answer these questions, this book examines how generations of Americans have experienced vanity and narcissism, loneliness, boredom, attention and distraction, anger, and awe, during two centuries of rapid technological change. In expressing and acting on their feelings, American men and women have shaped and reshaped their sense of self and the larger social order. Simple, ordinary, everyday acts and gestures—writing a letter, making a phone call, taking a picture, turning on the radio, texting, scrolling through Facebook—may seem private, insubstantial, ephemeral, and unworthy of historical inquiry. However, their social and emotional effects are cumulative and have had large-scale and very public consequences.

The research methods we used for tracking these changes were multifold. We read letters, diaries, and memoirs from the nineteenth and twentieth centuries, as well as psychological and sociological studies. We tried to capture the diversity of the past by studying the experiences of pioneers, enslaved people, early factory workers, businessmen, soldiers, teachers, and clergy. To study our own era, we interviewed, between 2014 and 2018, fifty-five Americans, sometimes speaking with them more than once over the course of the project. These were extended conversations, each generally lasting between forty-five minutes and two hours. Initially we conducted most of our interviews in our home state of Utah. We employed the "snowball" technique, getting referrals for interview subjects from acquaintances, and then having those subjects refer others to us. To gain access to the voices of late adolescents and young adults, we used the Weber State University Psychology Department's Human Subject Pool. We also posted notices on social media sites, and then arranged to meet volunteers in person. Additionally, we traveled out of state to conduct interviews.

In the end, our fifty-five subjects hailed from thirteen states and the District of Columbia. Roughly 60 percent of those we interviewed lived in Utah, though many of these had migrated from other states. While our interviewees came predominantly from the middle class, they represented a range of occupations. Our pool included a trucker, a nurse's aide, nurses, a building contractor, a schoolteacher, a school principal, social media marketers, software developers and entrepreneurs, a police department employee, retired military personnel, a speech therapist, a tax preparer, an ecologist, a nanny, an insurance adviser, a writer, a mental health counselor, lawyers, politicians, retirees, and students, among others. They ranged in age from eighteen to eighty-seven. The sample skewed white, but 20 percent of our interviewees were African American, Asian American, or Latinx.

We generally began the interviews by asking our subjects about their digital media habits, and then presented them with the questions that had prompted the book. Then we would let the conversation become less structured and go where it might. We made efforts not to "lead the witnesses," for we were eager to hear a wide range of opinions and experiences.

Admittedly, our interviews were shaped by our own social realities. First, a good portion of the time we did our research here in Utah. While neither of us is a native of the state (we moved here twenty years ago from—as they say here—"back east"), half of our interviews took place with Utahans simply because that is where we make our home. We wanted to get a pulse on the life of our own community.

Some people derisively dismiss Utah as "flyover country," suggesting that we are rubes who are out of touch with the intrepid capitalism that pervades Silicon Valley, Silicon Alley, and other high-tech regions on America's coasts. In reality, Utahans are one of the most urbanized populations in the nation. And our dynamic high-tech scene, which ranges for one hundred miles along the western edge of the Wasatch Mountains, easily earns us the moniker "Silicon Slopes." When it comes to silicon-oriented industries, our residents are as forward-looking and as cosmopolitan as our coastal neighbors (even if our cuisine might still be lagging a few years behind). Anybody who says otherwise simply does not know where the internet was born or that storied companies like Qualtrics, WordPerfect, Novell, and Adobe have made their home on the Wasatch Range.[11] Given this environment of innovation, we thought there was much to discover in our own backyard.

We did not want to limit our pool just to Utah, however, so to expand the scope of our study, we traveled to other states for interviews, to get more varied voices and a sense of national trends. In the end, in addition to native Utahans, our pool included people who hailed from Arizona, California, Colorado, Delaware, the District of Columbia, Florida, Hawaii, Illinois, Iowa, Minnesota, Nevada, New Jersey, and Ohio.

Our findings suggest that there are some broad new emotional patterns emerging in the United States, but also substantial variation, both in the past and in the present, for not all people use or have used technology in the same way. First, access to technology varies considerably. Ninety percent of Americans use the internet—some via their phones, some via computers. In 2018, 65 percent of Americans had broadband service at home.[12] Seventy-seven percent owned smartphones, a figure that is fairly consistent across racial and ethnic groups.[13] There is greater variation visible across income groups, with only 67 percent of those earning less than $30,000 per year possessing smartphones, while more affluent groups own them in greater numbers.[14] The greatest spread in smartphone ownership is between age groups. While 94 percent of those aged eighteen to twenty-nine own the devices, only 46 percent of those over sixty-five do.[15] Finally, there are some who have access neither at home nor in their pocket, for 15 percent of the nation's population does not have either smartphones or broadband.[16]

Not only do people have differing access to technology, but they also have varying modes of emotional expression. American emotional styles are shaped by individual personality, age, economic status, race, and gender. This book cannot possibly trace every distinct emotional style that is emerging in the digital age, but it does highlight broad trends affecting a wide swath of Americans, particularly those in the middle class.[17]

This new style is increasingly visible in daily life. For instance, every day in the United States, millions of people take selfies and upload them to social media. For many, the internet has turned the world into a stage. These developments, which impel them to think constantly about self-presentation, frequently provoke worries about narcissism, the subject of Chapter 1. Such worries, while often traced to the appearance of digital cameras and social media sites that allow pictures to be posted for free, actually have a far longer history. Similar anxieties arose in the nineteenth century, as Americans fell in love with photography, became prolific letter writers, and began to study themselves in newly affordable, mass-produced mirrors. These technologies offered novel opportunities for self-presentation and self-scrutiny; they also heightened self-consciousness and worries about sinful vanity. Victorians, then, seemingly had their own selfies. Chapter 1 explores the uncanny similarities between the digital selfies of today and the Victorian selfies of yesterday.

But the book also highlights some important differences. For centuries, self-regard and vanity were considered folly—mortality was an ever-present reminder of humans' limited powers and worth, and these beliefs kept self-promotion in check. That sense of vanity, however, has gradually faded into the background, as the language of sin has been abandoned, as moral strictures on vanity relaxed, and as new technologies—from the daguerreotype to the mirror to the selfie—emerged and hastened these transitions. In the wake of these changes, many Americans today feel it is acceptable to have a less humble sense of self than people did a century ago, and they no longer fear God's wrath should they celebrate themselves and their accomplishments. A sign of this is the recent estimate that, on average, a millennial will take more than 25,000 selfies over the course of his or her life.[18] There is still some alarm about such trends, but now it is couched in psychological terms rather than moral ones. Rather than being vain or vainglorious, the self-obsessed today are labeled as narcissists. This new terminology, along with the marketing campaigns of technology companies that profit off of personal aggrandizement, make many more forgiving of self-promotion and high self-regard.

Yet contemporary Americans, as they use social media, have created a curious new form of narcissism, which, while still selfish, is far more sociable than the mythic Narcissus who stared only at himself. The way that modern individuals get affirmation today is not by merely contemplating their own images. They also look for the likes and retweets of their friends and followers. Modern narcissists, in order to satisfy their need for self-affirmation, depend on the accolades and validation of others much more than in the past.

As they post selfies and wait anxiously for likes, many find that this culture ultimately offers neither a strong sense of self nor a rewarding sense of community. Because modern narcissists depend on the approval of others for their self-esteem, their self-reliance and independence are not robust. And even when they do feel validated, the ephemeral nature of tweets and posts and the fast pace of social media mean that the affirmation does not last long, and it does not bring an enduring sense of security and contentment. Likewise, the online communities they look to for validation often turn out to be fickle, built on weak bonds that lead to anxiety as often as they give affirmation. As Alta Martin, a twenty-eight-year-old university student told us, social media was often "more hurtful than helpful."[19] Camree Larkin, a twenty-two-year-old student who liked to post selfies, confided that sometimes social media left her feeling "distant from people. If that's my only interaction with people, I feel like there's a huge gap between us."[20]

Many of the Americans studied here longed for something more, however, and that longing often led them to worry about their bonds with friends and families. These anxieties are visible in ongoing discussions of loneliness and solitude, the subject of Chapter 2. With their phones and computers, a growing number of Americans expect to be constantly connected to others, and when this is not the case, they are uneasy. Fears of loneliness have increased, because celebrities, doctors, therapists, self-help gurus, and tech companies with a vested interest in drumming up concern about the feeling have turned it into a formal malady with purchasable cures. As a result, it is no surprise that there has been a proliferation of articles and studies devoted to loneliness and the ways that modern technologies have affected the condition.

These rising anxieties about loneliness reflect a shift in how Americans imagine the ideal self. Whereas today many long to live up to an ideal of a hyper-social person, happily networked into a larger community, earlier generations were more tolerant of loneliness and endured solitude with fewer misgivings. For example, in 1861, Lucy Larcom, a teacher at Wheaton Seminary in Massachusetts, confessed to her diary that she anticipated a long "year of lonely labor." But, she continued "I shall not be alone; I shall feel the sympathy of all the good and true. . . . I shall quiet my soul in the peace of God."[21] Nineteenth-century Americans often saw virtue and value in solitude. They frequently idealized the independent, rugged individualist (even if they rarely lived up to this vision) and did not believe they needed to be in touch with strangers from across the nation or across the globe in order to stave off loneliness. In fact, nineteenth-century Americans were known to chop down telegraph and telephone poles, for they felt them to be bothersome nuisances and intrusions.[22] Their expectations for social connection were far more modest than our own, and loneliness was, to them, a less worrisome condition. However, today, in an age when technology leaders like Mark Zuckerberg celebrate digital connection to others as a fundamental "human right," being disconnected makes an increasing number of Americans uncomfortable and anxious.[23]

As the tolerance for being alone declined, and solitude came to be regarded as a symptom of loneliness or oddity rather than a source of strength, Americans lost some of the conceptual tools that enabled them to justify and celebrate occasional moments of disconnection. In the process, they also lost a resource that sustained their independence. Solitude is a hard sell—it resists being commercialized or packaged. In contrast, the networks that contemporary Americans often turn to in order to stave off loneliness are commercialized, advertising instant sociability, and raising expectations for easy fellowship and friendship—expectations so high, that they are impossible to meet, yet so alluring that they still lead people back, again and again, to check Facebook just one more time.

Being bored and unoccupied also distresses the people we interviewed, although in different ways than in the past. Boredom may seem inherent to the human condition, but, in fact, it only emerged as a word in the middle of the nineteenth century. As Chapter 3 demonstrates, the drudgery that Americans experienced 150 years ago felt different than the boredom of today. In agrarian America, monotonous toil was burdensome, but free laborers conceived of it as morally redemptive. Indeed, in 1890, a popular book in America was the short work *Blessed Be Drudgery*, written by William Channing Gannett, a Unitarian minister.[24] As workers moved into factories, however, labor lost many of those redemptive qualities, and the tedium that accompanied it began to be experienced as a newly invented feeling: boredom. Those feelings were exacerbated by the rise of leisure time; new inventions like the phonograph, telephone, and radio; and the emergence of a therapeutic culture that pathologized boredom. A new view of human nature gradually emerged: to be fulfilled, one needed constant diversion and variety. The result of such views is that Americans have become more sensitized to boredom and increasingly intolerant of it. Today, in an era when people can summon the world on their phones, few seem willing to be unoccupied and deal with empty time. Many believe that they need to be constantly stimulated, engaged, and connected; it is not sufficient to merely sit quietly. As thirty-three-year-old Adam Kaslikowski, a technology entrepreneur living in Southern California, told us, "I think it is a shame that every spare moment is generally taken up by reaching for your phone and looking through it, specifically what I'll call bottomless apps like Facebook and Instagram, where you can just scroll forever and there is no endpoint. We are consciously saying goodbye to the ability to sit in silence and just be with yourself."[25]

The story of boredom parallels the story of loneliness. As Americans have become less tolerant of boredom and seek out constant engagement in an effort to avoid it, their independence has also suffered. In the nineteenth century, a range of thinkers, from Emerson to Nietzsche, observed that boredom, or a remove from communal entertainments, often spurred people to be creative.[26] The idea that boredom might be productive, might catalyze imagination, has largely disappeared, however, and instead it is now regarded as a useless feeling to be avoided at all costs.

Given the growing unwillingness to sit without outside stimulation, many today also worry about their ability to concentrate and pay attention. Chapter 4 takes up this concern. Such worries have been stoked by a rash of studies on concentration, multitasking, and distraction, including one which concluded that contemporary Americans have shorter attention spans than goldfish.[27] The belief that focused, single-minded concentration should be the preferred mode of mastering information developed over the course of the nineteenth century. As telegraph wires were sprawling across continents and oceans, and railroads connected distant cities, Victorian Americans began to feel that the world had become more complicated, and many became convinced that to understand it they needed to cultivate attention and the art of concentration. Yet as demands for focused attention increased in the nineteenth century, prominent medical authorities suggested that concentrating might overstrain mental workers' delicate brains. There seemed to be too much information to absorb and assimilate, and too much concentration could cause illness.

Today, in contrast, the reverse is true: the inability to concentrate is deemed an illness, for in the twenty-first century, in the midst of a rapidly expanding information economy where the amount of data multiplies each year, Americans feel they can know more and do more than ever before. They regard their phones, computers, and sometimes even their Ritalin pills as neuro-enhancers that can give them infinite mental power. They no longer believe there are limits on what they can know. And with these tools they also often believe they can have greater mastery over their minds than ever before. For instance, in the middle of a work project, in a doctor's waiting room, a boring meeting, or a tedious lecture, many pick up their smartphones as a way of reclaiming and asserting that their attention is their own. The phones emit a siren song, however, for while they allow their users to redirect their attention away from environments that are constraining it, as often as not, the smartphone itself recaptures this attention through commercialized distractions and entertainments. Nevertheless, contemporary Americans dream that with technology they can infinitely expand their mental powers and achieve true mastery of their minds.

As a result of their growing faith in their own unbounded cognitive powers, many have come to feel less awed by the forces of the universe, nature, and technology. The decline of awe is the subject of Chapter 5. Americans once regarded new inventions like the telegraph as the astonishing creation of divinity, believing that they had harnessed supernatural powers. Amazed by their technologies, they sometimes worried that their new machines might be crossing divinely established lines, taking power from the gods, and inappropriately aggrandizing the self. Those who were less fearful were no less awed, and some hoped they might be able to telegraph to heaven and communicate with God, the dead, or both. By the early twentieth century, psychologists, physicians, and sociologists began to dismiss such awe as a primitive emotion, one that modern individuals should no longer experience. The result of this shift is still visible today, for contemporary Americans use a more secular and tempered language to describe their relationship to their technology. They may be amazed by their new inventions, but that feeling reflects an awe at what humans can achieve, rather than a thrill at the prospect of something that lies beyond their powers. Their sense of the grandeur of the universe and its forces has diminished as their sense of the self has grown.

Recently, however, in response to the perceived decline in awe, the self-help industry and the psychological profession have begun to think of ways to resurrect it. While this certainly has implications for the development of rewarding personal lives (for who wants to go through life without being awestruck at least once in a while?), it is about more than just the self. This is because in its more powerful manifestations, awe helps people appreciate the larger world and communities of which they are a part, thereby upsetting and resetting their commitments to themselves and to others.

For modern psychologists, awe promises to be a social glue that unifies Americans at the same time that it enriches individual emotional lives. These promises are also taken up by Silicon Valley capitalists who hype the marvels of high technology and its power to connect people. Although many continue to believe in these promises, they have begun to ring hollow as Americans contemplate the economic inequalities and social injustices spotlighted by American populism and the #BlackLivesMatter and #MeToo hashtags. By many accounts, there is more social division and incivility rather than less. So if some psychologists continue to harbor hopes for awe, it is unclear, based on the amount of anger that Americans currently express, whether feelings of wonder and awe are really, in any significant way, tempering social discord.

The anger that contemporary Americans are today expressing suggests that limitlessness is not evenly distributed.[28] If some fortunate people feel and celebrate the beneficent effects of being connected, others still feel constrained. Chapter 6 details how American attitudes about anger have changed across time, and, in this case, how they have changed in parallel with transformations in the workplace. Much more so than in the past, the gears of modern corporate capitalism are oiled by workers who have learned to keep their anger in check. The anonymity of the internet provides a new venue where those passions can be released—but, our interviews suggest, only by some. Chapter 6 details the changing experience of anger and how that feeling has been shaped by the development of a corporate work ethic, the internet and social media, and the awesome but imperfectly realized promises of high technology.

Americans' emotional lives have changed as their society, culture, and technologies have changed. Today many are less humble about themselves and their powers, and also less awed by the world around them. They expect constant stimulation, connection, affirmation, and activity, and they are disappointed when these expectations go unmet. This new conception of self breeds a sense of power and also a sense of disappointment.

This book traces the development of a new and troubled American self, torn between individualism and community, selfishness and sociability, caught between the dream of limitlessness and the reality of limits. It explores how that self has been shaped by technological change and how it in turn is reshaping the social world.

A Note about Technological Determinism

Because this book examines the relationship between technology and emotions, it is worth asking whether technology can shape emotions. Conventionally there are two ways to address this question. So-called technological determinists argue that technology plays a central role in shaping human experience and that it is possible that machines are rewiring the way people feel.[29] This has been implied in the titles of well-circulated articles like "Is Facebook Making Us Lonely?" and "Is Google Making Us Stupid?" Composed in this fashion, the titles suggest that technologies might be driving their users to behave in certain ways.

In the other camp reside observers who find technological determinism problematic and who try to avoid attributing such power to machines. Scholars of this ilk often argue that emotions remain relatively static across time and that what is really happening is that long-standing feelings and instincts are being expressed through new technologies. Thus, for example, in *It's Complicated,* danah boyd provides an incisive study of teens and their use of social media. But while she recognizes that social media has certain features that "make possible . . . certain types of [teen] practices," she concludes her book with a static view of human psychology, writing that

> teens are as they have always been, resilient and creative in repurposing technology to fulfill their desires and goals. When they embrace technology, they are imagining new possibilities, asserting control over their lives, and finding ways to be a part of public life. This can be terrifying for those who are intimidated by youth or nervous for them, but it also reveals that, far from being a distraction, social media is providing a vehicle for teens to take ownership over their lives.[30]

Never mind that the whole category of the "teen" is a constructed concept that came into existence only in the early twentieth century.[31] The underlying idea is that human nature is fixed, and humans will ceaselessly find new ways to use technology to meet perennial and innate emotional needs.

In contrast, this book suggests that human nature and emotions are not static categories; instead they change subtly as a result of shifting economic orders, vocabularies, ideologies, theologies, and technologies. When interpreting and describing the relationships between technology and emotion, we admit that humans express feelings through their tools. Yet because these feelings are shaped and reshaped by the environment in which they are expressed, they are not stable, unchanging referents across time. We cannot say, for example, that Americans have always been lonely and that they merely express the same sentiment in new ways with each passing technological innovation, nor can we prove that the absolute incidence of the feeling has increased or declined. Instead, we argue that the experiences of loneliness, boredom, narcissism, attention, anger, and awe have changed across time, and that technology, as well as economics, culture, and social life, has had a role in changing inner life. Technology alone does not determine feelings, but the larger culture, of which it is one part, undeniably shapes them.

The History of the Emotions

As should be clear by now, we do not treat emotions as static concepts. It is a central argument of this book that they have changed throughout history. Most of the feelings and inner states we examine have become important categories in Americans' lives only in the last few centuries, and, even in that time, they have changed profoundly in meaning and experience. All of them are intimately linked to the rise of individualism and have been affected by the emergence of modern technologies. Earlier generations, while they worried about their internal states of mind, did not worry in the same ways or about the same things, and did not even use the same words to understand their feelings.

To study these shifts in Americans' inner lives requires that we take seriously the idea that emotions and mental states have a history; they are not solely biological but also the product of culture. That perspective—that feelings are, at least in part, historical artifacts—is increasingly accepted by historians, sociologists, anthropologists, philosophers, and some psychologists.

This proposition may provoke questions, for most contemporary Americans have grown up thinking that the feelings that seemingly well up from inside of them are wholly biological in origin. They seem elemental and concretely real, not only because one can genuinely feel lonely, or bored, or awestruck, but because they appear to offer an accurate reflection of the world around us. Like many, we were raised on this assumption. Only gradually, as we aged, did we begin to understand that feelings can be altered not only by making changes to one's environment but also by describing them with different words or making sense of them with different stories.

This is the case not merely for individuals but for societies as a whole, and the goal of the history of emotions is to uncover the shifting meanings and experiences of feelings across cultures and eras. Historians of the emotions take as their starting point the idea that feelings are not constant categories across time and space, but that instead they vary; they are not strictly the product of biology, but instead are also shaped by culture and society.[32] For a time, this notion might have seemed heretical, because for several decades prominent psychologists maintained that there were "basic" emotions such as fear, disgust, anger, happiness, distress, and surprise, which could be found across the globe.[33] In recent years, however, a growing body of research disputes this idea, and a number of psychologists are gravitating instead to conceptions of emotions as culturally variable. One example is the theory of "constructed emotion." That theory is based on the idea that feelings are not discrete, preexisting "entities," situated in a particular location in the brain, but that instead humans may have pleasant or unpleasant sensations that vary in intensity; they categorize, name, and understand these sensations based on past experience and cultural systems.[34] The research behind the theory of constructed emotion suggests, therefore, that there are not universal feelings across the world.

The work of historians, anthropologists, and sociologists supports this claim. Over the last several decades, scholars have documented a wide variety of emotional experiences, demonstrating, for instance, that some of the feelings that people take for granted today have not existed in other times or places.[35] The words and phrases that contemporary Americans use to describe their feelings do not map perfectly onto other cultures' or generations' inner lives, for many societies have a far different emotional vocabulary, stocked with feelings that have no ready equivalent in our own tongue. Indeed, even the concept of "emotion" itself does not exist in many other languages, and it appeared in English only in the seventeenth century, taking on its modern meaning in the nineteenth. While emotion might seem like a natural category of experience, it too is an historical artifact of rather recent vintage.[36] Further, although Americans and Europeans have long assumed that feelings and rationality stood in opposition to each other as inherently different mental processes, the work of both historians and neuroscientists is undercutting that assumption, suggesting that there is no clear divide between what is emotional and what is rational.[37]

Emotions, then, are not merely unregulated, unmediated outbursts untempered by thought; instead, the culturally specific words and categories people use to understand and describe feelings actually affect, shape, and hone them. The terms that are used—from happiness to grief to anger—carry with them connotations, value judgments, and expectations about how these states are expressed and displayed. As a result, in the process of identifying and naming feelings, emotions are given form.[38] Because words and their meanings can shift profoundly over time and across space, emotional experience does not hold steady.

Within a culture, there are varying emotional norms and rules that differentiate individuals' inner experiences as well. As this book will make clear, even today, as a new emotional style is coming to dominate online social life, there is still substantial variation, for not everyone has been subject to the same emotional rules or has been entitled to express the same feelings.[39]

It is worth tracing these changes because it helps to explain how the American emotional style developed, how modern personalities took shape, for personalities are not "natural" or inevitable but are instead the product of history and culture. How people feel—and how they feel about themselves—reflects, and in turn shapes, larger social values. Inner, private experiences are related to shared, public ones. The shifting ways people choose to express emotion, how they cope or flee from feelings, shape what and who they are individually and collectively.

These changes are important, for over the last two centuries, as Americans have debated, defined, and redefined loneliness, narcissism, boredom, attention, awe, and anger, they have also, on a subtler level, been debating long-standing questions about their relative commitments to the needs of the individual and of the community. When Americans have worried about loneliness, they have also been asking, "How much social connection does it take to be a fulfilled person?" When they have fretted about vanity, they have also been considering: "How much outside affirmation is required to achieve a virtuous sense of self?" When they have reflected on the best way to channel their attention, they were also wondering, "Do we need constant stimulation or focused attention to best realize our potential? Do we think best when we're alone or together?" When they have considered awe, they have often been puzzling over the question "Should our sense of self be small or large? Should we expect to be awed by our own powers or those of a vaster universe?" When angry, they have often asked whether expressing the feeling will unify or divide.

In an age when constant technological innovations promise to augment human capacities, Americans are consumed by the question of whether these tools enhance or degrade their lives and their humanity. As they wrestle with humility and hubris, connection and disconnection, stimulation and solitude, they have been defining what it means to be an emotionally fulfilled person in the digital age. Today it means never being lonely, always being engaged and affirmed by others, being unconstrained in anger, and able to multitask and apprehend everything. Left out of this new emotional style is a recognition of limits.

CHAPTER 1

From Vanity to Narcissism

Type "Is the internet making us more narcissistic?" into Google and you will get back countless links to popular and academic articles that attempt to answer the question. Ask your friends whether they take photos of themselves with their phones, and you will likewise get a range of replies: some will admit to taking selfies, while other will denounce the practice as narcissistic.

Against this backdrop, commentators like Jean Twenge and W. Keith Campbell, in *The Narcissism Epidemic,* and David Brooks in *The Road to Character,* have roundly condemned narcissism as a destructive behavior. Others, like Michael Maccoby in *The Productive Narcissist,* celebrate narcissism as a trait often found in effective leaders. Narcissism's defenders emphasize that narcissists have the confidence and high self-esteem to lead effectively, while critics highlight narcissists' overinflated sense of self, their willingness to take foolhardy risks, their selfish disregard for other people's welfare, and their need to have their accomplishments affirmed by others.[1]

These differences, and the fact that clinicians have at times misapplied the diagnosis, have led some psychologists to suggest that narcissistic personality disorder (NPD), which was first listed in the *Diagnostic and Statistical Manual of Mental Disorders* (*DSM*) in 1980, be removed as a formal diagnosis from the fifth (and current) *DSM.*[2] Others have suggested that the term "narcissism," because of its protean meanings, be avoided altogether.[3] In spite of these misgivings, NPD is still in the *DSM.*

Despite the word's ambiguities, the Americans interviewed for this book seemed to have a clear sense of the term, for popular understandings of narcissism are much more uniform than academic ones. For the lay individual, the core of narcissism is embodied in the myth of Narcissus. According to the ancient Greeks, Narcissus was so taken with his own reflection that he stared at it day in and day out until he perished. People might disagree about the ancillary qualities associated with this behavior, but all of our interview subjects acknowledged the self-absorption inherent in narcissism. Yet they also recognized how this core aspect of narcissism has been rendered anew in the digital age. Where the ancient Narcissus stared at his reflection in a pool of water, the twenty-first-century Narcissus looks at himself as he appears on a Facebook page, in a selfie, or in a blog post. And in a significant departure from the original myth in which Narcissus saw only his own reflection, the denizens of the digital era look at their selves through the "likes," "favorites," and comments they receive from the readers of their social media feeds. Modern narcissism, then, differs from earlier conceptions, because it has become both more common and, in an unexpected way, also more communal, as people attempt to use images of themselves as a way of reaching out to others.[4] While self-promotion may be perceived as inherently individualistic, to contemporary Americans it is also reflective of communal commitments, born out of a need to connect.

Given these tensions, twenty-first-century Americans are uneasy about how they should present themselves to others. They are not, however, the first to have anxieties about the amount of time and attention they should devote to crafting and presenting their images and identities. Earlier generations worried as well, but they used a vastly different language, born of a different moral landscape, to assess their problems.

Americans once believed that vanity and excessive self-regard were sins. They feared being too proud or arrogant lest they run afoul of religious rules that preached humility before God. They believed that as humans they were inherently flawed and finite; they therefore must be always mindful of their mortal limitations. They should not celebrate themselves and their accomplishments, for to do so would be to lose sight both of their own smallness and of God's greatness.

Today, many Americans are far more tolerant of self-promotion. They have few moral worries about presenting a grand sense of themselves to the world. In polished, perfected updates, tweets, and selfies, they celebrate their strengths and try to minimize their flaws and failings. While technology alone did not cause this change, it has undoubtedly abetted it. Key cultural transformations in the nineteenth and early twentieth centuries, including the rise of letter writing, the development of photography, the mass production of mirrors, and the repudiation of religious teachings that portrayed humans as inherently flawed, slowly accustomed individuals to public self-presentation and self-promotion. With the advent of digital media, these habits of self-promotion have become even more widespread and accepted. As a result, Americans' sense of their own importance has grown and their sense of humility has diminished. Yet, paradoxically, the rise in self-promoting behavior also reflects an ongoing, often anxious, longing for connections to others.

Traditional Views of Vanity, Pride, Egotism, and Narcissism

Contemporary Americans' concerns about self-presentation differ markedly from that of earlier generations. When eighteenth- and nineteenth-century men and women discussed excessive self-regard, they did not use the word "narcissism." That term was invented by the physician and social reformer Havelock Ellis in 1898 and only gained popularity after Sigmund Freud began to use it, most famously in his 1914 publication, *On Narcissism.*[5] Before these psychological labels emerged, Americans thought of high self-regard as a sin, and they used the words "vanity," "pride," and occasionally "egotism" to describe the trait. It was only in the early twentieth century, as sin became a less common way to think of the behavior, that those words gradually declined in significance and were transformed in meaning, and the language of moralists was replaced with the labels of professional psychologists.[6]

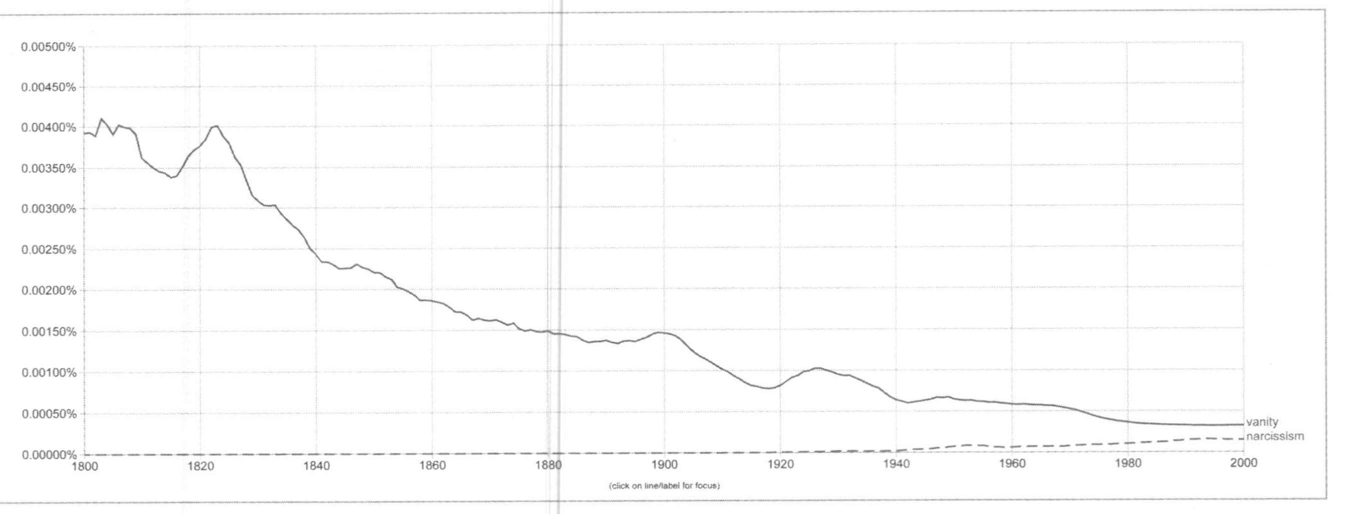

FIGURE 1.1 Google Ngrams, although they have biases, track how frequently words are used in a database of five million books published between 1500 and 2008. Ngrams produce provocative charts that support some of the historical trends described in this book. This graph charts the declining use of the word "vanity" and the rising use of the word "narcissism." (The "vanity" data also reflect the new use of the word as a term for a bathroom cabinet with mirror.) *Google Ngrams: http://books.google.com/ngrams*

For centuries in Europe and in America, there was significant pressure to conquer self-absorption, which was considered morally dangerous behavior. In the Judeo-Christian tradition, too great an obsession with the self could lead to sin—specifically to pride and vanity. Today both pride and vanity are rather innocuous terms. Americans are told they should feel proud of themselves, their towns, their work; likewise, they may encounter the word "vanity" as a fancy name for a sink with a mirror. But these words have been defanged, and the contrasts between their original definitions and their modern ones serve as potent reminders of their changing moral valence.

Vanity originally meant "vain, futile, or worthless." However, when directed toward a person, the word could also signify "the quality of being personally vain"; having a "high opinion of oneself"; or displaying "self-conceit and desire for admiration."[7] Although at first glance the meanings might seem quite distinct, there was a causal relation between them—it was futile and vain to have a high opinion of oneself, for all individuals were mortal and would fade away. Then too, all humans carried within them the flaw of original sin, and so were imperfect. Vanity and self-regard were useless, for individuals were never as good as they hoped or thought they were.

The Dutch Vanitas paintings of the seventeenth century gave a physical form to this conception. Painters depicted symbols of worldly aspiration—armor, books, portraits—alongside powerful reminders of their transience, such as skulls, candles with guttering flames, and rotting fruit, to remind viewers that human desires, titles, and glories ultimately meant nothing. Life was short, and worldly ambitions and desires ephemeral.[8]

Generations of authors penned equally telling verbal descriptions of the sin. In the seventeenth century, the Protestant writer John Bunyan offered a rich definition of vanity's meaning—and a highly critical one. In *The Pilgrim's Progress,* he famously named one of the stops "Vanity Fair," describing it as a place where one could buy "honors, preferments, titles, countries, kingdoms, lusts, pleasures . . . silver, gold, precious stones, and what not." The fair was populated by "cheats, games, plays, fools, apes, knaves, and rogues, and that of every kind."[9]

FIGURE 1.2 Harmen Steenwijck's *Vanitas,* circa 1640, represents the Vanitas genre, which emerged during the sixteenth and seventeenth centuries. The paintings reminded viewers of the transience of human life and attainments and of the imminence of death; they also served as a warning against vanity and pride in worldly achievements.
Wikimedia Commons, https://commons.wikimedia.org/wiki/File:Harmen_Steenwijck_-_Vanitas.JPG

Bunyan's point was that all human aspirations for distinction and entertainments were empty and sinful, unworthy of those intent on getting to heaven. Pride, a sin closely related to vanity, received equal attention from religious writers.[10] Andrew Jones, a seventeenth-century divinity student, for instance, labeled it "the Devil's Disease," in his 1699 tract, *Morbus Satanicus. The Devil's Disease; OR The Sin of Pride Arraigned and Condemned.*[11]

The warnings about vanity and pride came to America during the colonial period. Throughout the seventeenth and eighteenth centuries, ministers again and again reminded their congregations about the sinfulness of self-regard and the transience of life, using the word "vanity" in both its senses. As Solomon Williams declared in a 1752 funeral sermon,

> *Our Days on Earth are as a Shadow* . . . it is a *vain* Life. A Shadow is a thin airy Image, which has no Substance in it; . . . It flies from you, and runs away as fast as you pursue it; you find Nothing in it. . . . We are for a great Part of our Lives deceived with vain, and empty Hopes of some Comfort, which we never obtain; Abundance of the Pleasures of Life are in our Imagination, and chiefly the Fruits of our own Fancies.

Instead of harboring such fancies, he intoned that "we should live in an humble sense of our meanness."[12] This bleak view, of both human worth and human aspirations, this belief in life's ephemeral, shadow-like nature, was widespread in America—and surprisingly long-lasting.

Later generations of ministers continued to inveigh against the sin of vanity in all its varieties. In a series of sermons, Timothy Dwight, a Congregational minister and president of Yale, identified vanity as a "gross sin," intimately connected to a host of evils, including covetousness, ambition, prodigality, and sensuality.[13] Churchgoers could hear such sermons throughout the nineteenth century. Delia Locke, a settler in California, reported that at a funeral for a neighbor, the minister, "Mr. Stewart[,] preached the sermon upon the vanity of earthly things."[14]

And warnings against conceit, vanity, egotism, and pride proliferated in more popular forms as well. Early nineteenth-century readers in the United States might avail themselves of warning tales such as the 1810 story *The New Instructive History of Miss Patty Proud; or, The Downfall of Vanity: With the Reward of Good Nature,* or the 1818 volume *The History of Caroline: Or, a Lesson to Cure VANITY,* which told the story of a young girl who wanted a new fine dress. She was so pleased with it she decided to send invitations to "all her acquaintances in order to enjoy the inexpressible pleasure of being gazed at. . . . [S]he would walk backwards and forwards before them, like a peacock, and seemed to consider herself as empress of the world, and they as her vassals." However, her friends "discovered her intention in thus bringing them together, which was only to shew her fine clothes, and they were therefore resolved to mortify her vanity," which they did, by leading her through muddy fields so that her fancy frock was ruined.[15]

Yet alongside these warnings against self-love, pride, and conceit, and these stark reminders of life's futility and the pointlessness of vanity, there gradually developed a somewhat contradictory belief in the power of self-fashioning, a belief based on a more exalted vision of human potential. It was born of Renaissance humanism, which suggested that it was worth cultivating the self and developing individual skills and strengths. The literary scholar Stephen Greenblatt notes that "in the sixteenth century there appears to be an increased self-consciousness about the fashioning of human identity as a manipulable, artful process." He suggests that while medieval Christians had been encouraged to model themselves in imitation of Christ, Renaissance culture afforded them a freer hand, allowing them to create more secular identities.[16]

Acting in this tradition, George Washington famously copied down rules about self-presentation so that he might make himself into an honorable gentleman. Others of his generation read the *Letters of Lord Chesterfield,* which offered advice on how to mold one's identity in order to respectably withstand the scrutiny of the outside world. This genre expanded in the nineteenth century, offering middle-class Americans—and those aspiring to the middle class—advice on how to best present themselves in an increasingly urban and anonymous society.[17]

But even these guides, which put the focus squarely on the public presentation of self, and were devoted to developing particular kinds of behaviors and character traits, nevertheless still counseled against undue vanity, self-promotion, or pride. An advisor to young women asked her readers in 1849: "Has it ever happened to any but myself, to listen to I, I, I, in conversation, till, wearied with the monotony of the sound." She then reminded her readers, "We may be assured there is nothing so ill-bred, so annoying . . . as this habit of talking always with reference to ourselves."[18] Such advice applied to all sexes, as another guide noted: "In all conversation, nothing is so generally displeasing as egotism: it is highly censurable as a ridiculous mark of vanity."[19]

So while guides taught individuals to think about their selves, they also taught that they must not appear unduly focused on them. Self-fashioning was acceptable, but individuals still needed to avoid immoral vanity. It was a fine line that separated sin from appropriate modes of self-presentation.

Popular fiction of the era likewise underscored this message. The English novel *Vanity Fair*, written by William Makepeace Thackeray, was widely read on both sides of the Atlantic. Its title referred back to Bunyan's *Pilgrim's Progress*; its subtitle was *A Novel without a Hero*, because all of the characters had defects.[20] Louisa May Alcott's *Little Women* included a chapter entitled "Meg Goes to Vanity Fair," which described her short-lived efforts to enter fashionable society, the vanities she engaged in (including frizzled hair, powder, and "fuss and feathers"), as well as her ultimate renunciation of them.[21] Too great a focus on self could lead to sharp condemnations.

Supposedly the nineteenth century was the age of individualism, but many Americans displayed a profound discomfort with self-promotion and self-preoccupation. Ralph Waldo Emerson famously declared the nineteenth century to be the "age of the first person singular," yet many others thought that was a bad thing, not a good one.[22] In the age of the self-made man, newspapers nevertheless roundly condemned individuals who were too egotistical and self- absorbed. In 1838, a Mississippi paper, *The Ripley Transcript*, carried a column entitled "Likes and Dislikes." One of the items in the latter category concerned self-absorption: "I dislike to hear a man talk much about himself, his horses, his dogs, or any thing else belonging to him.—It's too egotistical."[23] In 1844, the *Baton Rouge Gazette* railed against those who were "so swelled up with the idea of their own consequence," and reminded readers that "egotism is looked upon as a *vice* and not a virtue."[24]

In short, in the nineteenth century, one could easily be condemned for egotism, vanity, and pride—for dressing too fancily or with too much affectation, for talking overmuch about oneself, for seeking titles or glory, and for overestimating one's powers. Self-presentation was a tricky thing, requiring modesty and self-restraint.

Yet while there were social pressures that worked to suppress unseemly or excessive promotions of the self, there were also new encouragements for self-presentation, which first developed in the nineteenth century and came into full bloom in the twentieth.

New Technologies, the Self, and Vanity

In the nineteenth century, as new technologies made communication faster and less expensive, Americans created rituals for sharing themselves through words and pictures. The rise of a national postal system facilitated the easy exchange of letters. The invention of photography allowed people of all classes to obtain images of themselves, which they could share with others. And the mass production of mirrors offered another way for individuals to refine their public identities. All of these technologies represent the antecedents of modern modes of self-presentation. They are the ancestors of the selfie, the text, the Instagram account, the Facebook post. These older technologies affected how nineteenth-century Americans regarded and displayed themselves to others, and they continue to influence modern Americans as well.

Letters

When new ways of communicating emerged in the nineteenth century, they excited enthusiastic commentary as well as anxieties about self-presentation. One key transformation was the "postal revolution" of the 1840s, which lowered prices for letters and improved their delivery. It enabled an increasing number of Americans to present themselves in writing to friends, family, and the larger world.

Yet just as there is a digital divide today, there was something of a postal divide in the nineteenth century, especially in the antebellum period. During the first half of the century, Americans corresponded at wildly different rates—for instance one historian calculates that in the 1850s, a North Carolinian sent an average of one and a half letters each year, while a resident of Massachusetts sent ten per year. Overall, city dwellers sent more than farmers did. New Yorkers each averaged thirty letters each year, while Bostonians held the lead with forty-one.[25] Not surprisingly, the literate corresponded more than the illiterate, yet letter writing was still possible even for those who did not read or write, as they could ask or hire others to pen letters for them. Before the Civil War, enslaved people had far less access to the post than whites or free blacks, but some slaves nevertheless found ways to avail themselves of it. Historian David Henkin notes that despite high rates of poverty and illiteracy, "slaves and free blacks used the pot to a striking degree." And gradually, over the course of the nineteenth century, more and more people were incorporated into the postal network, leading Henkin to observe, "The best evidence suggests . . . that in 1820 most Americans did not engage directly in any form of interactive, long-distance communications network, while by 1870 most of them did."[26]

This rise of popular access to the post meant a growing number of individuals had to learn how to put forth their ideas and their images, and still conform to rules about humility, for they lived in a culture deeply concerned about the vanity of life, in every sense. Indeed, their concern with the vanity—and the frailty—of life was evident in the first lines of the letters they exchanged, for many sensed their mortality acutely and this shaped their motivations and communications. For at least the first two-thirds of the nineteenth century, writers often began their letters with a ritualized affirmation of their existence. Again and again, writers reassured their readers that they were alive. "My Dear Brother,—You see that I am still in the land of the living," wrote Asahel Grant to his brother in 1841.[27] A Union soldier, Henry Clay Bear, wrote his wife in 1863, "I can still say I am in the land amongst the living."[28] Letters offered their writers a means of declaring their continued presence on earth and in the lives of their friends and families. The hope was to receive a response that would validate that existence and affirm the bonds of affection and social relation through words.

Within the letters themselves, correspondents poured forth other concerns as well, writing of their feelings and thoughts to friends and families scattered across the country. Some wondered how much of themselves to disclose. There was plenty of advice for them. Model letter books proliferated, instructing writers on how best, and most genteelly, to present themselves to those with whom they corresponded.[29] According to such guides, writers who wished to appear genteel were supposed to worry about the appearance of their stationery, ink, and handwriting, not to mention the moral value of their words.[30]

Indeed, some commentators suggested that there was almost a moral duty to write family and friends—the failure to correspond was a sign of self-absorption. Sarah Josepha Hale, editor of the popular women's magazine, *Godey's Ladies Book,* condemned the "selfishness" of those who claimed they did not have time to write friends and family, reminding readers that the "dear ones at home" were anxious to hear the news of those who traveled far from it.[31]

Because advisors portrayed letters as generous gifts one could bestow on friends and kin, there were fewer fears that they might incite egotism. Letters were governed by different rules than face-to-face social relations. Indeed, many Americans believed the self should be very present in these missives. Whereas vanity and egotism were unwelcome when people interacted in person, etiquette advisors suggested that in letters such egotism was acceptable.[32] In 1857, *Littell's Living Age* carried a feature, "Letter Writing and Letter Writers," which encouraged correspondents to write about themselves: "What but egotism should there be in a letter, if you care a fig for the writer? What other capital can be put out to such interest, if he interests you at all, as his own capital I?"[33]

Despite such encouragement, some letter writers nevertheless worried that they were talking too much about themselves. Jenny MacNeven, visiting relatives in Washington, D.C., wrote to her mother in New York City, describing her daily activities. In an 1840 letter, she wondered if she was dwelling over much on herself. She ruefully noted, "I suspect my organ of egotism (is there such an one?) will develop itself largely during my visit," for she had already sent a three-page letter in the morning and was sending a second one later in the day to tell her mother of her excursions and "all about myself."[34] Although she worried her letters might make her seem too self-absorbed, she continued to write frequently, and with great detail, about her time away from home.

The same fear of dwelling too much upon the self struck Allen White, of Emporia, Kansas. While courting his future wife, he wrote to her of his genealogy. He then noted, "You did not ask me for my History but I concluded to volunteer trusting that you will not deem it improper or egotistical in me."[35] While he worried that he might seem boastful and vain, he nevertheless regaled her with a description of his ancestors. And other Americans, writing on a range of topics, did similarly, filling letters with details of their daily activities, family accomplishments, worries, and wishes. Few seemed concerned about how many times they used the word "I" or whether they were talking about themselves overmuch.

As nineteenth-century Americans mailed letters back and forth, exchanging missives filled with the details of their lives, they helped to inaugurate a new age of communication about the self. Before the government reduced postal rates, letter writing had been the province of elites. As mailing letters became more affordable, a growing portion of the American population was able to craft a sense of self through words and to broadcast that self to a network of friends and family across the nation.

Photographs

Often enclosed in the letters they sent were pictures. The earliest photograph, the daguerreotype, was created in 1839 in France and brought to America's attention by Samuel Morse, the inventor of the telegraph. Daguerreotype studios quickly spread across the United States. While originally conceived of as an art form better suited to capturing landscapes than faces, photography quickly became a popular way of having a portrait taken.

What was novel about the daguerreotype was that it democratized portraiture. Previously the ability to possess a likeness of oneself had been something only the wealthy enjoyed, for commissioning a painting was expensive. Middle-class people, immigrants, and workers could not afford such an indulgence. Very rapidly, though, photography became inexpensive enough that it was in reach of the many. The abolitionist Frederick Douglass celebrated daguerreotypes for enabling "the humblest servant girl, whose income is but a few shillings per week" to have "a more perfect likeness of herself that noble ladies and even royalty . . . could purchase fifty years ago."[36] Whites, free blacks, the poor, and the rich could now possess portraits of themselves.

According to the estimates of the scholar Robert Taft, there were two thousand daguerreotype studios by 1850 in America; by 1853 they had captured three million images of Americans, at a time when the nation's population was twenty-three million.[37] Photographers were in particular abundance in newly settled regions, such as California during the Gold Rush, for many who had migrated there were eager to document themselves and their lives for family back at home.[38] Rather quickly, photographers accommodated this desire by offering images at relatively reasonable prices. While at first a daguerreotype sold for as much as $5 for a 2 ¾ × 3 ¼ picture, prices fell to roughly half that by the early 1850s, and soon thereafter it decreased even more, with photographers offering a very small image for as little as 25 cents.[39]

While the opportunity to sit for a photo excited many, it also excited some alarm. For instance, historian Richard Rudisill noted that in Germany, the newspaper the *Leipziger Anzeiger* warned against the sinful nature of the new technology: "Man is created in the image of God, and God's image cannot be captured by any man-made machine. . . . For man to contemplate bringing into being a machine . . . is comparable to considering bringing about an end to all creation. The man who begins such a thing must [arrogantly] consider himself even wiser than the Creator of the World."[40] As the editorialist noted, the camera seemed to be surpassing human limits and taking on transcendent qualities. Perhaps this was upsetting the divine order; perhaps it would lead humans astray. In America, there were worries about the moral implications of the daguerreotype, but none quite as dramatic as those printed in the *Leipziger Anzeiger*.[41]

In fact, some American commentators in the nineteenth century vested hope in photography, believing it might bring moral improvement rather than decline. Perhaps the daguerreotype might cure people of vanity, for, in contrast to artists who painted portraits and always tried to enhance the beauty of their wealthy subjects, cameras made no such efforts or allowances. An early report on the new medium suggested, in fact, that the daguerreotype "will never do for portrait painting. Its pictures are quite *too* natural, to please any other than very beautiful sitters. It has not the slightest knack at fancy-work."[42]

Others believed that this new lens might capture individuals' true character, moral strengths, and weaknesses, in a way that traditional portraits could not. Somehow the camera would be able to freeze an expression that revealed the inner worth of the sitter. A journalist opined that the images allowed for "the successful study of human nature. . . . Daguerreotypes properly regarded, are the indices of human character. . . . There is a peculiar and irresistible connection between one's weaknesses and his daguerreotype: and the latter as naturally attracts the former as the magnet the needle, or toasted cheese the rat."[43]

Many photographers saw this moral purpose as fundamental to photography. Marcus Aurelius Root, a prominent photographer of the nineteenth century, described the potential of his "heliographic art," writing that the photographer should strive to capture "that expression which reveals the soul within,—that *individuality* which distinguishes *this* from all human beings beside." To capture a true picture of the soul, the photographer must create an environment in his studio that set his patrons at ease.[44] Another photographer suggested drugging subjects in order to make them look natural, arguing that "a good dose" of laudanum "will effectively prevent the sitters from being conscious of themselves or the camera, or anything else. They become most delightfully tractable, and you can do anything with them under such circumstances." But he admitted that "the procedure entails a certain loss of animation."[45]

Yet while photos were supposed to produce authentic portraits of their subjects' inner souls, photographers like Root also revealed the great care and attention they took to create such pictures. They offered advice on what types of clothing were most flattering to particular figures and what angles were best to offset facial and bodily imperfections. These were studied images they produced, and while many claimed photos could reveal the inner soul, they also stood as testament to the great effort individuals went to in order to cultivate particular selves.

For instance, an analysis of gold miners' daguerreotypes revealed that even before they had headed west, many young men had their photos taken with all the imagined trappings of California life. James Ryder, a daguerreotypist whose studio was in Cleveland, Ohio, reported on the eager young men who trooped in for a picture: "Many who came for likenesses brought with them outfits to be shown in their pictures . . . [such as] tents, blankets, frying pans in which to cook bear meat, buffalo steaks and smaller game by the way, and to wash out their gold on reaching the diggings."[46] These early photographs recorded aspirations and fantasies more than current realities.

Some worked at crafting their images in other ways, hoping to hide flaws or to present themselves in ways that differed markedly from reality. Photographers testified to the fact that many of their subjects were obsessed by their portrayals in a rather disturbing way. The midcentury moralist T. S. Arthur described "a stout, fat lady who would like to be made a little smaller" in her photo, and a "lean one" who "desires full handsome bust and round plump arms, as she is just now rather 'thinner than common.'" There was much a photographer could do: "in fact, nearly all peculiarities of person that tend towards deformity may be modified by a skillful artist in the arrangement of his sitter—though he cannot help cross eyes nor make a homely person beautiful."[47]

Another magazine pointed out the foibles of many portrait sitters who worked to cultivate pretentious and not always accurate poses and images. In an age when literacy was not universal, the ability to read was a mark of status that some were eager to display in their portraits. Subjects sometimes were therefore disposed to a "*literary* weakness. Persons afflicted with this mania are usually taken with a pile of books

FIGURE 1.3 George Northrup, circa 1858. Though Northrup lived an adventurous life, traveling as a scout throughout the Minnesota Territories and trading with Native Americans, he never became a gold miner. Nevertheless, he chose to pose with mining equipment in this daguerreotype. *Courtesy of the Minnesota Historical Society*

around them—or with the fore-finger gracefully interposed between the leaves of a half-closed volume, as if they consented to the interruption of their studies solely to gratify posterity with a view of their scholar-like countenances—or in a student's cap and morning robe, with the head resting on the hand, profoundly meditating on—nothing." A young lady, for instance, who showed little interest in books, nevertheless "appear[ed] in her daguerreotype to be intensely absorbed in the perusal of a large octavo. What renders the phenomenon the more remarkable is, that the book was upside down." Others displayed a "*musical* weakness which forces a great variety of suffering, inoffensive flutes, guitars, and pianos, to be brought forward in the company of their cruel and persecuting masters and mistresses." The author reported that one woman, who was tone-deaf, nevertheless insisted on posing with "a sheet of Beethoven's most difficult compositions, in her delicate, dexter hand. Some amusing caricatures are produced by those who attempt to assume a look which they have not." Women were not the only ones with pretensions. "Timid men, at the critical juncture, summon up a look of stern fierceness, and savage natures borrow an expression of gentle meekness. People appear dignified, haughty, mild, condescending, humorous, and grave, in their daguerreotypes who manifestly never appeared so anywhere else."[48]

Other photographers provided a wider array of props to their clients, in order to help them create attractive, if misleading, images. Indeed, some photographers made it part of their service to provide their clients with clothing appropriate for portraits, as an early African American daguerreotypist, Augustus Washington, did in his Hartford, Connecticut, studio in the 1850s.[49] Others offered even more finery and luxurious accouterments to patrons of modest incomes so that they could appear affluent in their portraits. A photographer whose studio was in a poor neighborhood explained his booming business: "It is because I throw in the clothes with the picture that I get along. . . . You can come in here, for instance, in the dress of a tramp, and I will photograph you in the dress of a millionaire." In doing so, he believed he was bringing joy to his clients and to those who viewed their deceptive images. He asked the interviewer to imagine he was a

> young immigrant girl, and as yet you are not prospering here, for you don't yet know the language. But to the old folks in the home country, in order to cheer them up, you write that you are doing wonderfully well. . . . Then you see my studio, with its sign in seven languages . . . , and in a half hour I transform you from a poor girl in calico to a belle in a low-necked silk dress and white

> satin slippers, a diamond aigrette in her hair and a collar of pearls about her white throat. Consider the joy of the old folks, of the whole village, when that belle's picture arrives. To create envy, as well as to create joy, my costume department is employed.[50]

Some photographers went beyond props in helping their subjects indulge their vanity by creating images of dubious veracity. Just as modern-day photographers might rely on Photoshop or Instagram and Facebook filters, nineteenth-century portraitists did as well, devising ways to take photos of people that sometimes falsified their traits. An 1892 report detailed the financial successes of a Chicago photographer who began to rake in profits once he realized that women with thick ankles and big feet, or, as a journalist termed it, "large, misshapen hoofs," would prefer to be pictured with "small and loveable trotters." So he fashioned shapely, false limbs that could hang from the inner seams of the sitter's dress. Sometimes, alas, the sitter's real feet crept into view, resulting in her appearing to have four feet, but those who succeeded in tucking back their own (often with the aid of a rope) were quite satisfied with their portraits.[51]

Observers often chalked up such deceptions to vanity and the desire to appear better than one actually was. However, it was not just those who resorted to visual tricks who were considered vain, for some photographers suggested that the whole photographic economy depended on vanity. "If it wasn't for vanity and fashion New York photographers would drive a poor trade," suggested a newspaper article in 1879. The article quoted a society photographer who declared, "Vanity, sir, vanity. It's all vanity. . . . But as I told you before, there isn't usually much trouble in getting anybody [to sit for a photo]. This world is full of vanity; full of it."[52]

And as photographic technology advanced, new ways to reproduce and circulate images of the self proliferated, to the alarm of some observers. For instance, the *Nashville Patriot* reported in 1860 that it had become possible to print a daguerreotype portrait on porcelain: "One rather vain gentleman" was "having an entire dinner service made with his ugly face" printed on it.[53] By the 1880s, papers were reporting on "postage-stamp photographs," describing the tiny images as "the latest contribution of science to vanity."[54]

Long after photography was invented, many still continued to worry about its moral value. Such concerns were evident in the early twentieth century, more than sixty-five years after the daguerreotype's debut. "A press photographer asked Goldwin Smith, the well-known author, to pose for a number of pictures. Mr. Smith declined. . . . Such things, he said, smacked of vanity, and vanity was a fault that he desired to avoid."[55]

Yet alongside these concerns about vanity grew other perspectives on photographs and their moral value. The most photographed American of the nineteenth century was Frederick Douglass, who believed photographs to be a powerful weapon in his effort to build a just and free society. Pictures might communicate to white Americans a new vision of what African Americans were and could be. The many images that circulated of him offered a counterpoint to the crude caricatures of slaves and free blacks that abounded in the popular press. They also stood as a rebuke to the photographic images of slaves, created by white scientists such as Louis Agassiz, who hoped to use them to catalog physiological differences that he believed distinguished the races. Agassiz's photos portrayed the enslaved subjects as scientific specimens of an exotic race. In contrast, Douglass, as well as other African Americans who sat for portraits, offered the nation a far different vision.[56]

When he visited the photographer, Douglass admitted to feeling self-conscious and concerned about his appearance, noting that "conceitedness and vanity of personal appearance" were common and shameful traits. "And yet, it may be doubted if any man ever sat for a picture . . . without being conscious of more or less of that girlish weakness."[57] While concerned with how he looked, he was even more concerned about the way his image represented African Americans, free and enslaved. He observed that whites sometimes described African Americans as "black and ugly."[58] He countered, "I am black, but comely." He doubted, however, that white artists using brush and pencil would accurately capture that truth in their renderings.

FIGURE 1.4 Frederick Douglass was the most photographed American in the nineteenth century, according to John Stauffer, Zoe Trodd, and Celeste-Marie Bernier. *Wikimedia Commons, https://commons.wikimedia.org/wiki/Frederick_Douglass*

He noted, "Negroes can never have impartial portraits at the hands of white artists," for when a painter set out to portray an African American, racial stereotypes "exercise[d] a powerful influence over his pencil."[59] Perhaps photographic portraits would offer more accurate and flattering images, for he believed that photography might be a powerful means for capturing and conveying truth and reforming the republic.[60]

Sojourner Truth, the prominent African American abolitionist and women's rights activist, saw a similar liberating potential in photographic portraits as well, sitting for at least fourteen during the 1860s and 1870s. In the portraits, she presented what historian Nell Irvin Painter describes as the "the image of a respectable middle-class matron," wearing fashionable, well-tailored clothes, sitting in a parlor setting, often with flowers and books nearby. Painter notes, "Truth could not write, but she could project herself photographically. Photographs furnished a new means of communication—one more powerful than writing. They allowed Truth to circumvent genteel discourse and the racial stereotype embedded in her nation's language." Truth spread this image of herself by selling small copies of her portraits to supporters for 33 cents and larger images for 50 cents.[61] For those African Americans with the freedom and finances to obtain portraits of themselves in the mid-nineteenth century, their images often served both as representations of self to give to friends and family as well as arguments for racial equality for the wider society.[62]

Douglass and Truth believed that photos had the capacity to combat stereotypes and racism; other nineteenth-century observers attributed to photography a different kind of social power. They believed that photos might counter individualism and build community. Marcus Aurelius Root, the prominent nineteenth-century daguerreotypist, portrayed his art as building bonds between individuals, offsetting selfishness, sustaining love. "By heliography, our loved ones, dead or distant; our friends and acquaintances, however far removed, are retained within daily and hourly vision." Photographs might serve as tokens of communal values and familial bonds. Indeed, so powerful were photographs that Root claimed they offset the individualistic tendencies of

Americans that separated one from another. He wrote, "In this competitious [*sic*] and selfish world of ours, whatever tends to vivify and strengthen the social feelings should be hailed as a benediction." He believed that photography kept memories of loved ones fresh. As a result, he asked, "How can we exaggerate the value of an art which produces effects like these?"[63]

In practice, Americans often used photographs to connect with each other. For while photos might be taken as tokens of vain individualism, they were often put to use to build community and to strengthen social bonds.[64] The mails were full of them, for Americans exchanged them at a rapid clip, especially as they became more affordable. In fact, Henkin estimates that they were among the most common enclosures in nineteenth-century envelopes.[65] Men and women often had themselves photographed, sometimes out of vanity, but also because they were eager to share these images. Sometimes it was to show off; but other times photos were used as a weapon against time and distance, a way of maintaining memories of people long gone or far away, a means of updating scattered kin on new births, spouses, and even deaths.

One distinctive use of the photo that nineteenth-century Americans embraced was as a weapon against mortality.[66] In capturing an image, there might be some victory over death. Harkening back to the earlier notion of vanity and the Vanitas genre of painting, Americans tried to freeze time, capture faces, as part of their recognition of life's brevity and death's inevitability. In some ways, these photos were symbols of vanity in its traditional sense, because men and women, in having their images taken, recognized the vainness of life—that is, its shortness and futility.

The ultimate expression of this impulse was the posthumous photograph, designed to provide one last image of a loved one. These abounded in the nineteenth century and were a common part of a photographer's trade.[67] The bereaved could purchase a picture of a departed family member for as little as two dollars. One historian suggests that Americans commissioned these images more often than they did pictures of births or marriages, while another notes that in the

1840s and 1850s there were more photos of dead children than living ones.[68]

When families did spend money on photographs of the living, they often did so with a recognition that death lurked close by. Delia Locke, a New Englander living in California, was happy when in 1859 she had her children's pictures taken. It was not just because she was a proud parent, but because she also feared her children might not long survive: "How sad must those parents be, who are suddenly deprived of the pleasure of gazing upon the faces of their dear ones, when they have no likeness or picture to remind them of their dear features. When the loved one's face can no longer be seen by us, what a privilege to gaze upon a portrait, which we always considered so nearly resembling him or her."[69] When her sister Hannah's son died, Locke was aware that his mother lacked photos of him. She wrote of his funeral, "Hannah had only a short service at the house, then we took him to the grave and sang a hymn while they buried him. Hannah has no good picture of her dear boy."[70]

Indeed, photographers often made their images' death-defying qualities an explicit selling point. Playing upon the traditional conception of life as a shadow, many a mid-century photographer used the refrain "Secure the shadow, while the substance yet remains."[71] An overtly morbid ad from photographer Andrew Baldwin put the case more baldly, telling his fellow photographers they should "induce the old, the middle-aged, and the young to go to the nearest Picture-Gallery BEFORE the Great Destroyer comes."[72]

This theme struck a chord with those who saw in photography a weapon against death and the ephemeral nature of life. They indulged in vanity in its most morbid sense. A poem that accompanied the daguerreotype of Nancy Jane Wright, of Brook County, Virginia, dated 1852, read:

> When I am dead and in my grave and all my bones are rotten
> when this you see remember me
> when I am quite forgotten
> Six months from this where I may be

> no human tongue can tell
> perhaps I may be far far away
> never to return again[73]

Nineteenth-century Americans not only regarded photos as a way to stave off death and allay fears of mortality, but also considered them as tools to link the living. In the first half of the nineteenth century, Americans moved more than during any other period; photos served to counteract feelings of displacement.[74] Arozina Perkins, teaching in New Haven, Connecticut, at some distance from her family in Massachusetts, recorded in her diary in 1849 how she regarded images of her kin: "I have taken my treasure of daguerreotypes from my trunk, and opened and grouped them on my table. I talk to them, and look at them, till I fancy they are real, and almost expect their lips to move in answer to my questions."[75]

Abby Mansur, living in California far from her family in New Hampshire, again and again begged her sister to send her "miniature" images of those she had left behind. Finally, in December 1854, more than two years after she'd first asked for portraits, they arrived in California. Her reaction showed just what early photographs meant to Americans, how they embodied other values than vain individualism and reinforced deep sentimental bonds:

> Dear Mother and Sister and Brother
> I seat myself to answer your letter and also to thank you for your kindness in sending those miniatures, Hannah[.] I cannot describe to you with pen and ink how thankful I was to receive them[;] I liked to have cried myself to death over them[;] Horace cried[,] our partner cried also. . . . [W]e was crying for joy.[76]

These were the last images she saw of her family, for she died eight months later in California.

Professional photographers in the nineteenth and early twentieth centuries tried to attract customers by emphasizing the social and familial value of photos and dismissing any suggestions that portraits might be a sign of vanity. They worked to convince readers that purchasing portraits of themselves was not bad or self-indulgent, but socially and psychologically beneficial. In 1919, the Heyn Elite Studio wooed men and women with these reassuring words: "it isn't a sign of vanity for one to visit a photographic studio often—it's a sure sign that one has awakened to a realizing sense of obligation to family, friends and society."[77] A studio in Klamath Falls, Oregon, made a similar appeal, promising, "It isn't vanity which causes you to have your photograph made, but a sincere desire to visualize your real self for the pleasure it will give your friends and loved ones."[78] In this telling, photos were a link in the chain of sociability and a sign of virtue rather than vice.

For Americans living in the nineteenth and early twentieth centuries, photos served many purposes. They offered opportunities to strike self-aggrandizing poses, but also to fight for racial equality, and to sustain bonds with far-flung family and friends, and with earlier generations who had passed away. They may have sometimes been reflections of vanity, but they were also symbols of other feelings, relationships, and values. They were presentations of self, designed to build bonds and connections.

Yet many Americans' expectations about photographs began to change in the last decades of the nineteenth century. It was then that cameras became inexpensive and easy to operate, and as a result, photographic culture expanded in the United States. When the Kodak camera was patented in 1888, it became far easier for Americans of modest and middling means to take pictures of their families, their vacations, and their adventures. The most famous Kodak camera, the Brownie, created in 1900, sold at a brisk pace. Consumers purchased 150,000 the first year it was available, and had bought one million Brownies by 1906.[79] Families came to take for granted the opportunity to document their own lives without the help of professional photographers. The scholar Nancy West has demonstrated that, in the process, new expectations about what were appropriate subjects for photos developed. Images of the dead fell into disfavor, as did the rationale that one should photograph the living in order to have an image once they died. She argues, "Before the Kodak burst upon the scene, Americans were much more willing to allow sorrow into the space of the domestic photograph." After its emergence, "Kodak's advertising purged domestic photography of all traces of sorrow and death." The medium, at least in the hands of the populace, came to focus on capturing happy moments rather than sorrowful ones.[80]

This dovetailed with a new cultural imperative that shaped many Americans' emotional lives in the twentieth century—the need to appear cheerful and smiling in almost all moments of daily life. Over the last century and a half, men and women in the United States have been subject to increasing pressure to appear cheerful as a way of broadcasting their success in life. In a competitive society where individuals are supposed to be in control of their lives and destinies, the public display of happiness signals that one is doing well and has things under control.[81] A very visible sign of success and self-mastery is the smile, which became de rigueur in American photography by the 1940s. During the nineteenth century, many had avoided smiling in portraits, for artistic conventions suggested that the "grin was only characteristic of peasants, drunkards, children, and halfwits, suggesting low class or some other deficiency." During the twentieth century, however, misgivings about the smile were replaced by a new faith in its power to reflect social status and success.[82]

Consequently, as cameras came to be mass-marketed commodities, and as ebullient smiling faces came to be the expected subject of photos, many abandoned old fears about vanity and came to believe they should document themselves and their lives in exquisite and singularly happy detail. As a result, over the course of the twentieth century, personal photos came to focus more on joyful moments—on family gatherings, vacations, new possessions. Some of the pathos, sorrow, reality of life was edited out.[83] Photos lost their connection to the vanity of life—that is, its shortness and futility—and instead became items through which one might indulge vanity in its other, more prideful, sense.

Mirrors

So too did another consumer object often connected to vanity and the presentation of self: the mirror. In his study of nineteenth-century manners, John Kasson observes that the increasing number of mirrors was "indicative of the habitual self-regard and preparation for public roles that etiquette advisers demanded." In the eighteenth century, looking glasses had been relatively costly, and families, if they did own a mirror, generally had just one. "By the mid-nineteenth century, however, mirrors were a standard feature of middle-class homes, spreading in many cases to virtually every room. On walls, over mantels, on hallstands, sideboards, and cabinets, mirrors of varying size and expense ornamented the best rooms. . . . [O]ther mirrors appeared on bedroom and dressing room walls, wardrobes, dressing tables, and chests of drawers." But they were not mere pieces of furniture. Instead, Kasson speculates, they "played an important role in heightening a theatrical sense of the self. They were not only an indispensable aid to new standards of grooming. They taught users to appraise their images and the emotions they expressed frequently and searchingly, anticipating the gaze of others."[84]

By the late nineteenth century, such opportunities for self-scrutiny were available to a growing number of Americans, for mirrors were being mass-produced and priced to suit a range of pocketbooks. By 1897, consumers could buy a 10-by-10-inch mirror for 50 cents, and an even larger 16-by-16-inch looking glass was a mere $1.35.[85]

As mirrors became increasingly affordable and spread across the land, concerns about their moral meaning still lingered. One late nineteenth-century editorialist noted that "since looking glasses came in—personal vanity has been enabled to indulge itself at a more moderate cost than formerly."[86] An early twentieth-century writer asked, "Was anything more entirely provocative of human vanity ever invented than the many-sided shaving glass?"[87] Such worries about the morality of mirrors were also evident in the mourning practices of many American religious groups: bereaved families often covered their mirrors or turned them to face the wall when there was a death in the

FIGURE 1.5 While mirrors were becoming more affordable in this era, they were still morally controversial. "All Is Vanity," by the American artist C. A. Gilbert, circa 1892, warned against self-admiration and reminded viewers of their mortality.
Chronicle of Word History / Alamy Stock Photo

household.[88] To gaze upon oneself in the midst of grief was to engage in sinful, self-indulgent vanity and to ignore the ephemeral nature of life.

Yet, despite such misgivings about mirrors, a growing toleration for them was developing in the United States. In 1871, a newspaper columnist, Mrs. F. F. Victor, offered a defense of looking glasses. She noted that many believed mirrors signified "personal vanity, and that a man or woman who consults one is sure to be stigmatized as 'vain.'" But, Mrs. Victor maintained, this was a narrow, "prejudiced" view of the subject. "It is quite as often the timid and doubtful person who refers to his or her mirror for an opinion, as the fop or the coquette." The looking glass allowed people to see themselves through others' eyes. Consequently they no longer needed to "go into society ignorant of our defects, or unconscious of our excellencies of appearance." On the whole, mirrors offered a gift and carried with them a duty: the "duty is to show us where we may improve ourselves." In dismissing moral concerns about the mirror, she also put a different face on vanity, noting that "the fear of encouraging vanity in others is a mean fear. Vanity does far less real injury in the world than silent detraction."[89]

Many editorialists took similar tacks by the turn of the century, believing that a healthy awareness of how one looked was all for the good. An 1894 editorialist suggested that despite the fact that females were often condemned for vanity, men were "as vain as women" and were particularly prone to staring at their reflections in shop windows or pocket mirrors. Yet was this a bad thing? The author thought not. His goal was merely "to show that vanity lives alike in both sexes." He argued that self-scrutiny, a concern with appearances, and the study of reflections all arose from "a desire to look as well as possible in what one wears." Such a desire should be "commended instead of cried down," for it was "our duty to look as well as we can."[90] Vanity, here, was recast as a sign of carefulness, refinement, and virtue, and mirrors were a tool for improvement.

Certainly, those who encountered looking glasses for the first time remembered the heightened self-consciousness they brought—a self-consciousness that was not always the product of vanity so much as embarrassment. Mary Paik Lee, born in 1900, emigrated from Korea as a young child, and she first saw her full reflection only as a teenager: "There was a bureau with a large mirror on top. My first look at myself in the mirror was shocking. I had never seen a full view of myself before. At home, Mother had a very small pocket mirror that didn't show more than a few inches of our faces. I didn't even know what I looked like, and I had to admit I was a strange-looking thing, our family's main goal was to earn enough money to buy food to feed all of us. There was never any left over to do anything about our appearance."[91]

The self-consciousness that Lee experienced, and the mirrors that enabled it, were increasingly expected in twentieth-century America. Individuals were supposed to know how they looked to others and to care about their appearances. By viewing themselves in photos and in mirrors, they could see themselves as others saw them and then could work to refine their image, with little moral censure. Indeed, that mirrors were becoming widespread and that the vanity they were suspected of provoking was becoming a less worrisome condition was evident in the renaming of mirrors as "vanities." Merchandisers first began to refer to pocket mirror sets that would fit into a purse as "vanity mirrors" at the turn of the century.[92] Eventually, by the 1930s, the mirror that sat above a dresser or sink came to be called a vanity as well.[93] Although the concern with vanity as a vice did not entirely disappear, by the early twentieth century there was far less anxiety that it was sinful, or that it reflected the shallow worldly concerns of the unregenerate, who were unaware of life's brevity and the terrible fact of mortality. Now individuals could easily indulge in it with the help of a renamed mirror—they could stare at themselves in their vanities.

The Acceptance of Vanity and the Rise of Self-Esteem

Indeed, by the early twentieth century, vanity, self-love, and self-absorption had come to seem innocuous and, to some, even beneficial. Once regarded as futile and sinful inner states—since the self was mortal and ephemeral, and self-love a distraction from the divine—they now were accorded a new value. As a correspondent in the *London Mirror,* whose words were reprinted in American papers, declared in 1913, "In my view, conceit is a splendid thing for the future race. Conceited people make better and more careful fathers and mothers than men and women who have little pride or faith in themselves."[94] Vanity, long a sin, was now a social good.

Why did this reevaluation occur? First, many Protestant denominations throughout the nineteenth century had gradually but steadily moved away from the concept of original sin and inherent human frailty and flaws, and had put increasing emphasis on humans' potential for goodness. This widespread trend was only magnified by new theologies that arose in the late nineteenth century. For instance, with the emergence of a popular religious movement called "New Thought," many Americans were exposed to the idea that positive thinking might cure their ills and make them satisfied with their lives.[95] The movement's exponents suggested that self-regard was a desirable trait, a position famously encapsulated in the teachings of the French psychologist Emile Coue, who promised that with just one sentence of self-affirmation, people could find true happiness. Across America in the 1920s could be found men and women repeating the words "Day by day, in every way, I am getting better and better."[96] While such self-confidence would have been condemned as vain or prideful a century before, it was increasingly accepted as a sign of psychological well-being.

Another powerful force for change was the growing prominence of psychologists, most notably Freud and his followers, whose theories gradually spread across the United States. They suggested that pride, vanity, conceit, and egotism were natural and inherent to the human psyche. And here an important transition in the language began to occur. While Americans never stopped using the word "vanity," a new term gradually and slowly gained popularity in the twentieth century: "narcissism." Havelock Ellis, the British physician and sex researcher, created the term in 1898; Sigmund Freud began to use it in 1909, and he wrote an essay devoted to it in 1914.[97]

Freud's interpretations of the human psyche seemed to naturalize human striving and vanity; his discussions of narcissism made self-love appear normal, at least at some stages of life. Freud argued that "being enamored of oneself" was widespread, and maintained that all experienced that feeling to some degree. He pictured it as a normal developmental stage that most people transcended as they matured. If they did not, their narcissism might be a pathology. Yet elsewhere in his writings, he also suggested that the narcissist was one of several normal personality types.[98] In this rendering, self-love was not inherently sinful; instead it was a natural part of human psychological makeup.[99]

While the popular literature of the early twentieth century rarely made direct reference to the intricacies of psychologists' and psychoanalysts' theories, their opinions gradually reshaped traditional understandings of human nature. Many Americans began to think of humans as aggressive, competitive, and individualistic, motivated by innate drives and instincts, including narcissism, self-assertion, and egotism. A 1912 bon mot, reprinted across the country, reflected this new understanding. A child asked his father what an egotist was. "'An egotist,' the father answered, 'is a man who thinks he is cleverer than anyone else.'" The boy's mother then qualified that statement: "'No, my son, that is not quite right. An egotist is a man who *says* he is cleverer than anyone [else]—every man *thinks* he is.'"[100]

Finally, new economic structures affected prevailing views of vanity. Corporate office culture of the early twentieth century emphasized the importance of self-promotion and the development of a likeable, winning personality for success in business. Part of the secret to advancing in the corporate economy was to subtly broadcast one's own strengths, while offering ample praise and little criticism of others.[101] Appealing to the vanity of coworkers, clients, and bosses would help one advance. As Dale Carnegie reminded his readers in *How to Win Friends and Influence People*, "Remember that a man's name is to him the sweetest and most important sound in the English language." Likewise, he counseled that individuals should "always make the other person feel important."

Success and advancement seemed to rely on an economy in which vanity and praise were the currency.[102]

Vanity was perhaps even more essential to the consumer economy. As many an editorialist, sociologist, and economist argued, it often seemed to undergird consumer spending. Increasingly, advertisers made explicit reference to it as they pitched their products, selling not just mirrors, but also fashion, cosmetics, clothing, and photographic services.

The growing acceptance of the word was visible in the use of *Vanity Fair* as the title for a new fashion magazine in 1913. While both Bunyan, in *Pilgrim's Progress,* and Thackeray, in his novel, had used the phrase in a pejorative way, by the early twentieth century, *Vanity Fair* was not a marketplace of sin to be assiduously avoided, but a parade of fashion to be joined.

Vanity Fair magazine was only one of many businesses to employ the word in a new, and generally more benign, sense. Velveola Souveraine Face Powder offered its customers a free "Vanity Box," with the purchase of its powder, and illustrated it with a picture of a woman looking contentedly at herself in the mirror.[103] Styleplus clothing appealed to men with the headline "Has Your Real Vanity Been Tickled Lately?" It then queried, "Have you had a suit of clothes recently that just please you? About once in ten times you get such a suit. . . . This is an appeal to your common sense as well as to your pride."[104] Here, pride and vanity were traits that individuals should indulge, not repress.

By 1927, photographers embraced the term. Kodak began to market to women the Series III Vest Pocket Vanity Kodak in five colors: Sea Gull grey, Cockatoo green, Redbreast red, Bluebird blue, and Jenny Wren brown. Whereas in the past, photographers and their subjects had sometimes worried about accusations of vanity, by the 1920s, neither the term nor the behavior excited moral anxiety.[105]

These ads used vanity to sell particular goods, but many believed that vanity underlay the purchase of not just a few select products but almost all consumer spending, whether or not it was articulated explicitly in advertising copy. In 1911, the *El Paso Herald* celebrated the "virtue of conceit," noting, "Were vanity to cease suddenly from among us, society would go all to pieces and half the commercial world might shut up shop." The economic implications were grave: "Vanity sets millions of men and women at work in spinning and weaving ornamental fabrics," as well as making other elaborate accessories. The author speculated that if people repressed their vanity, "tailors would find the greater part of their occupation gone. . . . The cult of the waistcoat, the tie, and sock would disappear. . . . The girl who now buys 18 hats a year and gowns in due proportion would content herself with two of the second and possibly four of the first." Beyond its economic value, vanity was also an important aid in social life, for it made people craft appearances that would please others. Consequently, vanity stood "midway between virtue and vice," and the author declared that it was a quality that came to everyone "quite naturally." While the trait could be overcultivated and taken to extremes, "In the well-balanced nature vanity is one of the most useful qualities."[106]

By 1936, the *Delineator Magazine* was declaring "vanity is sanity." Appraising the widespread displays of vanity, the magazine concluded they were a good thing. "Doctors . . . know that a vanity case is a sure sign, not of a scatterbrained, neurotic, self-interested and designing little minx . . . but of a healthy, generous woman who loves her fellow beings and likes to look attractive far more for their sake than for her. There's a lot more than vanity in those cases. There's health. There's sanity." In fact, those who showed no vanity were the worrisome individuals, for their lack of interest in appearance and the opinion of others often signaled larger psychological problems.[107]

Consumers likewise seemed to have few qualms about the new vision of vanity. They purchased vanity sets, vanity mirrors, and *Vanity Fair.* Annie Maude Dee Porter, an affluent and religious resident of Ogden, Utah, was a regular reader of *Vanity Fair* magazine, which she mentioned repeatedly in her diary.[108] In 1938, she also reported that she had gone to "the Vanity beauty shop. . . . Had a shampoo, paper curl & manicure."[109] The word had slipped into common parlance as an unexceptional term. It had lost much of the moral freight it had once carried, and no longer provoked concern.

By the 1950s, a letter writer to the *Christian Century* was giving an all-out defense of the trait. "It is incontestably true that a good supply of vanity helps to lift and keep a person up to the level of good manners—a good level on which to live, and a high one too." He celebrated the "little dressing table where the noble wife and mother can look into the mirror when she is applying lipstick. . . . It is called a 'vanity,' a sound piece of furniture. Every home should have one. . . . [V]anity can be a spiritual force. There are many times when a new tricky little red hat will do more for a woman's spiritual morale than a dozen prayer meetings. Honest!"[110]

The idea that a new hat could offer spiritual succor was a direct renunciation of long-standing warnings against vanity. Indulging in purchases and cultivating an attractive appearance were not harmful behaviors; instead they were laudable. Whereas earlier generations had seen vanity as a flaw and a sin, one that was ultimately futile in the face of death, mid- and late twentieth-century Americans encouraged self-love and admiration as essential to a good life—an attitude still very visible today.

It was in this context that psychologists also deemed a related trait, "self-esteem," an increasingly important characteristic for Americans of all races and classes to develop. The phrase had been coined at the end of the nineteenth century, but it took root in modern theories of psychology only in the mid-twentieth.[111] As the idea gained currency, Americans were advised to develop self-respect and self-love rather than to see themselves as inherently flawed and degenerate.[112]

Simultaneous with this trend was the suggestion by psychoanalyst Heinz Kohut, a onetime follower of Freud, that self-esteem and narcissism could be healthy personality traits. He even suggested that some individuals might benefit from being more narcissistic rather than less so. Kohut admitted there was a pathological narcissism that grew out of childhood neglect, which left adults feeling inferior and in need of validation. But there was also a healthy narcissism, a kind of self-confidence that helped one function effectively in the world, a quality that should be nurtured.[113]

By the mid-twentieth century, then, many Americans had become far more receptive to displays of vanity than their forebears. These transitions were driven by new technologies that made the presentation of self easy, affordable, and commonplace; new theologies that redefined sin; new psychological conceptions of the healthy self, and the new imperatives of an emerging consumer culture fueled by vanity. Yet while the moralists of the nineteenth century no longer held sway, in the late twentieth century a new generation of social critics and psychologists was evaluating these transitions, and not always in a receptive light.

The Rise of the Therapeutic Society and the Growing Concern with the Self

David Riesman's 1950 book, *The Lonely Crowd,* which focused on the insecurity and self-consciousness of mid-century Americans, ushered in this new wave of criticism. Riesman's central argument was framed around two cultural types, which he termed "inner-directed" and "other-directed." Inner-directed Americans navigated through life by consulting precepts they had internalized from their parents or older generations. In contrast, other-directed types responded to their peers and to the messages of mass media. Riesman, who was not averse to the use of metaphor, referred to inner-directed types as people guided by an internal gyroscope while other-directed types were guided by radar. In Riesman's view, the "inner-directed" person had given way to an "other-directed" type who was much more apt to follow the shifting norms and fashions of society. Riesman's readers could find additional examples of this personality type in works like *The Organization Man* and *The Man in the Grey Flannel Suit* in which the protagonists made their livelihood working for corporations. Unlike the fabled American yeoman, the 1950s corporate executive depended on the company for his daily bread. As a result of that dependency, he used the dictates of the corporation and his peers as a reference point for personal action.[114]

Even though Riesman was not writing a polemic against other-directedness, by the 1960s, with the rise of a youthful counterculture, many Americans held such personality traits suspect. For those involved in the era's protests, the other-directed type was conflated with conformism, corporatism, and phony glad-handing. The reactions against it ranged from shirking traditional institutions and joining communes to rallying to Timothy Leary's cry to "turn on, tune in, drop out." Many hoped their efforts would revolutionize American politics and redress the racial, gender, and class injustices that in their view were perpetuated by the stodgy conformity of their elders. And to their credit, real political change did occur that benefited groups previously marginalized or oppressed.[115]

Like many rebellions, however, the outcomes were not as monumental as their leaders hoped, and, according to social critics like Tom Wolfe, Christopher Lasch, and Richard Sennett, the resulting frustrations triggered a turn inward. Instead of finding redemption through the politics of the commune, by the 1970s, a growing number of Americans turned to therapy and the homilies of self-help for guidance. The therapeutic imperative impelled people to discover their inner selves and to use whatever they discovered there as a polestar for action. While this behavior certainly departed from the other-directedness described in Riesman's work, social critics vacillated as to whether to celebrate or condemn it. Among the celebrators was Tom Wolfe, who described the change in a well-circulated 1976 article titled "The 'Me' Decade." In Wolfe's view, society had taken a distinctive turn away from a focus on the needs of the collective (the "we") to the needs of the individual (the "me"). For most of US history, Americans had seen themselves as part of an intergenerational collective. But during the "me decade," that communal ethos had been eclipsed by a newfound concern with the self. As Wolfe explained:

> Most people, historically, have *not* lived their lives as if thinking, "I have only one life to live." Instead they have lived as if they are living their ancestors' lives and their offspring's lives and perhaps their neighbors' lives as well. They have seen themselves as inseparable from the great tide of chromosomes of which they

> are created and which they pass on. The mere fact that you were only going to be here a short time and would be dead soon enough did not give you the license to try to climb out of the stream and change the natural order of things. . . . For anyone to renounce the notion of serial immortality, in the West or the East, has been to defy what seems like a law of Nature. . . . And now many dare it.[116]

In Wolfe's account, individualism was undergoing a renaissance. He claimed that modern Americans now shirked the responsibilities that their parents had felt obliged to shoulder. Instead, they pursued a more libertarian and inner-directed ethos and challenged the teachings on vanity that nineteenth- and early twentieth-century moralists had preached. The fact that human lives were short had led earlier moralists to conclude that it was vain to invest overmuch in the individual self. But Americans in the 1970s began to draw different conclusions. The fact that life was short did not mean that it was insignificant. Instead, it meant that one should extract as much enjoyment out of one's brief time on earth as possible.

Wolfe was not the only prominent social critic to reflect on changing attitudes toward the self in the 1970s, and his article, famous as it was when it appeared in 1976, soon had to compete with Lasch's *The Culture of Narcissism,* published in 1979.[117] Lasch contested some of Wolfe's conclusions, arguing that in fact, the conformism of the 1950s and the communitarian experiments of the 1960s had not yielded to individualism, because people, for the most part, still depended on corporations and state bureaucracies for their livelihood and for social guidance. As he explained in *The Culture of Narcissism,* Americans "surrendered" their "technical skills to the corporation," and consequently the household had lost its traditional functions. This was part of a larger trend that he termed "the atrophy of older traditions of self-help," which left individuals feeling incompetent and "dependent on the state, the corporation, and other bureaucracies." Lasch then honed in on the psychological implications of this trend, arguing that

> narcissism represents the psychological dimension of this dependence. Notwithstanding his occasional illusions of omnipotence, the narcissist depends on others to validate his self-esteem. He cannot live without an admiring audience. His apparent freedom from family ties and institutional constraints does not free him to stand alone or to glory in his individuality. On the contrary, it contributes to his insecurity, which he can overcome only by seeing his "grandiose self" reflected in the attentions of others, or by attaching himself to those who radiate celebrity, power, and charisma. For the narcissist, the world is a mirror, whereas the rugged individualist saw it as an empty wilderness to be shaped to his own design.[118]

Lasch and Wolfe were part of a larger stream of conversation about narcissism in America, for as they were writing, psychotherapists and psychiatrists also were paying new attention to the condition. While they did not agree on its symptoms or its value, its causes or its cures (or even whether it needed to be cured), the psychoanalysts Heinz Kohut and Otto Kernberg helped to spur interest in narcissism.[119] That interest ultimately culminated in the inclusion of a new mental illness—Narcissistic Personality Disorder, or NPD in the *DSM* in 1980. During this era, various narcissism scales and inventories were developed, designed to gauge whether individuals possessed the defining cluster of traits of the true narcissist.[120]

As it made its way into psychological diagnoses, the concept of narcissism also continued to make inroads into popular culture. Thousands of news articles appeared on the psychological condition in the 1970s and 1980s. For instance, in Ohio, the *Lima News* declared narcissism to be "on the upswing" and went on to report, "Some of the country's top social scientists are calling narcissism the major disorder facing Western society today. Analysts say patients with traditional Freudian neuroses are being replaced by the New Narcissus—a self-centered, self-serving creature so preoccupied with himself and his problems that he has no energy left to relate to others. . . . Many protestors of the '60s now concern themselves with nothing more selfless than the state of their Cuisinarts. People pay more attention to wearing the right clothes today, and disco dancing is a singularly narcissistic activity."[121]

If the trends that were being reported in the 1970s and 1980s seemed relatively new, by the end of the century they had solidified into persistent habits. As Americans, and particularly those in the middle class, increasingly gazed inward, they became more and more interested in psychology and its popularized self-help teachings. As Eva Moskowitz argued in her 2001 book, *In Therapy We Trust,* late-century Americans had begun to "turn to psychological cures as reflexively as they once turned to God."[122] And in the early twenty-first century, Moskowitz saw salient evidence for this gospel everywhere: from the publication of best sellers like *Prozac Nation;* to the popularity of daytime TV talk shows like *Oprah*, *Ricki Lake*, and *Geraldo*, devoted to emotional issues; to the rise of prescription medication like Zoloft, Prozac, and Paxil to treat mental depression; to the rapid growth of twelve-step programs and the tripling in the number of disorders that had been codified in the *DSM* between 1980 and 1990.[123]

Compounding her worries, the therapeutic culture was no longer confined to Americans' brick-and-mortar lives. It migrated online with the proliferation of internet support groups, which by Moskowitz's count exceeded five thousand even as far back as 2001.[124]

Narcissism in the Digital Age

As if to confirm the concerns of these earlier critics, a plethora of articles have appeared in the twenty-first century, with titles like "Is Facebook Really Turning Us into Narcissists?" or "How to Spot a Narcissist Online" or "Does Facebook Turn People into Narcissists?"[125] This expanding body of commentary gives further credence to the possibility that the patterns of behavior, which Wolfe, Lasch, Kohut, and Kernberg described as narcissistic in the 1970s, continue today and have even expanded.[126] And that possibility seems likely, because over the course of two centuries Americans have become far more accepting of self-promotion and have narrowed the range of behaviors that they condemn as vain. Today, they seem to have few fears about celebrating themselves, their accomplishments, and their possessions.

One key difference between the narcissism of Lasch's era and contemporary patterns is that in the 1970s and 1980s, many were turning inward, worried about their psyches and their inner lives. They may have shared their concerns with a few friends or family members, but such self-scrutiny was largely private and invisible to others. In contrast, today those who are concerned with themselves—whether it be their internal states or their external appearances—have ample opportunity to broadcast their concerns, their self-images, and their identities to the world through the internet. Self-absorption has become more public and more visible.

Some suggest that the internet itself has heightened narcissism, for it has offered unprecedented ways to scrutinize, glorify, and advertise the self, and to do so far more publicly than ever before.[127] For instance, Adam Kaslikowski, a tech entrepreneur living in Venice Beach, California, noticed how many people were promoting and marketing themselves on the internet. He told us, "I've been fascinated at the development of the concept of a personal brand in the last five, ten years. That didn't exist before. . . . Now everybody knows the phrase 'personal brand.'"[128] Katie Schall, a social media manager for an artisanal cheese company in Ogden, Utah, likewise believed some facets of the internet were "just really braggy."[129] Michelle Braeden, a Utah high school teacher and mother, observed that the internet was making her students feel like they were on constant display: "I don't think they ever get a break from like being pressured. . . . They're always on some sort of stage. . . . I hear the kids saying, 'Oh, I got this many retweets' or 'I got this many likes,' . . . so they have a constant reminder of if they're accepted or if people . . . are following them."[130] The stagelike nature of the internet and the pressure that it imposed on people to perform and promote themselves were also noted by nineteen-year-old Cooper Ferrario (who, perhaps not coincidentally, had recently become a marketing major). He described his own and his peers' worries. They constantly wondered, "Should I post this? How will that be perceived?"[131]

Twenty-seven-year-old Anna Renkosiak, who worked at a museum in the Quad Cities of Illinois and Iowa, sensed this same increasing pressure to perform, although she believed it was a burden that girls shouldered especially: "Girls growing up now . . . they always have to be . . . on; they always have to look pretty, 'cause you never know if you're going to take a selfie."[132]

But even while many are becoming accustomed to presenting and aggrandizing themselves through Twitter, Instagram, Facebook, selfies, and blogs, this does not mean that concerns about narcissism and excessive regard for the self have disappeared completely from public conversation. After all, in spite of the fact that Adam believed that having a personal brand was a "smart" thing to do, he did not take selfies or post much: "I'm not the guy who takes pictures of my life."[133] And while Katie believed that online interactions were "braggy," and Michelle noted that the internet felt like a stage on which people were pressured to perform, in their own lives they were trying to resist these trends. Katie, for example, lived in a house with spectacular views of sunsets over the Great Salt Lake, but she had never posted a picture of those sunsets to social media, even though some of her guests had.[134] Similarly, a Utah retiree, William Speigle, allowed that some people liked to be center stage, but he had no "need to be in the limelight." The secret for him to be liked was "to keep my damn mouth shut."[135] This was one of the reasons he had quit Facebook. The internet made it far easier to present one's self and one's images and ideas, but many still worried a great deal about how much of their selves they should make public.

Summing up these anxieties, Emily Meredith, a nurse living in Oakland, California, noted, "I just think if you step back from the whole thing, . . . it's just really strange to be posting pictures and liking each other's pictures . . . so I think I'm embarrassed about my own participation in it. . . . You feel like you've engaged in this like weird bragging . . . or narcissistic enterprise . . . and I just cringe at that."[136] Americans post and tweet and upload, trying to project the image of a perfect self, but they often worry about it as they do so, for the presentation of self is an activity that is still policed, even if the contexts in which it is monitored have shifted. Today, physical mirrors and so-called vanities are no longer controversial artifacts as they were in the nineteenth century, and neither is the therapist's couch as it was in the 1970s. The new battleground is the internet—and the areas where the sins and virtues of narcissism are still heavily debated are in the realms of selfies and social media, blogs, Fitbits, and GoPros.

Selfies, Social Media, and Seeing Oneself through the Eyes of Others

The explosion of selfies, which by some counts may exceed a million taken every day, has triggered concerns about narcissism, as has the rise of social media more generally.[137] The classic selfie is a youthful woman doing a duck-face in front of a mirror or with a selfie stick. It's an image that has been widely mocked, for the production and consumption of selfies remains contested terrain.

A 2014 Pew study found that 55 percent of the millennial generation (born between 1981 and 1996) shared selfies. The millennial rate for posting selfies was more than twice that of Generation X (born between 1965 and 1980). Baby boomers (born between 1946 and 1964) posted even more rarely.[138] When we interviewed Kent Winward and JulieAnn Carter-Winward of Ogden, Utah, we could see some of the generational differences that Pew described as well. Sometimes their children, who were in their teens and early twenties, would include them in pictures, but Kent and JulieAnn said they generally avoided selfie taking. JulieAnn described her daughters as "selfie queens." Her son had similar proclivities and had recently asked her to buy him a selfie stick while they were shopping at Staples. JulieAnn had refused, in keeping with her and Kent's desires not to indulge their children's interest in selfie culture.[139] In contrast to JulieAnn, forty-two-year-old Christine Fulmer, also of Ogden, Utah, posted selfies and liked doing so: "What's cool is . . . you can pick . . . the best part of yourself . . . to put up there." Yet she ultimately acknowledged that "it seems rather narcissistic."[140]

Kent, JulieAnn, and Christine were in their forties and early fifties and were more reticent about selfies than those ten or twenty years their junior. In contrast, twenty-two-year-old Camree Larkin liked taking photos and posting them, noting that she really enjoyed sharing pictures that depicted weddings, or hiking and camping trips, or other events that were "really gorgeous." And, she noted, "I like to look good in my pictures. Especially if I'm going to post online."[141] College student Skyler Coombs likewise admitted to taking selfies fairly frequently, but he posted them less often.[142]

Yet even those who had seamlessly adopted the use of selfies into their day-to-day lives still were critical of it, and posted with an eye toward moderation lest they appear conceited. Most admitted there was a line that separated appropriate posting from overposting, modesty from self-promotion. They were constantly crafting their images, and part of that self-fashioning involved limiting the number of self-portraits they shared. However, defining the boundary between healthy self-promotion and overpromotion or narcissism was a difficult enterprise. K., a college freshman, said that while he took selfies, he never posted them because

> I'm not really so much to post . . . personalized pictures. I felt like that's kind of like, um, excuse the term . . . a douchebag sort of thing to do. You take a selfie and post it up just to like make yourself look cooler than the rest. . . . I really don't want to do that. . . . People taking pictures in the mirror. . . . That's kind of annoying 'cause you see it so often. So I don't want to . . . flow into that crowd.[143]

He wanted to distinguish himself from those who "are constantly, daily talking about themselves or constantly posting things. Like you'd see people with pictures up to the thousands. . . . That's so much time on taking selfies and some of them . . . look exactly the same as another one. It's like what's the point? . . . It seems kind of narcissistic."[144]

Some worried about coming off as self-absorbed. Eddie Baxter, a twenty-nine-year-old social worker, at times felt like he was "kind of bragging," even if he did not intend to. He had just returned from a trip to the Galapagos. While there, he confided, "I was blowing up my

Facebook with a whole bunch of amazing photos and just giving thanks. . . . My post was not 'Ha ha, look at me, I'm in the Galapagos.'" Instead he was trying to "share" his experience and gratitude. But he was not sure how those images ultimately were received: "depending on the viewer, they're going to decide whether or not I'm being braggy."[145] Camree, while acknowledging the pleasure she felt when her pictures were liked, was at the same time worried about how others perceived her. As she put it: "I definitely can be vain. . . . I can be way more obsessed about myself and horrified if someone else realized or thought that I was obsessed with myself."[146]

This self-criticism was on display when twenty-eight-year-old Alta Martin, a radiologic science major, talked about her selfies. For Alta, the desire to moderate selfie taking was motivated by the fact that she imagined others would get "really annoyed with it."[147] She herself felt this way about some of her friends' online habits, particularly one stay-at-home mom who posted far too many pictures of her kids, a habit typical of what another interviewee, Katie, called "mommy over sharers."[148]

Emily, the nurse in Oakland, and also a mother of two sons, recognized that posting pictures of one's children could be a form of boasting, and she worried about doing it excessively. She talked critically about the "natural parental impulse of . . . sharing your joy and delight in your children" and the downsides of doing it on the Web. As she put it, "It's something about how public it is. . . . There's a vulnerability in how much you love your children . . . and want other people to appreciate them . . . and then there's . . . just like a kind of bragging that is annoying and off-putting." She believed there was something "coercive" in making others applaud your child: "That's the cringy part."[149] The people we interviewed wanted to share their images and their updates but did not want to overdo it. Being able to judge how much was too much was difficult, however, and left many feeling uneasy.

"I got to a point where it was like I was pursuing likes," recalled Gregory Noel, a twenty-eight-year-old college admissions counselor, who ultimately deleted his Facebook account precisely because of how uncomfortable it made him feel about himself. When he was an active Facebook user, he remembered thinking, "What can I post or what can I do to . . . get this certain amount of likes, to get this certain amount of recognition . . . for the thing that I posted?"[150] Like Gregory, many people as they posted were acutely aware of their audience, conscious that they were being observed. Yet even if it made them feel self-conscious, they often found it impossible to stop posting.

The need to display their lives, their successes, their happiness—a cultural necessity ever since smiles had caught on in snapshot photography in the early twentieth century—had become acute in the twenty-first century. If one wanted to be popular, to be a social actor, to be relevant, one needed to post and promote, to project a polished, flawless image, despite the anxiety it might cause. Behind such posts lurked the hope that one could find a true and lasting sense of worth, a permanent feeling of acceptance.

People were eager to share glamorous images of themselves, but no matter how proud or pleased they were with their pictures, they still often worried about how they were perceived. Social media might be a way to gain admiration but it could also make one feel profoundly vulnerable. Eddie constantly second-guessed "every single post" that he created. He explained, "I want to sound educated in my post." When we met him, he was completing his master's degree in social work, but not so many years earlier, he told us, "I was a high school dropout and a . . . single teenage father." He winced when he saw old posts from that period of his life on Facebook's timeline, for he was acutely aware of spelling or punctuation errors he had made long ago. Now that he had graduated from college and was completing graduate work, he confided, "I want to sound educated. I want to be articulate myself. So I overthink a lot of my posts now. . . . I'm afraid to really share too much."[151] Anna, who used social media for her job as well as in her personal life, noted, "You put yourself out there and you're completely exposed." She admitted that the feeling of exposure limited how often she posted. "I'm not one to want to be out there so that's why I don't post that often. . . . If I put that out there what are you going to think?" She continued, "I don't want to put myself out there for that potential ridicule . . . for nobody to like it . . . for it to go completely unnoticed." Social media offered a way to find affirmation, but when people did not respond—or responded in the wrong way—it was demoralizing.[152]

Although many tried to predict their friends' responses to their posts, they realized their ability to do so was imperfect. They might not be able to gauge how their followers would react to an update or a picture. For that reason, younger people prescreened their Instagram and Snapchat posts with an intimate circle who served as a jury to decide if something was worthy of posting to a larger audience. For instance, Hailee Money, a college student who had spent her early teenage years in Aurora, Colorado, but now lived in Utah, admitted, "Every time I post something I'm making sure . . . I even ask people, . . . 'Is this good for Instagram?' Like I don't want to put something on there that people might think doesn't look good." She thought this was necessary "because people see you differently than you see yourself."[153]

Once they did post, people fretted about what their friends' responses—even the positive ones—actually meant. They plumbed their depths for subtexts. What were their followers' true opinions? Could one trust the likes and comments that one received? After all, while people might respond positively to a post, it was not always clear that they were being honest and forthright. L., from San Bernardino, California, thought that social media sometimes falsely inflated self-esteem: "More often than not people will post positive remarks, instead of like 'that's not the best picture of you.' So it's kind of . . . [a] false sense of like [a] compliment."[154] In a vicious cycle, the very unreliability of feedback heightened self-consciousness and self-scrutiny, and that often increased anxiety—leading people to post repeatedly with the hope that they might receive the affirmation they sought. But a lasting sense of validation proved elusive.

Not only did people sometimes distrust online responses to their own posts, but they also questioned the accuracy of others' photos. Emily noted that people she knew often called Facebook "Fakebook." Images—from the "curated," glamorous selfie to visions of perfectly contented family life—were deceptive. She planned on having talks with her children about "the fact that what's on Facebook—not just the photos, but everything that's posted—is curated. . . . It's only people's most wonderful experiences, and everything's great, and the whole Fakebook idea. . . . Nobody puts up . . . 'I screamed at my kids and they're driving me crazy or whatever. . . .' People only present . . . a polished version of their life. You know like, 'We baked cookies and then we danced.' . . . [Not] 'my kids are on the computer all day and then I screamed at them.'"

In her home on the northern border of Oakland, the polished selves people tried to present on social media, were, paradoxically, nontechnological. So even though she and her friends were using digital media to present themselves, the image they were trying to project was distinctly low-tech. As Emily put it, "Nobody says . . . my kid spent three hours playing *Minecraft*. . . . It's always about the cupcakes they made or the . . . wonderful drawings they did." These parents wanted to present the most "wholesome" and "most nontechnological . . . version of themselves." She parodied that instinct: "Oh! My children spent hours building this fairy fortress out of leaves in the backyard and aren't they amazing . . . and here's the picture of it posted on social media."[155]

Everyone we spoke with was keenly aware of the ways people edited their identities and lives on social media. Consequently, they eagerly looked at each other's beautiful and glamorous images, but often with considerable skepticism, realizing that what they saw was not a realistic portrait of their friends' lives. Thirty-three-year-old Diana Lopez, the community outreach coordinator for the Ogden Police Department, observed that when people posted on social media, "They show all the beautiful, showy parts of their lives and it kind of seems like they need that validation, they need those likes, they need followers. . . . People aren't always genuine, . . . aren't always expressing what's really going on—for the sake of likes." She explained why they did this, noting, "Everybody wants to put their best foot forward; the problem is that . . . everybody can always put their best foot forward all the time, when it comes to social media."[156] Despite their distrust of the accuracy of their friends' posts and their inability to read how other people responded to their own, they continued to upload them and to self-consciously fashion their online identities through such images.

In fact, often they seemed to live their day-to-day offline lives with the goal of appearing happy, confident, and glamorous online. Camree admitted that sometimes she decided what to do based on the photo opportunities the activity might present. "I do take pictures in mind thinking that I'm going to share this with other people." On occasion, the thought of pictures would be in her mind as she made plans: "I do think sometimes . . . about doing something . . . and posting it later." Visions of future online responses to selfies could shape one's daily actions.[157] People moved through their lives with an online audience in their heads.

Living with that audience in one's head and then posting for them might create what Emily termed "a feedback loop," which "creates more need for us to get more outside affirmation." She speculated that as people became accustomed to such affirmation, they "need more of it. . . . They post things in order to get it. . . . You start to need it." She realized, however, that for young people, such posting and liking was the only way to have an identity in America. It might be cringeworthy to one generation, but essential to another. Wondering aloud, Emily asked, "For young people . . . do they exist without a . . . digital presence? . . . Is it even narcissism as we understand it if it is just a requirement for existing . . . in order to have a presence in their world?" The Web provided affirmation of one's identity and worth—affirmation that people longed for and yet were simultaneously embarrassed to admit they needed.[158] Social media heightened self-consciousness and magnified social need.

From the sheer ubiquity of selfies and the admissions made by our interviewees, it's clear that Americans, and especially youthful Americans, use images to seek validation and to deal with insecurity as never before. On the one hand, they want to broadcast images of their successful lives; on the other hand, they do not want to look desperate for approval or overeager while doing so, for that would undermine the effect. They need social validation but they often fear making that need public by posting too much. And while the controls on vanity might not be as strong as they once were, some degree of policing oneself and others is still evident. Contemporary Americans use social media and post selfies to advertise their success and to celebrate themselves, but they do it with anxiety. They may not fear vanity as a sin, they may not worry about offending God by thinking too much of themselves, but they do worry about losing face on Facebook. Individuals' concerns about appearing narcissistic are not so much concerns about maintaining a particular ethical position so much as an interest in shaping other people's perceptions of them. They fear social, not divine, sanctions if they cross the line. Although they are focused on the self, they are acutely aware of the community they are posting for, whose regard they value and long for.

Politicized Selfies

Selfies can convey other longings as well. Some selfie posters believe their portraits have meanings that go far beyond personal self-promotion and that their images could in fact be deeply political. Rather than being a sign of vanity, they contend that a selfie can communicate values, critiques, and commentaries and is not inherently a symptom of self-absorption. For example, the researchers David Nemer and Guo Freeman demonstrated that, in Brazil, the selfie was often accompanied by text or mise-en-scène that called attention to dangers and social pathologies that selfie takers were facing in their poor, often crime-plagued favelas. Their images intentionally captured not only their faces but also their trying living conditions. The pictures were about more than self-promotion; they were also social commentary.[159]

Feminists have made similar arguments about the political power of selfies. In November 2013, the journalist Erin Ryan published an article titled "Selfies Aren't Empowering. They're a Cry for Help" on Jezebel.com. Ryan argued that when women took selfies they were actually succumbing to sexism and that selfies are "a reflection of the warped way we teach girls to see themselves as decorative."[160] Many women of color disagreed, arguing that a deeper political meaning could be found in their portraits. According to media scholar Minh-Ha

Pham, readers soon began to protest Ryan's piece with the hashtag #feministselfie, paired often with a selfie.

> @OHTheMaryD: "I'll be damned if an able-bodied . . . White woman is gonna make me feel bad for taking a damn selfie. HELL NO! #feministselfie"
>
> @bad_dominicana: "selfies are the only place I see women like me. Unlike whites, I don't have entire industries made in my image. #feministselfie"
>
> @bad_dominicana: "taking time out of my day to admire myself in the midst of constant antiblack misogynist degradation, #feministselfie pic.twitter.com/7mgZOfU8dg"
>
> @FireinFreetwn: "Need a White woman to come teach me about feminism. This is my cry for help #feministselfie pic.twitter .com/EkRkwGyw5Y"[161]

In Pham's reading, these tweets demonstrated how selfies might be used for purposes of social activism, and suggested that the significance of posting a selfie depended in part on how the groups one was affiliated with were represented in the mainstream media. As Pham puts it, "Feminist selfie tweets . . . are not simply digital forms of self-regarding. They are a decentralized mode of political action."[162]

The political use of selfies has also been evident in the online responses to recent killings of unarmed black men. After police in Ferguson, Missouri, shot Michael Brown, many criticized the media coverage of the shooting by posting selfies of themselves with the hashtag # IfTheyGunnedMeDown. They displayed side-by-side portraits, offering images of themselves wearing graduation caps and gowns, on the one hand, and partying, on the other. Their point was that news outlets often portrayed African American victims of police violence in a negative light, showing pictures that made them look tough and potentially dangerous, rather than choosing images that depicted them as upstanding citizens.[163] A century earlier, Frederick Douglass had become the most photographed man in America in an effort to change perceptions of African Americans, and modern activists were using the selfie for the same purposes. These were and are explicitly political uses of selfies. Such portraits function as social criticism rather than as a means for mere self-aggrandizement.

And even when selfies are used for self-promotion, they do not represent quite the same brand of narcissism visible in the ancient Greek myth. To be sure, there are similarities: Narcissus acted on his self-love by looking at his reflection in a pool of water. Similarly the counterculture used therapy, and modern narcissists count the number of likes on their Facebook feeds. Yet these similarities elide important differences. Where Narcissus gratified himself by sundering his relationship with others, the modern narcissist depends on maintaining social connections with others. For contemporary Americans, staring at oneself is not enough, because on social media, one exists only in the eyes of others. For selfies to succeed requires a response from one's followers. Selfies and social media might be about promoting the self, but in the end, it is a socially constituted self.

Blogging

The same tension between the desire for individual glory and the hope for community support is visible in other online venues. By 2012, an estimated thirty-one million Americans blogged.[164] Bloggers, some of whom have called the blogosphere the "narcissystem," have on occasion voiced doubts about their activity, as has a wider public.[165] As Claire Tanner notes in *Vanity: 21st Century Selves,* since bloggers do not have to subject their writing to editorial review, they are ostensibly at liberty to post on a whim, to post on anything they like, and to indulge in extreme forms of self-gratification. Free from the constraints of the press, these liberties naturally open up the possibility for bloggers to be self-referential, introspective, and entirely oblivious to the concerns of a potential audience. Likened at times to vanity presses, blogs, of course, do have attributes that cater to vain instincts. And yet, as Tanner notes, in practice, these are tempered by the fact that bloggers are rarely writing just for themselves but instead seek to capture the interest of a wider audience. To that end, they will refer and link to other blogs and often (although not always) solicit comments from readers. These practices turn what would otherwise be a one-way, self-absorbed act of broadcasting into a more interactive activity, where the self is forced to engage, consider, and respond to opinions expressed by others.[166]

Kent, who had maintained a number of real and pseudonymous blogs, admitted, "There's a little bit of built-in narcissism just to be an author period. You can't be an author without just a titch of narcissism." JulieAnn, who, like her husband, also blogged, agreed: "Only if you think you are not shit will you write something and call it good."[167] But while Kent and JulieAnn allowed that there might be a little narcissism in blogging, neither thought it was the best way to describe the activity, since to be successful required connecting with others. JulieAnn explained:

> You have to be completely unafraid of self-examination. You have to be completely shameless in your self-examination. But at the same time . . . in order to be a good writer you have to have a level of compassion and empathy with the world around you. . . . That's really what it's about. You got to access the deepest darkest parts of yourself but at the same time you have to be connected to the world around you because you can't just write what you know; you also have to write what other people know.

Kent agreed, noting, "The reason it's not narcissistic is because when you write, . . . if you are doing it right . . . you write and you say, 'Who is my audience? Who am I trying to reach?' . . . When you keep the audience in mind then I think it works really well." He continued: "If you are writing or participating in social media and thinking solely about your audience and connecting with them, then the narcissism goes out the window because you are not thinking about yourself."[168]

Another blogger, Tamsin Howse, came to the same conclusion after posting about her looks and being accused of narcissism by a commenter on her blog. She posted in response, "I am, and always have been, a bit of a narcissist. But I wonder if all bloggers are the same. We all write about ourselves and our lives. . . . All bloggers are narcissistic when you consider that simple fact." Yet, like Kent and JulieAnn, she believed successful bloggers had to do more than think and write about themselves. As Howse puts it, "I have a suspicion there are very few bloggers out there who would manage to be successful by admiring their own attributes."[169]

Indeed, some bloggers ran into trouble when they were perceived as doing this. Adrienne Shubin, who blogged at therichlifeonabudget .com, faced online accusations that she was a narcissist. An anonymous reader had posted, "Have you heard of Narcissism, and its nasty big brother Narcissistic Personality Disorder? . . . Have you wondered ever if some of it might apply to you? . . . Almost all your blogs follow a similar shape—me, me, me, me, me, me oh, what do you think about me/ my style?" Shubin addressed the criticisms on her blog, writing,

> Although I do not agree with the commenter, I do acknowledge that my confidence has grown over the last nearly two years of blogging. . . . Does that make me a narcissist? I would have to ask a professional to really know for sure. But based on what I have read about it and what I know about myself, I think it makes me this: A self-confident, middle-aged woman who has found joy and contentment through writing and sharing interests, including fashion, style and general lifestyle topics, with other women all over the world.

She claimed that blogging had also given her the opportunity to make new friends and helped her master the fear of exposing herself to the world.[170]

Like Shubin, many bloggers claimed that critics misunderstood what they were up to. They were not interested in self-love so much as validation from a community they cared about. For many, there seemed to be a longing for connection rather than the mere celebration of self. Vanity might be the initial impetus driving people into blogging. And bloggers at first might be impelled to be self-referential and focus exclusively on themselves. But the desire to popularize their blogs often led them to consider their potential audience and to express themselves in ways that engaged and sustained the attention of others. Blogging might share characteristics of the therapeutic culture that Lasch railed against, because it encourages introspection. But unlike therapy, blogging is open to something more political since it impels people to risk having their ideas shot down and criticized. As the writer and journalist Scott Rosenberg puts it,

> The opportunity to write has always been available to anyone with an education, but until very recently the opportunity to publish one's writing has been rare and prized. By making that opportunity nearly universal, blogging has changed something fundamental in our culture. Writing in public lets you discover what you really think by attempting to set your thought in prose and "putting your views at risk" (Christopher Lasch phrase) in front of strangers. Now anyone who wants a taste of this experience can have it.[171]

JulieAnn, who had shut down some of her blogs because of stalkers and trolls, also recognized that writing could be risky. She confided,

> I guess if you only want to exist in an echo chamber where you only want everyone to agree with what you have to say and you only want praise and you only want good things coming your way then I guess that could be considered a form of narcissism. But that's a little different than when you write for an audience. When you write for an audience you are not afraid to be controversial. But you're also not afraid to be compassionate and to be able to say your . . . piece in such a way that if somebody does have a beef with it, you're open to discussion.[172]

In the end, many bloggers recognize that their blogs have the potential to be narcissistic. But it is not self-evident that online forms of communication are any more narcissistic than more traditional ones like book writing. In fact, because books are not as interactive as blogs and are impervious to live critique, they might sometimes be considered even more self-indulgent.[173] In contrast, bloggers write in the hope of generating online, interactive discussions with their readers. Like those who take selfies, the mark of a successful blogger is one who connects to an audience.

The Quantified, Documented Self

Social media platforms and blogs are not the only recent technologies to make Americans turn inward (as they also, in the process of sharing, paradoxically turn outward). In 2014, 3.3 million wristbands like the Fitbit or Nike's Fuelband were sold.[174] By 2017, Fitbit reported that there were more than 25 million people using their product, which is marketed with the slogan "Know Yourself to Improve Yourself."[175] These "fitness trackers" encourage users to explore who they are, not only through writing or digital images posted to social media, but also through the measurement of their physical activity and biorhythms. Moreover, since Fitbit and Fuelband also facilitate the automatic posting of this activity online, they provide another venue, albeit less verbal or visual, for the expression and measurement of self.[176]

Gary Wolf, perhaps the most zealous proponent of this sort of self-examination, began calling this form of personal digitalized measurement "the quantified self." In addition, he spun up a website with an eponymously named domain (e.g., quantifiedself.com) and wrote a manifesto titled "The Data Driven Life" for the *New York Times Magazine* in 2010. As Wolf saw it, modern fitness trackers allow users to know themselves in unprecedented ways. Since time immemorial, humans had been trying to uncover a "truth buried at another level." But in Wolf's view, past approaches had limitations because individuals' memories were "poor," humans were "subject to a range of biases," and their attention spans were short. With fitness trackers on the other hand, many of these limitations were superseded. Not only did the new technologies allow users to "take stock of ourselves" in unprecedented ways, but it was "the self we ought to get to know."[177]

In a parallel initiative, some Americans have taken advantage of the falling cost of digital memory and digital recording in an attempt to

FIGURE 1.6 GoPro's "Be a Hero" campaign encourages consumers to record and share their outdoor adventures. *Courtesy of GoPro*

log as much as possible about their daily lives. A semblance of this activity is on display in the feed of a super-tweeter or very active user of Facebook. Likewise, GoPro, which allows people to record their activities during their athletic exploits with a helmet-mounted camera, advertises "This is your life. Be a Hero," promising new ways to present glamorous live footage of oneself to one's followers.

But so-called life logging takes this to the next level. Gordon Bell, who pioneered the practice in 1999, puts it this way:

> Each day that passes I forget more and remember less. I don't have Alzheimer's or even brain damage. I'm just aging. Yes, each day I'm losing a little bit more of my mind. By the way, so are you. What if you could overcome this fate? What if you never had to forget anything, but had complete control over what you remembered—and when? Soon, you will be able to. You will have the capacity for Total Recall. . . . [S]oon you will be able to record your entire life digitally. It's possible, affordable, and beneficial.[178]

Between quantifying the self and life logging, Wolf and Bell offered contemporary Americans the prospect of learning about themselves and remembering themselves in new ways. But whether they represent social progress and a more profound understanding of the self is another matter, since, as the business consultant and blogger Lindsey Rainwater suggests, they might lead toward "self-obsession" rather than to "self-awareness."[179]

Rainwater's distinction between self-awareness and self-obsession captures the tension many Americans are wrestling with. They want to know themselves but worry that in the process of doing so, that is all they will know. Their vision will be narrowed, and they will be able to see only themselves reflected in their phones and on their computer screens. Changing views of vanity and narcissism, psychology and economics, have had a role in narrowing many Americans' vision. But these changes have also been fostered by the development of technologies that allow people to dwell more intensively on their inner selves and the way that self is projected to others.

Contemporary Americans are, as the writer Nicholas Carr argues, in an age dominated by "technologies of the self."[180] Technologies have, at times though, also originated from and bred more communal sentiments.[181] As bloggers, selfie takers, social media posters, and Fitbit wearers will testify, technologies of the self may foster a turn inward, but it is not usually accompanied by an abdication to a pure and unadulterated form of vanity or narcissism. These devices of the self are still often used for the purposes of social connection, for the whole point of posting a selfie, a blog post, or a Fitbit score is to get a response from friends, family, and followers.[182] For many, these audiences are essential to sustaining their expanded sense of self.

People post, hoping that an audience is watching and liking what they see. They are eager to project an appealing and attractive identity to their followers, that is (usually) consistent with who they are offline.[183] That hope of presenting a credible identity has evolved since the advent of the internet. In 1993, the *New Yorker* published a cartoon with the caption "On the Internet Nobody Knows You're a Dog." The cartoon, now semifamous with its own Wikipedia entry, emblematized Americans' early hopes for how identity might be nurtured on the Web. Because the Web was perceived as a virtual sphere, separate and distinct from offline lives, it appeared to invite Americans to experiment with different identities and to take on whimsical personae. It encouraged users to develop online personalities without worrying whether they jibed with their offline ones.[184]

In the intervening twenty years, however, the lag between online and offline selves has diminished. In the wake of Edward Snowden's revelations about government surveillance, and recent disclosures of Facebook's data-sharing tendencies, Americans are cautioned to take greater care of their online selves. Those identities might at one time have floated freely in the cybersphere but they are increasingly associated with corporeal selves.

Because of this development, Americans are ever more mindful of their online selves. And here, in this space, surveillance and narcissism start to meld. The narcissist wants to be watched. And while others may not want to be, they are warned to imagine that they are being watched anyway.[185] The result is heightened self-consciousness, an increased desire to perfect one's online identity and image. Adam, the tech entrepreneur living in Southern California, observed this tendency in himself. "This is a worldwide audience. . . . You have to be very careful about what you put on here. And as long as you're being careful and therefore editing, well, then why not edit it in a certain direction? . . . You can slice [your posts] . . . to make yourself look like an extreme athlete or a homebody or anything in between. . . . As long as I'm editing, I can edit it in my favor."[186] Q. had a similar feeling about being online: "I very much feel like I've got somebody looking at me constantly. That changes the way you actually do things, if you feel that way."[187] The belief that they were being watched made them self-consciously shape and fashion their identities, for both Adam and Q., like other contemporary Americans, were acutely aware of the eyes on them, whether they belonged to friend or foe. The need for a polished, flawless self was therefore all the more pressing.

That sense of being watched, and having to craft an identity for others to consume, affects what people post, and their very sense of self. The imperative to prepare for such scrutiny is being inculcated at a young age. Warren Trezevant, a software designer in Oakland, California, advised his adolescent daughter, "Just think that every time you hit Send, anyone in the world will read this. . . . So if somebody in Germany reads it, what will they think of it? If the government reads it, what do they think of it?" Between the scrutiny of friends and family, and the surveillance of the government, modern Americans are being conditioned to feel acutely self-conscious, for they rarely feel free of the gaze of others.[188]

The feeling or burden that life was a public performance did not start with the Web. After all, as Shakespeare famously proclaimed in *As You Like It*, "All the world's a stage." And the idea has earlier roots, as well, with religious people believing that God was watching their every move. The Web, however, has added new complexity to the problem of being observed. It forces Americans to be increasingly mindful of the selves that they are now, as well as the selves that they might become, and to contemplate how those selves are reconstituted through the opinions of others.

Vanity has changed in another way as well. Whereas previous generations were taught that life was fleeting and the mark that one made on the world was ephemeral, contemporary Americans are warned that what they say or do will last forever on the internet. It is, supposedly, permanent. Whereas once God judged humans and condemned them for thinking too much of themselves, for having a false sense of their own permanence or importance, today Americans feel judged by their friends, followers, and the anonymous eyes on the internet. It is difficult to say which is more frightening.

Americans have become more involved with technologies of the self, immersed in the so-called narcissystem. As a result, in a world where their identities are not ascribed to them, but instead are something they believe they can and should fashion, narcissism comes knocking.

Thus, in the digital age, we all have become narcissists. But unlike the original Narcissus, we are obsessed by the self as it is mirrored by the community, and the technologies in which it is reflected. This obsession, in turn, cranks up our anxieties. One might think that anxiety would breed more humility than hubris. But, paradoxically, in modern America, such is not the case. Our sense of our own importance—and our longing to broadcast that importance to others—has grown.

CHAPTER 2

The Lonely Cloud

For over a decade, pundits have asked whether social media and the internet are making Americans isolated and alone. The discussion began in the 1990s but gained new intensity as a result of Stephen Marche's 2012 essay in the *Atlantic,* which asked "Is Facebook Making Us Lonely?" Marche told the sad story of Yvette Vickers, an actress and former Playboy Playmate, who had died alone in her Los Angeles home in 2010, but whose body was not discovered until many months later. In the years before her death, Vickers had withdrawn from family and friends and had conducted most of her social life with remote fans via internet sites. Marche argued that Vickers's death was emblematic of a larger cultural malaise. Either social media was making people lonely, or, as may have been the case with Vickers, lonely people were attracted to it.[1]

The article caught the public's attention, and was quickly followed by a spate of others, including Eric Klinenberg's "Facebook Isn't Making Us Lonely," sociologist Claude Fischer's "The Loneliness Scare," and Zeynep Tufekci's "Social Media's Small, Positive Role in Human Relationships."[2] And in the intervening years, the debate has continued to rage with partisans on both sides; for instance, psychologist Jean Twenge, in her book *iGen,* claims that young people born after 1995 not only endure significantly greater loneliness than millennials, but are subject to it in part because of their iPhone use.[3] While there has been little agreement on the issue, the debate turned a spotlight on the emotion.

News outlets, doctors, and celebrities have likewise given increased attention to the condition over the last few years. According to some reporters, America is experiencing a "loneliness epidemic," and its citizens are living in an "age of loneliness."[4] Estimates vary as to how many Americans experience the emotion, with findings ranging from 20 percent to 35 percent of the population.[5] In 2014, Oprah Winfrey developed a campaign titled "Just Say Hello" to help people combat the feeling.[6] Medical and mental health experts have likewise warned that loneliness not only is psychologically trying, but also has physical costs that are devastating, with the lonely dying earlier than the nonlonely, a point that the foremost psychologist of loneliness, John Cacioppo, made repeatedly in a series of publications and interviews on the topic.[7] In 2016, the *Washington Post* labeled the emotion a "lethal risk."[8]

While much of the discussion has focused on whether loneliness has increased or decreased in recent years as a result of digital technologies, posing the question that way obscures key issues. To really understand the present-day concern with loneliness, one needs to know how the social experience and meaning of the feeling have changed over time. For instance, the *Washington Post* recently declared that loneliness "carved 'caverns' in Emily Dickinson's soul and left William Blake 'bereaved of light.' Loneliness, long a bane of humanity, is increasingly seen today as a serious public health hazard."[9] However, such statements overlook the fact that nineteenth-century loneliness was not the same as twenty-first-century loneliness. Loneliness felt different—and meant different things—to earlier generations of Americans. Although many eighteenth- and nineteenth-century Americans experienced the feeling, there was much less cultural concern about it than there is today. Further, they had different, and somewhat more modest, expectations about the number of friendships and social connections they should have. Modern technology as well as new psychological theories left a decisive mark on Americans' inner lives, and they have reshaped loneliness from an unpleasant part of the human condition into what many regard as a dangerous social epidemic of the digital era.

The Solitude of One's Own Heart

The words "lonely," "lonesome," and "loneliness" entered the English language only around 1600.[10] This is not to say that earlier generations had not contemplated the problems of being alone—indeed, scriptural references to aloneness abound in the Bible, with the first appearing in Genesis, when God declared: "It is not good that the man should be alone; I will make him a help meet for him." He then created Eve.[11] Both in the Torah and the New Testament, there are also reminders that God's presence offered an antidote to the solitary nature of life.

Yet while humans had long been alone, within English there were no words for the feeling that is now called loneliness until the sixteenth and seventeenth centuries. Some have suggested that with the advent of Protestantism came a new cultural emphasis on loneliness. Sociologist Max Weber argued that Calvinism in particular, with its doctrine of predestination, isolated people and led to an "unprecedented inner loneliness of the single individual." During the Reformation, Protestants had repudiated the Catholic rituals that might allay fears of damnation. As Weber saw it, each person therefore "was forced to follow his path alone to meet a destiny which had been decreed for him from eternity. No one could help him. No priest. . . . No sacraments. . . . No Church. . . . Finally, even no God."[12] Individuals were on their own in life and in death.

As a consequence of this theology, most of the early Protestants who came to America believed that loneliness was part of the predestined order of the world and that those who suffered from it must resign themselves to it as the will of God. This was how David Brainerd, a missionary to the Housantonic tribe in New York, saw his situation. He complained to his brother in 1743, "I live in the most lonely melancholy desert, about eighteen miles from Albany," and he found his work "exceedingly hard and difficult." Yet such hardships were his burden to bear. He wrote, "Lord grant that I may learn to . . . endure hardness, as a good 'soldier of Jesus Christ!'"[13] This was also how Rebecca Dickinson, an eighteenth-century spinster, coped, as she struggled with feelings of sadness and isolation. Born in 1738 in Hatfield,

Massachusetts, Dickinson lived for decades in the house in which she had been raised. Again and again she complained in her diary of her lonesomeness. In September 1787, after making a visit to a local tavern, she wrote of the despair she faced when she returned home: "Came to this hous [*sic*] about half after Seven and found it dark and lonesome here." She knew she was supposed to submit to such circumstances, but found it difficult. "I walked the rome [*sic*] and Cryed my Self Sick and found my heart very Stubborn against the government of God." But she ultimately believed her aloneness was God's will, writing that although she did not like her solitary existence, "it is the will of god Conserning [*sic*] me[;] no other Place would Doe [*sic*] to Cure me of my Pride and to wean me from the world."[14] To be a devout and faithful Christian meant to accept hardship and suffering, loneliness and isolation, and to believe that such conditions ultimately would prove redemptive.

Dickinson was living in a small, tight-knit New England town, but she found that even in such a community she could still be lonely. The opportunities for lonesomeness only increased in the years to come, as towns grew larger and populations more scattered. After the American Revolution, as the new ideology of individualism took root in the nation, many felt pressure to adopt autonomous and self-reliant behaviors. Consequently, over the course of the nineteenth century, there was a growing cultural focus on the power and potential of the individual.

Some outside observers believed that because of this individualistic ideology, Americans were at greater risk of loneliness. Most famously, Alexis de Tocqueville claimed that rather than being linked by social and familial bonds, Americans were free agents, which made them both liberated and potentially lonely. He wrote, "Not only does democracy make men forget their ancestors, but also clouds their view of their descendants and isolates them from their contemporaries. Each man is forever thrown back on himself alone, and there is danger that he may be shut up in the solitude of his own heart."[15]

Like Tocqueville, many observers have portrayed individualism as innate to Americans. Over the past two centuries, there have been countless celebrations of the westering pioneer, and the self-sufficient person, but in reality, there were many in the United States who found it hard to be so independent and alone.[16]

As a result, loneliness was widespread in the nineteenth century. Men and women who could write frequently confided in their letters and diaries that they felt lonely, often using the then-popular term "lonesome."[17] Some were lonesome in idle hours, when family and neighbors were not around, and there was little to do. Others felt it acutely after a death in the family. Still others felt it when they moved or were moved away from home, a common experience, given that roughly 50 percent of the population migrated—or were forcibly transported—across state lines during the first half of the nineteenth century.[18]

Not all who migrated did so willingly, and enslaved people, who were often sold away from family and friends, frequently experienced a welter of feelings, including profound anger, sadness, and loneliness. Tom Jones, born in North Carolina in the early nineteenth century, regarded his lonely condition as the consequence of slavery. He recalled his feelings when his wife was sold away from him: "Oh, how lonely and dreary and desponding were those months of lonely life to my crushed heart! My dear wife and my precious children were seventy-four miles distant from me, carried away from me in utter scorn of my beseeching words. I was tempted to put an end to my wretched life." He saw his wife only once more when her new mistress was traveling in the neighborhood. He invoked his painful story of bondage, marked by separation, loneliness, and despair, to motivate opposition to slavery, and hoped that "the just God" would remember these incidents "in the last award that we and our oppressors are to receive."[19]

Other enslaved people found a balm for their loneliness in religious faith. Elizabeth, born in Maryland in 1766, was sent away from her family to a distant farm. On a rare visit to her mother, who lived twenty miles from her, she recalled that her mother "told me that I had 'nobody in the wide world to look to but God.' These words fell upon my heart with ponderous weight, and seemed to add to my grief. I went back repeating as I went 'none but God in the wide world.' . . . After this time, finding as my mother said, I had none in the world to look to but God, I betook myself to prayer, and in every lonely place I found an altar." Ultimately, her religious conversion filled her with "sweetness and joy" and made her feel less isolated.[20] After her religious conversion, Sally, an enslaved woman from North Carolina, no longer felt "friendless and alone," according to her biographer; a favorite song she liked to sing was the popular hymn that began "Jesus once was poor and lonely." She confided that it was her faith that had allowed her to "come through . . . [this] low ground o' sorrow."[21]

Unlike the enslaved, free whites had the liberty to choose whether to join or separate themselves from family and friends, and consequently, their experience of loneliness was somewhat different, for it was more often a consequence of their own decisions. Some believed it might be necessary to endure loneliness to get ahead in the world.[22] Like the enslaved, white Americans frequently coped with loneliness by relying on their religious faith, but they also invoked the rhetoric of unfettered individualism and the language of opportunity. Elisha Douglass Perkins, a gold miner, did both, as he explained how he endured loneliness while traveling to California in 1849. His party had just left St. Joseph, Missouri: "Now we were out of civilization & the influences of civilized society entirely, & cut loose from the rest of the world to take care of ourselves for a while. I confess to a feeling of lonliness [*sic*] as I thought on the prospects before us, & all we were separating ourselves from behind. Henceforth we shall have no society, no sympathy in our troubles, & none of the comforts to which we have been accustomed, but must work across these vast wild wastes alone." He was willing to make such sacrifices, for "Gold must be had & I for one am willing to brave most anything in its acquisition." To help him on his lonely voyage, he vowed to rely on "our own strength & his who takes care of us all."[23]

Throughout the nineteenth century, religious authors often counseled not only that God might offer succor to the lonely, but also that aloneness was part of the divine order. Although loneliness and solitude were not always equated, William Rounseville Alger treated them as synonyms in his 1867 book *The Solitudes of Nature and of Man; Or, The Loneliness of Human Life*. He wrote, "There is more loneliness in life than there is communion. The solitudes of the world out-measure its societies." He wrote of the "Solitude of the Desert," the "Solitude of the Prairie," the "Solitude of the Ocean," the "Solitude of the Pole," the "Solitude of the Forest," "the Solitude of the Mountain," and "the Solitude of the Ruin," before he turned to his main subject, "Solitudes of Man." Human solitude was more extreme than that encountered in nature—Rounseville described it as the "intenser inner deserts of mental and moral being." He believed that all individuals suffered loneliness because they were separate, and their inner lives remained hidden from their fellows.[24] An English religious writer enunciated much the same idea, putting it this way:

> We each of us are born into the Land of Loneliness. Our fathers and mothers belonged to that land, and their fathers and mothers belonged to it also. Every one in the world, by nature, belongs to it, and the strangeness of it is, that though there are so many millions of people living on the earth, each soul is a solitary soul; no one knows it thoroughly—it is hidden in its lonely cell.[25]

To many in the nineteenth century, loneliness was an expected part of the human condition.

Some theologians suggested that this condition was a result of the fallen state of man. Until Adam and Eve had sinned, there had been no loneliness; once they fell, however, they became separated spiritually from each other and from God. Loneliness was the result of sin and the manifestation of divine will.[26] In the nineteenth century, African Americans put such thoughts to music, singing the song "Lonesome Valley." The melody spread and became popular across the United States. The lyrics reminded singers that, as they journeyed through life and grappled with death, "You gotta walk that lonesome valley, you gotta go there by yourself, Nobody here can go there for you, You gotta go there by yourself."[27]

Yet despite the fact that loneliness might be God's punishment, there were some theologians who believed that God also offered the best cure for the condition. Faith might confer a sense of wholeness and communion and make loneliness abate. The congregational minister Amory Bradford explained, "Sooner or later everyone who lives long, knows what it is to be lonely in spirit. We try to make our lot endurable, . . . and earnestly seek to make it a blessing, but still 'a nameless longing and a vague unrest' is never long absent. There is but one medicine for this homesickness of the soul, and that is the realisation of the Unseen Companionship. The art of living alone is acquired with the art of living in the consciousness of God."[28]

More secular writers also addressed the condition, suggesting that true genius needed solitude for its expression. Ralph Waldo Emerson, perhaps the foremost philosopher of individualism in nineteenth-century America, articulated this perspective in "Self-Reliance." There he stressed the necessity of being alone at times, and he endorsed solitary experience, writing, "We must go alone. . . . At times the whole world seems to be in conspiracy to importune you with emphatic trifles. Friend, client, child, sickness, fear, want, charity, all knock at once at thy closet door, and say,—'Come out unto us.' But keep thy state; come not into their confusion."[29] In his view, to avoid social interaction, to revel in solitude, was the way to gain true freedom. Bradford made a similar argument in *The Art of Living Alone:* "Those who go ahead of their times, must always go alone. The great poets have seldom been men of the world. Divine music cannot be heard in the noise of a crowd. . . . Great men are like mountains,—the higher they rise the more lonely they become."[30]

Emerson's friend and neighbor Henry David Thoreau surely subscribed to a similar philosophy. His acquaintances sometimes said to him of his life at Walden Pond, "I should think you would feel lonesome down there, and want to be nearer to folks." He admitted he had been lonely early on in his stay at Walden, but that he had found solace in nature and solitude. Besides, he reminded his readers, "The sun is alone. . . . God is alone. . . . I am no more lonely than a . . . dandelion in a pasture." And, he claimed, all people were solitary to some extent: "I have found that no exertion of the legs can bring two minds much nearer to one another." Wise men should not live close to others, should not settle near the "post-office, the bar-room, the meeting-house, the school-house, the grocery, . . . where men most congregate," but should instead seek places where they could find their own inner meaning.[31]

Chopping down Telegraph Poles

The rhetoric of independence and solitude was far easier to espouse than loneliness was to endure, for despite his idealistic words and his ardent denials of loneliness, Thoreau often found himself walking back from Walden Pond into Concord in an effort to find sociability. He ate frequent dinners at the Emersons' house. (Some claimed that when he heard the Emerson dinner bell, he came "bounding out of the woods . . . to be first in line at the Emerson dinner table.") His mother and sister often visited him, and he them, and his mother, in fact, did his laundry. He walked into town to visit other friends as well.[32] He admitted in *Walden* that despite his words about self-reliance and solitude, "I think that I love society as much as most, and am ready enough to fasten myself like a bloodsucker for the time to any full-blooded man that comes in my way. I am naturally no hermit."[33] As a result, he sought out opportunities for companionship in village life, as did men and women across the nation.[34]

While such face-to-face fellowship was the most common way to allay loneliness, there were also new balms for the feeling in the emerging communications systems of the era as well. For instance, over the course of the nineteenth century, Americans came to rely on the US Postal Service as an antidote to social isolation. The expansion of the post in the 1830s and 1840s fostered a new outlook, making individuals feel connected to others far distant from them.[35] Through the mail, they conversed with distant kin and friends, in an effort to sustain a beloved community through pen and ink. And the post allowed them to move, with the assurance that they might assuage their loneliness and homesickness with letters. To some extent, the mail made it easier for Americans to strike out, to be rugged pioneers, alone on the frontier, precisely because it promised to make them less alone. It supported the myth of rugged individualism and independence even as it subtly undermined it by connecting people to each other.

Sara Stebbins, who had moved from Deerfield, Massachusetts, to Richfield, Illinois, wrote her mother-in-law a letter in early 1839. She began, "Feeling rather lonely I address a few lines to you hopeing [*sic*] that it will not meet with as much neglect as those I have written to the girls. If they knew how much pleasure it gives us to heare [*sic*] from home I think they would write us oftner. Would they once imagin [*sic*] how lonely we are . . . I know they would write."[36] Christian Miller and his cousin Benjamin Kenaga also found comfort in correspondence. They had both grown up in New York State; both had moved to the Midwest (Christian to Indiana and Ohio, Benjamin to Michigan). They stayed in touch throughout their entire lives, relying on letters for updates. A sense of loneliness was evident in their correspondence, as was a dependence upon the post to assuage it. In 1881, Christian wrote to Benjamin, "Dear Cousin . . . once wee [*sic*] were young but now we are middle aged. . . . [A]lthough once we were allmost [*sic*] in dayly [*sic*] conversation with each other[,] . . . by fate unavoidable . . . [we] have been separated and therefore obliged to carry on our conversations by the medium of pen & ink."[37]

While the mail offered some relief to the isolated, the telegraph, though frequently portrayed as a cure for loneliness, often did not live up to expectations. When the telegraph system was established, beginning in 1844 and expanding thereafter, there were extensive and euphoric celebrations of it, and many believed that God had ordained its creation as a way to join people and nations together. And in its early days, many Americans hoped it might relieve their loneliness. In 1852, Elizabeth Emma Sullivan Stuart, widow of the fur trader and explorer Robert Stuart, wrote to her son, "Oh, if I could only on Telegraph wires be seated in that dear little dining room between you & Kate, & *talk* over all things, how my heart would be gladdened."[38]

However, the high prices of telegrams did not allow for such easy intimacy. To communicate from New York to London in a brief, ten-word telegram cost $50 in 1866.[39] Sending a message cross-country was only marginally more affordable: In 1861, to send a telegram over the newly completed transcontinental wire cost between $5 and $6. Rates fell over the course of the nineteenth century—by 1867, it cost $1.09 to send a wire from New York to Chicago, and by 1900, 30 cents. However, in an era when the average daily wage of an industrial worker was $1.93, these prices were still too high for frequent use.[40] Consequently, throughout the nineteenth century, the telegraph's most regular customers were businesses and the wealthy.

As a result, despite the fanfare and hoopla that surrounded the telegraph in its early days, nineteenth-century Americans of the middle and working classes gradually tempered their hopes for the technology. While the telegraph system as a whole symbolized connections to the outside world, that connection was, for many, an abstract concept rather than a daily reality. Up through the early twentieth century, on a day-to-day basis, Americans of middling and modest means used the telegraph only rarely, especially when compared to the mail.

Because it was not affordable, nineteenth-century Americans often thought the telegraph to be of little use in ending their isolation and loneliness. It spanned the continent, theoretically connecting millions; however, people frequently just wanted to correspond with select friends and family. The telegraph might give access to the rest of the nation and the rest of the world, but if one did not know individuals in California or England, what good was that? As Thoreau famously queried, "We are in great haste to construct a magnetic telegraph from Maine to Texas; but Maine and Texas, it may be, have nothing important to communicate."[41] Some may have been lonely, but unlike contemporary Americans' perceptions of the internet, they did not necessarily regard telegraphic connections to a larger, foreign world as a solution to the problem.

In fact, in its early years, some actually resented the telegraph, finding it to be at best an irrelevant presence, and at worst a threatening intrusion into their social lives. According to one account, in the mid-nineteenth century, a telegraph mast "on the Iowa shore at Keokuk stood near the cabin of a crusty old pioneer" from Tennessee. The man "was much annoyed by the humming of the wire. One night during a high wind, the noise became insufferable to him; he sprang out of bed, seized his ax, and after an hour or more of hard chopping, laid the mast low." This was not uncommon in the early years of telegraphy: "In Indiana, not only were wires severed, but poles were felled. In Ohio it was a favorite trick to cut wires and leave them trailing across a road, where vehicles ran over them."[42]

The telegraph brought noise and trouble for many who were content with their lives the way they were. Some observers speculated the telegraph might even bring diseases into communities that had heretofore been safe from them. Perhaps, they suggested, the cholera sweeping the nation in 1849 might be spreading across the wires. Such worries, antipathies, and resistance suggested to one early scholar of telecommunications that these nineteenth-century Americans were "temperamentally not yet ready for the telegraph."[43]

That "temperamental" unreadiness may not have been a sign of backwardness so much as a sign of social expectations that were dramatically different from modern ones. These communities were isolated, yes, but that isolation may have been, in their residents' eyes, something of an advantage. Their known social world was largely intact, not scattered. Those with close relationships and strong communities felt little need for the telegraph—the people they wanted to talk with were in their immediate neighborhoods. And when they longed to converse with the growing number of family and friends who were at greater distance, they relied on the less noisy and obnoxious—and far more personal—medium of the mail. What good were wires that carried expensive messages, connected them with strangers, made noise, and possibly brought danger to their towns, congregations, and homesteads? Americans may have been lonely, but they did not always think the telegraph was a solution to their woes.

While promoters of the telegraph dreamed of global connections, some saw greater utility in local ones. During the 1860s, the Western Union Company attempted to build a telegraph line from San Francisco to Russia. The ambitious engineering project enlisted the help of indigenous people who lived along its route through British Columbia. Ultimately, however, the project was abandoned in 1866, and the Wet'suwet'en people repurposed the telegraph wires and poles to build

FIGURE 2.1 Hagwilget Bridge over the Bulkley Valley. The bridge was built by Wet'suwet'en people in British Columbia, using telegraph poles and wires left by the Western Union Company after it abandoned its plan to build a telegraph line from San Francisco to Russia. Photo circa 1904. *Courtesy of the Royal BC Museum and Archives / A-04067*

a bridge spanning a rocky chasm near their homes.[44] The telegraph materials, when refashioned into a footbridge, were more successful in strengthening communities, connections, and ties to others than they were when strung from poles to span oceans and continents. In strong, face-to-face communities, contact with the larger, unknown, outside world often seemed neither necessary nor appealing. To many in the nineteenth century, the idea that one needed a multitude of contacts across the globe was a foreign and unfathomable concept.

Similar reservations existed about the telephone. In its early years, many regarded it as an unwelcome and unnecessary presence. They objected to the noise it brought and the way its infrastructure intruded on their lives and properties. In 1888, the *New York Times* reported that a New Jersey judge had in essence decided that "a man who cuts down an obnoxious telephone pole in front of his premises because it is obnoxious cannot be convicted of malicious mischief."[45] Perhaps that inspired Henry Winters in Plainfield, New Jersey, who in 1889 cut down telephone poles near his property.[46] A Pennsylvania farmer, W. A. Zahn, cut down five poles in front of his land and threatened to cut down more in 1891.[47] In 1893, "a small-sized mob" gathered outside a pharmacy in Bordentown, New Jersey, "and cut down a pole erected by the Delaware and Atlantic Telephone and Telegraph Company."[48] In 1910, Mrs. F. O. Gorman of Montpelier, Vermont, was arrested after cutting down three telephone poles because she was "disturbed by the noise of the wires."[49]

Even those who did not cut down poles sometimes had limited interest or faith in the telephone as a cure for isolation. In his memoirs, muckraking journalist Mark Sullivan explained that each generation had its own expectations and pace in social life. "For each person there is a natural tempo. And this tempo is in many situations less rapid than the pace which modern life demands and modern machinery dictates." He remembered that his parents had little interest in modern means of communicating, writing, "Sometimes what we regarded as comforts for our parents were by them regarded as fantastic and unnecessary modernities. What we regarded as desirable ornaments they looked upon as wanton extravagances. . . . For years my brothers, concerned about my parents alone on the farm, wanted to put in a telephone so that we could call them up and talk with them. The innovation was resisted successfully until within a few years of my father's death, and was accomplished then only by some intimidation."[50]

Yet while Americans often initially resisted the telephone, most eventually came to embrace it. In 1880, four years after the telephone's invention, only one in a thousand Americans had a phone; thirteen years later, the number stood at one phone for every 250 people (and most of these early telephones were owned by commercial establishments). However, by 1907, there was one phone for every fourteen people, and by 1920, one-third of households owned phones.[51] As their popularity grew, communities often established their own small systems, for the phone excited hopes of local social connection and to some seemed to promise an end to isolation. In 1901, a columnist in the *Waterloo Semi Weekly Courier* reflected: "The phone is one of those things which a person does not appreciate until he has once enjoyed them. Independent of its value from a commercial and business standpoint, . . . it forms a social link which is almost invaluable to every resident of the farm house."[52] In 1904, another telephone promoter observed that the "telephone takes from the farmer's family its sense of loneliness and isolation," and argued that it would make the "pathos and tragedy" of isolated farm life "disappear." As Claude Fischer has shown, when people did begin calling, they generally used the phone to reinforce existing relationships rather than to create wholly new ones.[53]

Americans, then, found an array of social possibilities in these new technologies. Some continued to rely on face-to-face interactions; others used the mail. And many integrated the phone (and, as rates fell during the twentieth century, the telegraph to some extent) into their social routines.

Yet when individuals employed these technologies and techniques to offset loneliness, they did it at a far different pace and frequency than do modern-day Americans. A 1933 study, which analyzed national rates of mail, telegraph, and telephone use, estimated that per capita Americans had surprisingly few calls, letters, and cables, and that over time, this number grew only slowly. In 1907, Americans received per capita a local letter every eighteen days, and a nonlocal one every six; by 1927, the rate had picked up, with local letters arriving every nine days and nonlocal ones every four. Telephoning became more common—Americans received a local call every three days in 1907 and every one and a half days by 1927. Toll calls remained relatively infrequent: in 1907, the rate was one every four and a half months; by 1927, they were coming in every one month and ten days. Telegrams were extremely rare. In 1907, Americans received on average one telegram every eleven months and two days; by 1927 that rate had increased to one every six months and twenty-three days. Cablegrams from overseas were the most exotic form of communication. In 1907, Americans received them on average once every fourteen years; in 1927, every eight years.[54]

There is a danger in reading too much into these figures since they are aggregate numbers distributed across the population. They deceive, because in reality some Americans received letters and cables at far higher rates, and some at far lower, yet they nevertheless point to the fact that Americans in the past had different expectations about loneliness and sociability. Their expectations about contacts with the larger world were lower. Certainly they do not seem to have thought that constant contact with people farther afield would make them less lonely.

While at first blush, these figures might suggest that Americans had few social interactions, the reality is that they often had rich, locally oriented social networks with their neighbors, which required neither letters nor phone calls. Just as Thoreau had eased his loneliness by visiting Emerson, so millions of Americans visited their own neighbors, often at a breathtaking clip. For instance, in the 1880s,Martha VanOrsdol Farnsworth, a teenager living in southern Kansas, described her daily social life in her diary. During eighteen days in August 1882, she recorded interactions with 191 people in her small town.[55] During ten cold days in January 1896, Edward Rich, a Utah doctor, mentioned nearly forty face-to-face interactions with at least twenty-seven patients, acquaintances, friends, and family members (seeing some of them more than once), in addition to receiving letters, telegrams, and phone calls.[56] These patterns of local sociability helped people cope with loneliness. Meeting and greeting acquaintances on the street, attending lodge meetings, and chatting with neighbors at the store or post office offered chances for friendly interactions.

The Less Lonely League

But these patterns underwent dramatic changes in the late nineteenth and early twentieth centuries. In 1890, the US Census Bureau announced that the American frontier was closed. Thirty years later, the Bureau's tallies revealed that America had become an urban nation, with more than 50 percent of its population living in towns and cities. As Americans became city dwellers, they met more people, and some might have expected that they would encounter less isolation than they had on farms and in small towns. But these hopes were not always borne out. Some city dwellers concluded that metropolises might actually heighten loneliness, while on the other hand, some doctors, psychologists, and academics worried that these bulging cities offered too much socializing, and that the increasing number of interactions could actually cause harm. Whereas today Americans are worried about a "loneliness epidemic," a century ago some worried about the oversociability of the American people.

Doctors and psychologists in the late nineteenth and early twentieth centuries believed that, rather than resolving loneliness, the urban environment and modern conveniences were overstimulating people. Some connected the new inventions and the hectic pace of city life to the rise of neurasthenia, a nervous disease characterized by symptoms that included exhaustion, sensitivity to extremes of heat and cold, "early and rapid decay of the teeth," baldness, nearsightedness, indigestion, and "nasal and pharyngeal catarrh." The American doctor George Beard brought great attention to the condition in his 1881 book *American Nervousness,* believing it to be widespread among the nation's upper classes. He suggested its cause might be found in a host of developments in America, including the telegraph, the railroad, the printing press, the speed of American businesses, the rise of watches, the spread of democracy, the stress of electoral politics, and the excitement of American religious strife.[57]

Beard was not alone, for, as Stephen Kern shows in *The Culture of Time and Space,* an array of physicians and social observers in the United States and Europe believed that the effort to keep up with the fast rhythm of city life drained people of their energy and vitality.[58] Many doctors suggested that the best treatment for severe cases of this malaise was "the rest cure," which involved at least a partial retreat from society and an embrace of solitude and isolation.[59] That respite was necessary to maintain one's health, as the American philosopher Irwin Edman explained in 1920: "Just as men can be satiated with too much eating, and irritated by too much inactivity, so men become 'fed up' with companionship." He maintained that humans had a physiological need for some "solitude and privacy."

> "The world is too much with us," especially the human world. Companionship, even of the most desirable kind, exhausts nervous energy, and may become positively fatiguing and painful. To crave solitude is thus not a sign of man's unsociability, but a sign merely that sociability, like any other human tendency, becomes annoying, if too long or too strenuously indulged.[60]

He suggested that the stressful nervousness of turn-of-the-century America might be the result of excessive sociability.

> Much of the neurasthenia of city life has been attributed to the continual contact with other people, and the total inability of most city dwellers to secure privacy for any considerable length of time. In some people a lifelong habit of close contact with large numbers of people makes them abnormally gregarious, so that solitude, the normal method of recuperation from companionship, becomes unbearable. . . . But a normal human life demands a certain proportion of solitude just as much as it demands the companionship of others.[61]

In contrast to modern American commentators who worry that aloneness and solitude are dangerous, to Edman, they were natural and necessary.

However, if medical professionals and social critics expressed concern about the exhausting pace of social interaction in America's growing metropolises, others harbored different fears about urban life. They contended that modern cities were so vast and impersonal that it was hard for individuals to meet one another. As a journalist noted, loneliness "is coeval with the development of the modern city; in less populous communities, where 'everybody knows everybody,' it is not so acute."[62]

In the early twentieth century, isolated individuals across the nation banded together to fight off the feeling. In 1901, in Seattle, an African American writer implored his readers to create organizations for the lonely within their churches. He lamented the fact that newcomers might go to a church for months "and no one has ever asked them to come in and join their family circle and take even so much as a cup of tea." To remedy this, he proposed that "in all Negro centers there ought to be a woman's guild of Negro hospitality in every Negro church." Such guilds might address the "unutterable loneliness" that young men and women, new to the city, felt as they searched for community.[63]

A decade later, white citizens in the nation's growing metropolises began to form what they termed Lonely Clubs or Lonely Bureaus. These organizations took root in big cities like Chicago, New York, Detroit, Seattle, and Los Angeles to great media fanfare.[64] For instance, in 1911, papers reported that L. J. Wing, a prosperous but lonesome businessman in New York, had founded the Less Lonely League.[65] A few years later, in Chicago, a lonely woman wrote to papers about her plight. Her letter was the impetus to form a Lonely Club there, which quickly attracted 175 members.[66] The existence of such organizations suggested that one of the problems of the American city was that there were too many people all together rather than too few. The crowds, the mass of people, heightened isolation and nervousness. Counterintuitively, more people might actually mean more loneliness. Then too the city environment gave the illusion and seemed to promise that one could and should know scores of people. When one did not, when reality did not meet such expectations, this could cause discontent, sadness, and loneliness. Urban life heightened hopes for social connection and often dashed them as well.

That these Americans willingly and publicly referred to themselves as lonely and formed Lonely Clubs was an indication that the feeling did not yet imply a dangerous illness, a shameful neediness, or social maladjustment, as it does today. Instead, it was an unpleasant but unsurprising and unstigmatized condition. However, in other ways, the loneliness clubs anticipated many modern attitudes. First, they seemed to indicate that while the feeling existed, it ultimately need not be experienced and was not inevitable. One could fight it more aggressively than earlier generations had believed. Second, it suggested that the

CHICAGO'S LONELY CLUB HATCHES GREAT SCHEME; NOW IT ISN'T LONELY ANY MORE

FIGURE 2.2 In the early twentieth century, Lonely Clubs and Less Lonely Leagues sprang up in major American cities. Their founders hoped to fight the social isolation that many claimed was rampant in urban life. *North Dakota Evening Times*, May 22, 1913. *Library of Congress, Serial and Publications Division*

nineteenth-century palliatives for the condition—religious faith and resignation, a hope for reunion and connection at death—were losing traction. Indeed, Mr. Wing, the founder of New York's Less Lonely League, spoke directly of the inadequacy of religion and religious institutions in combating isolation. "Long ago I realized that New York was a mighty lonely place. . . . At first I thought that through a church I might meet congenial people. I attended one in Central Park, west, but it didn't take long to convince me that no friends were to be made there. I attended another church in Lenox avenue. The result was the same. Now I don't attend any church."[67]

If churches, once so vital to community life and integration, no longer fostered social relations, their old theologies also seemed to lose power over many Americans. Whereas Calvinism had fostered loneliness and implied it was inevitable, new forms of Protestantism suggested otherwise, for the idea that suffering and loneliness were divinely willed and to be expected no longer held as much appeal as it once had. In the early twentieth century, many churches and laypeople were influenced by the rising religious movement of "New Thought," with its message that individuals could perfect themselves and improve their conditions through positive thinking. This movement, which took off at the turn of the century, grew in strength and influence.[68] By mid-century, books like the Reverend Norman Vincent Peale's *The Power of Positive Thinking* were blockbuster best sellers, spreading the message that people could change their conditions through a positive outlook. As Peale wrote, "The fact is that popularity can be attained by a few simple, natural, normal, and easily mastered techniques. Practice them diligently and you can become a well-liked person."[69] In such a climate, the passive acceptance of loneliness seemed untenable. No longer were the lonely expected to resign themselves to the feeling. Instead they should take action to avoid it. And when they did feel loneliness, it was a sorry reflection on their own weak willpower and lack of initiative.

"You Can't Be Lonesome If You Own an Edison"

This message was reinforced by merchants. In the early twentieth century, salesmen hawking technologies from telephones to Victrolas to radios promised that these goods would banish loneliness. All one needed to do was purchase them. The ads reflected a new approach to selling that emerged in the 1910s and 1920s; rather than touting a product's physical specifications, advertisers presented their goods as cures for a wide range of psychological and medical maladies.[70] For example, Bell Telephone used the theme of loneliness repeatedly in the early decades of the twentieth century. "Telephone your friend and get an Instant Reply," Nebraska Bell promised in 1912, noting the phone "banishes loneliness and brings a feeling of comfort and security."[71] Bell suggested that parents install phones for the purpose of "Gathering Friends" for their children, noting, "A little girl can always get somebody to play with by using the Bell telephone. . . . There is no need to be lonesome with a telephone in the house,

Right in Your Neighborhood

EVERY little while some friend or neighbor has a Bell Telephone put in. If you have one, every new subscriber enlarges the scope of your personal contact. If you have not, every new telephone makes you the more isolated—cut off from the activities about you—"out of things."

Bell telephone service is the greatest of all modern conveniences. And it is more than a mere saver of time and distance. It keeps you alert; keeps you abreast of the rapid march of events—in your neighborhood and in the world at large.

You cannot be up with the times without the Bell Telephone.

For rates and other information regarding service, call the District Manager

The Central District and Printing Telegraph Company

BELL SYSTEM

FIGURE 2.3 By 1909, AT&T ads suggested to consumers that they would have more personal contacts if they owned a phone, and would be left out and isolated if they did not. *Courtesy of N. W. Ayer Advertising Agency Records, Archives Center, National Museum of American History, Smithsonian Institution*

because you can at least talk with your friends, even though they are far away."[72]

Some also raised the spectre of social exclusion to sell phones. A 1909 ad suggested that those who did not install a telephone might become lonely, as their neighbors shifted their conversations to the wires. "Every little while some friend or neighbor has a Bell Telephone put in. If you have one, every new subscriber enlarges the scope of your personal contact. If you have not, every new telephone makes you the more isolated—cut off from the activities about you—'out of things.'" The ad concluded its pitch with the dire warning that "you cannot be up with the times without the Bell Telephone."[73] Another ad from the same year reminded readers, "The Bell system brings eighty million men, women and children into one telephone commonwealth, so that they may know one another and live together in harmonious understanding."[74]

The ads also suggested that modern individuals needed to be connected to more people than previously, and that they risked social and business failure if they did not join the network. As a 1910 ad explained, a phone was "A Finder of Men." Americans had many friends, but did not always know where they were. With a phone, however, they might more easily locate them. It counseled, "Cities are larger than they used to be. Men know and need to know more people. Yet the need of keeping in touch is as great as ever. Without Bell service there would be hopeless confusion."[75]

Telephones were not the only products that promised to ward off loneliness. Phonograph merchants offered similar benefits. A 1905 ad proclaimed, "Buy an Edison Phonograph and you will never be lonesome."[76] A year later, the National Phonograph Company promised buyers that with a phonograph they would be "In Good Company," reassuring them that "you can't be lonesome if you own an Edison."[77] "Give Your Wife a Victrola," suggested the Corley Company in 1918. Directed to husbands, the ad noted that a housewife "spends long hours at home when you are not there, that are lonesome and dreary—a Victrola will enable her to pass these hours pleasantly."[78]

And radios too promised similar contentment. "Don't Miss the Miracle of Radio," lectured a 1923 ad, arguing that "the home without a radio receiving set is handicapped. Its outlook on life is circumscribed. Why should the neighbors enjoy all the happiness and fun?"[79] By the mid-1930s, Arvin Radio suggested that one could avoid ever being alone or experiencing even a moment of solitude, because their car radio "counteracts the depressing monotony of distance—keeps your spirits high. Instead of driving away from your friends of the air, you'll take them with you."[80] Phones, phonographs, and radio ads told Americans that loneliness could be easily cured with the latest technology; it was no longer necessary to simply endure it.

Some Americans testified that these technologies did in fact alleviate their lonesomeness. A listener to the early Massachusetts radio station WGCI wrote in 1923, "Hearing your voice as I do once or twice every day I feel almost a personal acquaintance with you. . . . I lost my dear old wife who had been my companion and pal for over 52 years on August 13th and I find much comfort from your broadcasting."[81] A woman from Brooklyn, New York, likewise confided to researchers, "I consider radio the most essential piece of furniture as thru months of loneliness (my husband being a tubercular veteran in a sanitarium) I have played it, sang with it, and kept my spirits up thereby saving my own reason and disposition."[82]

But the technologies' limitations were apparent on the pages of memoirs, diaries, and journals. Dorothy Johnson, born in 1906, recalled how radio changed domestic life in Whitefish, Montana, when it took off in the 1920s. It made solitude increasingly unendurable: "Listeners became addicts, so accustomed to having sounds of any old kind coming into the house that they were nervous when it was quiet. It was easier to leave the radio on and not listen than to bear the unaccustomed silence. For better or worse, the quiet, the isolation, the parochialism were gone."[83] By 1942, a reporter noted that Americans had become so dependent on the radio that they could no longer deal with any solitude or loneliness. She wrote, "I've nothing against radios. Indeed, I am very much for them, and have installed them myself in various parts of my home, including my car." But "I am very much against our hysterical need of constant noise and diversion as a means of escape from solitude." She explained, "Solitude is not a blight nor a nightmare. It is a normal and necessary part of our human experience, and no character can become . . . poised without large amounts of it."[84] Modern life, and particularly modern technology, had diminished individuals' capacity to accept aloneness.

Worse, many began to blame themselves when they felt lonely. Beginning in the twentieth century, Americans heard again and again that their isolation and social and business failures were due not to forces larger than themselves, not to the inherent sorrow of the human condition, not to divine will, but to their own failure to develop a likeable self. As historian Richard M. Huber notes, "Advice on how to achieve a radiant personal impression, which had been abundant before the 1930s, became a 'how to' mania after that decade. Through charm schools and correspondence courses, newspaper features and the radio, in books, articles, and 'hard sell' advertisements, there was plenty of advice on how to make yourself more personally and physically attractive to other people."[85] Success writers told aspiring businessmen that in the growing corporate economy, they should cultivate vast networks of friends and colleagues and strengthen business relationships through displays of friendly goodwill. Dale Carnegie, for instance, celebrated New Deal politician James Farley—who knew the names of fifty thousand people—as a model for others to emulate in their quest for success.[86]

For those who could not or did not want to master fifty thousand names, such advice could heighten feelings of anxiety and social inadequacy. As psychologist Rollo May wrote in 1953, "The feeling of loneliness arises from the fact that our society lays such a great emphasis on being socially accepted. . . . If one is well-liked, that is, socially successful—so the idea goes—one will rarely be alone; not to be liked is to have lost out in the race. . . . 'Be well-liked,' Willie Loman in *Death of a Salesman* advises his sons, 'and you will never want.'"[87]

Just as such views were spreading, a new word was taking root in the language in the late 1940s: "loner."[88] The word carried negative connotations, labeling those who sought solitude as abnormal. Well-adjusted individuals were supposed to be happy and sociable and seek out friends. Those who stayed by themselves and stood at the margins of society were worrisome and unsettling. They made their isolation manifest. Whereas nineteenth-century Americans would have understood the loner (though they would not have called him that), twentieth-century Americans saw him as maladjusted to the demands of a society that stressed sociability and the overt appearance of cheerful friendliness.

Taken together, these pressures made Americans who were lonely feel out of step with American culture. Perhaps they were responsible for their own isolation because they had not cultivated the right personality traits; or perhaps they had not purchased the right products to counteract their loneliness.

TVs, Cars, and *The Lonely Crowd*

These worries, doubts, and fears about social success and adjustment grew in the years after World War II. One very visible sign of this was the 1950 publication of *The Lonely* Crowd, written by David Riesman (in collaboration with Nathan Glazer and Reuel Denney). No history of loneliness in postwar America would be complete without referring to this book. Although the title was chosen by the publisher, its reference to loneliness underlined a central message: Americans at mid-century were sensitive to peer opinion and group-oriented, but this grouping compounded rather than alleviated loneliness.

The Lonely Crowd was often interpreted as an indictment of mid-century, middle-class shallowness, embodied in what Riesman termed "the other-directed person" who was acutely aware of the opinions of neighbors and colleagues. But unlike more caustic critics, Riesman saw virtues in other-directedness and held out hope that those virtues could be fused with inner-directed traits to create a new American, sensitive to others while still being autonomous. Yet the more indelible message from *The Lonely Crowd* was implicit in the title's reference to loneliness, as well as in its concluding passages:

> If the other-directed people should discover how much needless work they do, that their own thoughts and their own lives are quite as interesting as other people's, that, indeed, they no more assuage their loneliness in a crowd of peers than one can assuage one's thirst by drinking sea water, then we might expect them to become more attentive to their own feelings and aspirations. This possibility may sound remote, and perhaps it is.[89]

Riesman had not written *The Lonely Crowd* strictly as a work of criticism, but the book's portrait of the loneliness of other-directed individuals resonated with many. Ever attendant to the fashions, opinions, and trends of their peers, other-directed people were remarkably well socialized: they knew their friends and coworkers better than they knew themselves. However, their social bonds tended toward breadth rather than depth. And this lack of depth, in turn, often led to loneliness. Like Tantalus, the Greek mythological figure who eternally attempted to drink out of a pool of receding water, the America Riesman described was populated by people who sought to redress their loneliness through other-directed activities—which instead of assuaging the emotion, often compounded it.

Many hoped they would find an antidote to loneliness, and some looked for connection and conviviality in the suburban communities that were increasingly coming to represent the good life. Yet behind the facades of these new homes, many found aloneness rather than satisfaction. Whereas Riesman laid the blame at the feet of the other-directed, some, like Harvard sociologist Robert Putnam in *Bowling Alone,* have pointed the finger at the technological emblems of suburban life: TVs and cars.[90]

Television arrived in America's living rooms in the late 1940s and 1950s. By 1960, an estimated 88 percent of all households owned them, and by 1970, 96 percent did.[91] Some observers suggested that this was a positive development, for TV had the potential to unite people by offering them a common amusement.[92] And indeed, during TV's earliest days, viewing was a communal experience. Given the scarcity of sets, large numbers of friends, family, and neighbors would gather together to watch.

Pat Amino, a Chicago resident, recalled that her sister and brother-in-law were the first people she knew who had a TV. "So, every Friday, Saturday and Sunday, we would rush over there to look at this little TV. It must have been nine inches, right, in those days? About twenty of us are gathered around this little television, watching Lawrence Welk and different programs. It was fun."[93] Rainette Holimon, of Monmouth County, New Jersey, recalled similar gatherings in her neighborhood. "I remember when the first television came into the Black neighborhood. It was at the home of Dr. Carter, a Black medical doctor, who was very popular, and was very friendly. He had a huge basement, and whenever one of the Joe Louis fights occurred, he opened his doors to the people in the community. We just flocked to his house, those who could get in, just to see this television."[94] Televisions could bring together families and neighbors and seemed to be promoting sociability, at least in the eyes of some.

But others disagreed, saying TV was a destructive social force that left its viewers isolated. As TV sets proliferated, some observers suggested that they ceased to unite neighbors and instead isolated them. Bill Miller, a longtime resident of St. Johns, Oregon, in the Columbia River Basin, recalled how his town had changed. He had moved to St. Johns as a child, in 1924. He remembered the communal nature of town life when he was a boy and young adult.

> Well, everybody was trying to get along with everybody else, and you had your little Mom and Pop grocery stores there on the corner. Everybody left their houses open, toys out on the street, cars open, nobody stole anything because nobody had anything worthwhile to steal. They had what they called a five hundred club that they played, and they'd go from one house to another playing five hundred, or they go from one house to another doing Square Dancing—for entertainment—the older folks, my mom and dad.
>
> . . . And it was just wonderful. Everybody knew one another. Everybody knew one another's first name, the last name, and knew everything about everybody.

But he claimed such times were gone. And he put the blame on the isolating effects of modern household technologies. "The reason why you don't know your neighbors," he explained, "is because you got television, you got air conditioning and your homes heated so nice, you don't have to worry about going outside. You stay in there, and you see them at the store, and go 'Hi neighbor,' and that's about it. You may know their first name, but you don't know their last."[95]

Some contended that TV's isolating effects hurt not just neighborhoods but individual households as well. Even when families watched programs together, they might not actually be sharing much. As one critic put it, "It is true that the American family has been brought together in the evening by television, but the bringing together is much like that old genteel custom of building a family mausoleum and bring[ing] all the folks together that way. . . . With a television there isn't much more conversation than in entombment."[96] (Later generations of media theorists would claim such criticisms unfairly characterized TV watching as passive and stupefying, yet accurate or not, the critiques of television as antisocial and isolating continued—and continue—to circulate.[97])

TV's critics claimed that not only did television diminish family interaction, but that often family members did not even view it together. According to one 1975 tally, 60 percent of American households had at least two television sets, which the *Saturday Evening Post* claimed led to the "splintering" of the family. "Early articles about television are invariably accompanied by a photograph or illustration showing a family cozily sitting together before the television set. . . . Who could have guessed that twenty or so years later Mom would be watching a drama in the kitchen, the kids would be looking at cartoons in their room, while Dad would be taking in the ball game in the living room." The result of this was a less tightly knit family. The article lamented, "What has happened to family rituals, those regular, dependable, recurrent happenings that gave members of a family a feeling of *belonging* to a home rather than living in it merely for the sake of convenience, those experiences that act as the adhesive of family unity far more than any material advantages?"[98]

If television was perceived to be fracturing neighborhood and family life and thereby increasing opportunities for loneliness and isolation, the car brought other concerns. Since the 1920s, social critics had worried that the car might cleave families.[99] But those fears only grew in postwar America as more and more people took up driving. The car offered new ways to socialize; youth famously used cars and drive-ins as venues for dating and for associating with fellow teens. And cars could also help people get more easily to the Elks Club or to Christmas dinner with relatives. But in practice, autos encouraged people to live at greater distances from their work, from the places they shopped, and from the places they gathered for civic association. The lengthened commutes and the increased time that people spent in autos often came at the expense of other, more communal activities.[100]

As a consequence, the drumbeat of discussion about loneliness grew louder in the 1960s and 1970s. Just as *The Lonely Crowd* had fostered an initial scholarly interest in loneliness, another book magnified that trend two decades later. In 1970, Philip Slater published a best seller, *The Pursuit of Loneliness.* Slater placed the blame for loneliness firmly at the feet of Americans' individualistic tendencies and the technologies they used to support their independent lifestyles. He wrote,

> When a value is as strongly held as is individualism in America the illnesses it produces tend to be treated by increasing the dosage, in the same way an alcoholic treats a hangover or a drug addict his withdrawal symptoms. Technological change, mobility and the individualistic ethos combine to rupture the bonds that tie each individual to a family, a community, a kinship network, a geographical location—bonds that give him a comfortable sense of himself. As this sense of himself erodes, he seeks ways of affirming it. But his efforts at self-enhancement automatically accelerate the very erosion he seeks to halt.[101]

For Slater, these isolating technologies were embedded in suburban living and its associated affluence:

> We seek a private home, a private means of transportation, a private garden, a private laundry, self-service stores. . . . An enormous technology seems to have set itself the task of making it unnecessary for one human being ever to ask anything of another in the course of going about his daily business. Even within the family Americans are unique in their feeling that each member should have a separate room, and even a separate telephone, television, and car, when economically possible. We

> seek more and more privacy, and feel more and more alienated and lonely when we get it.[102]

Slater seemed to blame Americans for their loneliness, suggesting that if they were to veer away from their adherence to individualism they might be happier. This new view of the lonely, which blamed sufferers for their own condition, and which suggested that they were at fault for experiencing the emotion, was a dramatic departure from nineteenth-century understandings of the feeling.

If twentieth-century assessments of loneliness differed from nineteenth-century ones, so too did the suggested solutions. In *Democracy in America,* Tocqueville had described a people who had celebrated individualism but who at the same time sought to counteract that condition by participating in communal activities, such as civic associations and religious organizations. The Americans described in *The Pursuit of Loneliness* and similar works of the 1970s were supposedly equally individualistic. Postwar individualism was manifest in the production and consumption of cars, TVs, and other technologies of suburban living. But in the suburban landscape, there seemed to be no easy modern way to build community and cure loneliness, and as a result many looked for its solution in the psychology and self-help books.

A rapidly expanding array of books and articles were dedicated to the subject, as educators, journalists, psychologists, sociologists, and therapists of the era created a literature designed to offset loneliness.[103] For instance, some teachers believed they must prepare their students for lives filled with loneliness. Adele Stern, the vice principal at Paramus High School in New Jersey, weighed in on the subject in a 1973 issue of *Scholastic Teacher,* promoting the development of courses about loneliness and alienation aimed at teenagers. She argued that given the feeling's ubiquity, schools should help young people learn to cope with it.

> Since adolescents are concerned keenly with themselves as unique individuals, since they struggle with the notion of what kinds of persons they will become, since their actions and experiences daily define them existentially, they frequently suffer

> their loneliness silently and confusedly, not realizing that their peers and elders are or have been exactly where they are now. A mini-course on the literature of loneliness and alienation can give students insight into their own problems and some direction perhaps for their next behaviors.[104]

Stern envisioned a class in which students would explore the emotion through novels, films and poems. Course readings might include Albert Camus's *The Stranger,* Ralph Ellison's *Invisible Man,* which "touches on the nature of loneliness," as well as 1970s novels like *Jonathan Livingston Seagull,* in which the main character had "separated from his flock because he wanted to be braver and better than his peers." Through this literature, Stern hoped to prompt in her students "a recognition of situations which make people lonely," as well as the "ways loneliness and/or alienation may work positively for the development of the individual."[105]

Loneliness also made appearances in the popular press with eye-catching titles: "Chasing the Lonelies," "I'm Lonely, You're Lonely," "Christmas Can Be a Lonely Time Too," "The Lonely Youth of Suburbia," and "Seeking a Cure for Loneliness."[106] Martha Weinman Lear, writing in *Redbook* in 1980, proclaimed loneliness to be "a form of distress commoner than the common cold, and infinitely more upsetting. Yet we do not talk about it. We are somehow embarrassed. In the culture that invented back-yard barbecues, church suppers, potluck, popularity contests and block parties, who wants to admit loneliness? It is un-American. It does not exist." However, she believed, this was changing. "Recently, impressed by evidence that loneliness is its own creature, and that it lives with millions of Americans like a vicious household pet . . . experts have begun to study it."[107]

As Lear indicated, there was a new interest in the condition among academics, for by the end of the late 1970s, psychologists had begun to make loneliness a formal area of study and were touting its potential as a fruitful ground for research. In 1978, Dan Russell, Letitia Anne Peplau, and Mary Lund Ferguson developed a loneliness scale designed to measure who was lonely and how much loneliness they experienced. Users were asked twenty multiple-choice questions and once their responses were tallied, a loneliness score was reported. Loneliness was on its way to becoming a measurable and quantifiable psychological problem.[108]

The new interest in loneliness as a troubling psychological and social condition was even more in evidence at a 1979 conference on loneliness at UCLA, funded by the National Institute of Health and organized by Peplau and her fellow psychologist Daniel Perlman. Following the conference, the organizers put together a "sourcebook" on loneliness, which provided a rich compendium of loneliness theories, loneliness research, and loneliness therapies.[109] In their introductory remarks, Peplau and Perlman commented on the rapid growth in studies of the emotion since the 1930s. From 1932 to 1977, 208 professional publications on loneliness had appeared in English. But only 6 percent dated from before 1960, while 30 percent were published in the 1960s. The lion's share of studies had been published in the 1970s. As Peplau and Perlman proclaimed, "Today research on loneliness is flourishing." And they noted with satisfaction, "Now is a good time to be doing loneliness research."[110]

Although Peplau and Perlman were psychologists, sociologist Robert Weiss also attended the conference. Weiss had edited a volume on loneliness in 1973 in which he outlined a framework for studying the emotion.[111] He proposed that loneliness could be measured by reference to two social states: a person's attachment to a significant other (like a spouse), and the number of other social connections that the person might have. While Weiss was an esteemed expert at the conference, Peplau and Perlman proposed an alternative framework: in their view, loneliness was simply a "discrepancy" between the number of social connections people actually had and the number they desired.[112] Naturally the psychologists and the sociologists were unable to convince each other of the greater merit of their respective approaches. But the lack of closure, and the fact that both disciplines could claim the subject matter as their own, probably spurred further interest in loneliness studies.

As psychologists, Peplau and Perlman regarded the emerging loneliness research as shedding light on an emotion that had existed since the beginning of human time. But what was not said was how their work was very much an outgrowth of twentieth-century consumer culture. Behind the single-family home, the car, and the television stood a phalanx of building contractors, automobile dealers, and television salesmen who were more than happy to sell their wares to Americans eager to fulfill their dreams of the good life. But when their isolating side effects became manifest, consumers could then turn to the intellectual and therapeutic tonics offered by psychologists (like Peplau and Perlman), sociologists (like Weiss and Slater), or educators (like Stern).

Beyond the prescriptions of the "loneliness industry," Americans could also turn to technologies that were marketed as antidotes to isolation.[113] Communications companies continued to make appeals that played on people's fears about loneliness and their hopes for connection and friendship. A 1958 AT&T ad depicted a woman on the telephone in her living room with the night sky in the background. The ad promised that "a telephone call from out of town takes the blues out of the night."[114] A 1966 Bell advertisement featured a smiling bride with a headline targeting her parents: "When your children grow up and go their own way . . . keep them close by long distance. It's the next best thing to being there."[115] And, in 1979, AT&T rolled out its famous "Reach Out and Touch Someone" campaign.[116]

Writ large, these examples illustrate a story about the paradox of American progress and hedonic adaptation. As sociologist Eric Klinenberg suggests, more Americans are able to afford privacy and are consciously choosing to live alone.[117] And even with such living arrangements, Fischer argues in *Still Connected* that Americans are as connected as they have ever been.[118] But in spite of these trends, many still worry about loneliness. How can this be? One answer arises out of the loneliness frameworks that Weiss, Peplau, and Perlman had been promoting at UCLA in 1979. Loneliness may not be a condition that arises strictly from a paucity of social connections. Instead, following Peplau and Perlman, it may be impossible to overcome the emotion until people have bridged the difference between the number of social connections they actually have and the number they desire. If modern Americans really experience loneliness this way, then it is not enough to have more

FIGURE 2.4 In the mid-twentieth century, Bell Telephone celebrated the power of long-distance calls to assuage loneliness. *Courtesy of AT&T Archives and History Center*

options for social connection than ever before. What matters now, and has mattered since the advent of the American consumer economy, is whether people have as many friends and acquaintances as they believe they should, for the loneliness industry continues to ramp up expectations and accelerate hedonic adaptation.

FIGURE 2.5 AT&T's "Reach Out and Touch Someone" ad campaign was created in 1979, branding the telephone as a tool for maintaining friendship. It is one of the most famous American advertising campaigns of the twentieth century. *Courtesy of AT&T Archives and History Center*

From the Lonely Crowd to the Lonely Cloud

Modern technology and modern psychology have raised expectations for constant sociability and have made people worry more about loneliness. In an age when connecting is expected and purportedly easy, the failure to have fulfilling relationships feels all the worse. Psychiatrists

Jaqueline Olds and Richard Schwartz note that many of their patients feel lonely, and that the condition fills them "with deep shame."[119]

That shame comes because, supposedly, loneliness is easy to fix. For example, in 2014, *O, the Oprah Winfrey Magazine*, in a partnership with Skype, developed a campaign called "Just Say Hello," which promised a simple cure for loneliness. As the magazine's health editor proclaims:

> A quick hello, a warm smile, a friendly wave. These simple gestures can spark meaningful human interaction, which has proven to have a stunningly positive effect on your health and happiness. That's why *O, the Oprah Magazine* has partnered with Dr. Sanjay Gupta and Skype to get the world talking again. . . . We live in an era when we are more connected than ever. The average person's social media followers can total hundreds, if not thousands, yet many of us still feel disconnected. According to estimates . . . at any given time one in five people are lonely.[120]

To remedy that, the campaign suggested that people should greet strangers or reconnect with old friends and then post about the experience on Vine or Twitter. Oprah and company promised that through modern technologies like Skype and Twitter, and simple displays of good cheer, loneliness would be alleviated.

Similarly, Mark Zuckerberg, the founder of Facebook, helped create internet.org in 2013. That group's mission is to provide internet connectivity to the billions of people on earth who still did not have it. In a 2013 CNN interview, Zuckerberg said, "We just believe that everyone deserves to be connected and on the internet. . . . A lot of people think connecting five billion people is going to be challenging and it is, but I think it is one of the biggest problems of my generation."[121]

Without discounting the nobility of internet.org's underlying purpose, it is clearly in the financial interest of Facebook's founder to emphasize the human need for connection.[122] Moreover, the initiative also exacerbates anxieties about loneliness, especially when loneliness is understood as a discrepancy between the number of social connections that one currently has and the number that one desires. Of course, communication companies, whether old or new, have thrived on this discrepancy. AT&T exploited and called attention to it in the early twentieth century. But today's companies raise expectations further, suggesting that all one needs is goodwill and better infrastructure—better routers, faster connections—to make social life easy. This promise of frictionless social interaction often makes communications products enticing. For instance, Christian Wutz, the chief marketing officer for Tive, a wireless provider, explained that while he and his colleagues were not involved in marketing social media, their enterprise nevertheless benefited from the anxieties about sociability that Facebook, Instagram, and similar apps might create: "Do we capitalize on the anxieties? . . . Yeah, definitely. . . . The more people we have connecting, the more revenue we generate. . . . The stronger we can make the desire that you want to connect and check your things, then the more money we make. But I think we don't necessarily make people want access more; they already have it, we give them a means of getting that . . . [which] they haven't had in the past, so we make it easier for them to connect."[123]

Such longings for connection are not new. But it is only recently, with the zeal of Zuckerberg and internet.org as well as smaller tech companies, that increased connectivity has been turned into an entitlement—a desire, in other words, that tolerates no limits. For how much connection is enough? The implicit answer seems to be the more the better, a belief that writer William Powers calls "digital maximalism."[124] Such an outlook can lead to an unending quest for easier, faster connections and an ever greater number of social contacts, for the promise of the modern digital era is that once one has access to the Web and social media, friendship and sociability will be easy, and loneliness and isolation a thing of the past.

Triggers for Loneliness: The Quantity and Quality of Friends

Implicit in today's tech marketing campaigns is the idea that one can never have enough connections, enough friends, enough opportunities to reach out and meet new people. And it is clear that some

Americans—particularly young people—believe this and are trying to accumulate an ever-increasing number of followers and friends on social media. A study by the Pew Research Center offers a sense of these aspirations. The study finds that a significant portion of Facebook users have stunningly large numbers of friends. The average number of friends per user is 338, while the median number is 200. Fifteen percent of all users had more than 500 friends. The numbers varied greatly by age, with 72 percent of those older than sixty-five having one hundred or fewer friends, and 27 percent of those between the ages of eighteen and twenty-nine having more than five hundred friends.[125] Such figures indicate that definitions of what it means to be popular and sociable on the one hand or lonely on the other are in flux. An older generation clings to more modest expectations, satisfied with fewer connections. (Admittedly the modesty of the older generation's hopes may also reflect the reality of mortality, as friends die off, diminishing one's circle.) Meanwhile, the rising generation expects hundreds if not thousands of friends. While it is rare to see emotional norms change so dramatically, the twenty-first century offers one such opportunity, as exuberant hopes for sociability abound and as fears of loneliness simultaneously grow.

Our interviews corroborated these findings. Skyler Coombs, a college student, told us that his father felt he did not have enough friends and often felt lonely, especially when he compared his Facebook account to that of others. "My dad . . . sometimes he'll think he doesn't have a lot of friends. That's because . . . he's not . . . on Facebook and stuff all the time. . . . I think it does make him feel more lonely. . . . He thinks everyone has all these friends." Skyler suggested that when people studied other people's Facebook profiles, they would conclude "'oh, they have this many friends on Facebook, so they're obviously friends with all these people and they hang out with those people,'" and would then think "'why don't I have all these friends?' and kind of downgrade themselves and make themselves feel lonely."[126]

To avoid such feelings of social inadequacy, some, particularly the younger people we interviewed, believed they should try to accumulate as many online followers as possible. They were convinced that the more they had, the better, for those tallies would translate into an enhanced social life. They were often acutely aware of the number of their friends and followers. For instance, Cooper Ferrario, the nineteen-year-old marketing major, noted that "how many people follow you on Instagram, how many likes you get on your pictures . . . it affects your social ranking. . . . In high school where everybody knows everybody, . . . people know how many likes you get on your pictures; . . . people know how many followers you have; people are aware if you're someone who is active on Twitter or if you're not, and I would say that directly correlates with popularity. . . . In my experience, . . . there is a direct correlation. . . . The kids who are popular on social media, they're also popular in real life."[127] Chris Fowles, a twenty-three-year-old technology entrepreneur in Washington, D.C., said that while he worried about social media statistics less than he once had, he recognized that online popularity was "a part of our culture." He noted that there was "a threshold" of "Instagram likes or general . . . social interactions that you have to get," in order to achieve social success.[128]

In spite of these social pressures, many recognized that social media might be raising expectations and creating illusions about other people's social lives. Being on Facebook, Instagram, and Twitter had at first made them worry about the number of their online contacts, but their anxieties about such social statistics gradually diminished. For example, L. and her boyfriend Q. acknowledged that when they first became active on Facebook, they felt pressure to collect friends. But after a time that pressure eased: L. admitted that when she "was a first user," she worried about how many friends and followers she had. "I did feel . . . pressure to accumulate large groups of friends . . . online." But now, she noted, "I'm content with the community I have on there. So I don't seek to change it dramatically. Or try to improve it in any way." Q. said it took him "about a month" on Facebook before he stopped worrying about whether he had sufficient friends. At that point, he realized that he had the same friends online that he had offline "and I really stopped caring after that."[129]

Social media made many people worry about the quantity of friends, because it offered a tally, visible to all, of just how many connections they had. At first, a low number seemed to indicate a social life that was deficient and lacking. However, over time, many had come to worry less about the quantity of their online connections. They often did not speak with any precision about the number of friends they had on social media sites and often did not remember the last time they had counted. Yet while they might not have wanted more friends per se, they often displayed signs and intimations of loneliness in other ways. If the quantity of friends was not always a concern, the quality of friends and the way those online friendships translated into face-to-face relationships mattered a great deal.

The internet promised fast and painless sociability, but making true connections proved harder than expected. Ana Chavez, a twenty-three-year-old immigrant from Mexico, had grown up in a small, tight-knit ranching community in Michoacán. It was an impoverished, isolated settlement, and everyone shared possessions, from cars to a landline phone. That closeness and intimacy was impossible to replicate online, she said. "You can only get that by living with another person, not through social media."[130] Another transplant, eager for connection but not always able to find it on social media, was Christine Fullmer. A native of Philadelphia, she had lived for years on the East Coast before heading to the West. She had moved to Utah with her husband and children in 2013 and had been using social media to develop a new local network. As we sat in her garden, she told us that, at first, Facebook had given an illusion of friendship, making it seem like the world was populated with warm, welcoming, and familiar faces. Christine explained: "I think it gave me hope that I would quickly develop all these relationships, but then because everything was so dependent on social media, I became disappointed that that's all it was." Perhaps, in fact, social media had slowed down the process of developing close friends: "maybe it took so much longer than it normally would have because of this . . . almost lazy reliance on Facebook. It's a lot easier to just use Facebook than it is to actually take time to go see somebody." In Christine's view, talking on the phone would have been a better way to make connections. She remembered telling a friend that, on social media, "you don't get to know people and I just feel so shallow and

I just wish there could be more phone calls." She explained that on a phone call or in a face-to-face conversation,

> It's easier, it's more intimate, you get to know the person better. . . . I've been here for a year and a half. . . . I met so many people quickly, but the relationships were being built on Facebook, and to me . . . I'm not getting to really know that person; they're not really getting to know me from . . . posting cool articles on Facebook and sharing our reactions to them. You know it's not nearly the same level of connection. . . . I'm in speech pathology; the voice is crucial, the facial expressions are crucial, and you lose all of that.[131]

Some media theorists have argued that it's a mistake to distinguish between virtual and online friendships.[132] But whether or not there's an ontological basis for this distinction, for Christine it seemed tangible, and until she could actually turn her online friends into face-to-face ones, she felt lonely.

Triggers for Loneliness: Envy and Being Left Out

Like Christine, Hailee Money had moved to Utah from out of town. Only she had come from Aurora, Colorado, and whereas Christine was married and had two kids, Hailee was a teenager. The move, as she recollected, was full of loneliness and homesickness: "When I moved here . . . looking through my Facebook page would make me sad. . . . I would see . . . all the things they were doing over there and I felt like they had forgotten about me. . . . I didn't have any friends here." Just remembering that period in her life was painful. She said tearfully, "I don't really know why I'm crying. . . . So I was really sad at first. . . . I had a really hard time making friends and I would go home after school and look at Facebook and just wish that I was there so much. . . . It made me really sad. . . . But . . . when I was there I would look at my friends here. . . . Either place I was, I would be like 'oh, I wish I was there!'"

Hailee admitted that in her first months in Utah, "I hated looking at Facebook. . . .'Cause it made me feel really sad. So I told myself 'Don't look at Facebook anymore; that just makes you feel too sad.' So for a while I tried to not look at it." Hailee's description of her loneliness was the most poignant we heard. At one point she said that she even experienced "dread" at the prospect of logging in to Facebook. Fortunately, her feelings were ephemeral. As her connections in Aurora faded and her friendships in Utah grew stronger, her loneliness abated.[133]

Hailee's unease was compounded by distance. She measured her life in Ogden by reference to the lives of her friends in Aurora. In contrast, twenty-seven-year-old Anna Renkosiak, who lived in western Illinois, measured it by unfavorable comparisons to more proximate friends, some of whom had boyfriends and others who neglected to invite her along to social events:

> Before I met Justin and before I had . . . a boyfriend . . . you'd see people on Facebook and you'd . . . be sitting at home on a Friday night like "Oh, woe is me," and you'd see people being happy, people getting engaged . . . people out with friends . . . and that would make you lonely . . . but it's the comparison. . . . When I see things on Facebook like that, it's that comparison—they can have that, why can't I?

It was not just that Anna saw people with seemingly more full and rewarding social lives; it was that Facebook also kept her aware of what social gatherings she had not been invited to join. "Lonely is a good" way to describe her feelings, she said. "If I see . . . some of my friends out doing stuff and I'm not there 'cause I wasn't invited. . . . you sit at home, and, well, 'where's all my friends?'"[134]

L. recounted similar feelings of being left out when she logged in to Snapchat. "Your friends and your group of people will share some social event that they're all included in and you aren't a part of it, and that makes you feel a little left out."[135] The ability to know when one was excluded made loneliness harder to endure than it had been before these technologies emerged.

Some quit social media for this very reason. Eli Gay, the tax preparer living in Oakland, California, recalled learning from Facebook that an acquaintance was at the same concert she had attended, but then she realized her friend had not made an effort to meet up with her. "She never came and found me . . . and I felt really bad." That was painful, but it made her recognize that Facebook gave her too much information, which gave her "fodder" for constructing negative and not always accurate stories about herself and her relationships. "I think what I've realized is that I don't want to know that much. . . . We're not meant to have lenses into everything."[136]

Triggers for Loneliness: Counting Likes

The lenses that social media provided were many, and could offer provocative visions that might induce loneliness. They could serve as a way for users to witness friends participating in activities that they themselves were excluded from. They could heighten feelings of inadequacy in other ways as well. Individuals were acutely aware of how many "likes" they got on Facebook, Snapchat, or Instagram—how their messages and updates were received by a broader public. Just as Riesman's "lonely crowd" had been attuned to their peers and seemed to flourish or wither based on the opinion of others, the Americans we interviewed expended a great deal of energy wondering how others perceived them. When they received no responses to their posts, it could be devastating, as Alta explained: "I post stuff with them in mind and when I see they're not responding to it or liking it, it's like, 'Yeah, I know you're not following me anymore.' And then it causes just that more drama. . . . I'm too old for it."[137]

In an effort to avoid such scenarios, some young adults became very focused on posting only popular items. Cooper Ferrario, the marketing major, talked about the pressure his generation felt to craft the right image in order to be well liked. He knew many who asked themselves, "'Is this interesting enough? Do people care about this? Will people like this?' . . . and that determines whether they post it." Cooper made it clear that such questions were not frivolous, but had significant effects on their social standing and connections. "As a teenager," one's online image "definitely affects your social status."

Those who were mostly in their teens, twenties, and early thirties, eager to avoid loneliness and to find friends, had to scrutinize every word and post to make sure they were well received. They also lamented that they could not do the same in their offline, face-to-face interactions. Cooper noted, "I'm sure that social anxiety is increasing in my generation, [it] has to be, because we're so used to being able to communicate without having to say words or look at anyone's face." He admitted, "It's a mask, it's a barrier and it's protective."[138] Twenty-two-year-old Camree Larkin was grateful for that protection, for there were greater opportunities to edit oneself to make the desired impression. As a result, she found online interactions less risky than face-to-face ones: "I think there's more room for me to mess up when I'm with actual live people."[139]

But one might also draw opposite conclusions—that online editing made one less knowable, less approachable, more hidden. In the quest for popularity, young people crafted their words and Photoshopped their images, but by doing so they might be keeping others at a distance rather than building more intimate and genuine connections. Gregory Noel had noticed this. He had enjoyed using social media to keep up with his friends' and acquaintances' lives; however,after a point, he realized he was seeing only the "highs," but not the "lows," not the "entirety of their lives." It was after he disconnected from Facebook that he felt he knew his friends better, for he made an effort to call, visit, and text with them, to find out what was really going on with them. "Disconnecting has given me more of a reason to connect interpersonally with my friends."[140]

The Costs and Benefits to Sociability: Reactions Young and Old

At times during the interviews, some people seemed to manifest an unspoken desire to cut down on their social media use because they perceived that it might be increasing their loneliness and unhappiness. Hailee recalled a feeling of "dread" when logging on to Facebook; Alta observed that Facebook for her was often "more . . . hurtful than helpful." And a few returned Church of Jesus Christ of Latter Day Saints missionaries we talked with actually reported that their missions (which placed very strict restrictions on use of the internet) had made them more conscientious and mindful users. But on the whole, disconnecting rarely happened, because as poet JulieAnn Carter-Winward told us, posting and participating on social media seemed to many to be "a way out of loneliness."[141]

Some suggested that social media enriched and sustained existing relationships. Forty-eight-year-old Warren Trezevant, the software product manager from Oakland, California, believed that Facebook enhanced face-to-face interactions, for it gave him a "low-level knowledge" of his friends' lives, so that when they talked in person they could cut through all the old news and instead "engage . . . at a more current level."[142]

Elena Ellsworth, a retired nurse's aide in Colorado and a native of Hungary, believed social media had reduced her loneliness and her worries about her children, for she could check in on their doings on Facebook. She observed, "Definitely, it's connecting me with my children, any time I want to . . . and definitely the loneliness is a lot smaller than was before. . . . Well, before, like for three days, my children don't call, you know, they don't care about me, of course, I am old, you know. . . . they are busy." But once she was on Facebook, she knew what they were up to: "I find out more about their lives than I did before. . . . The phone, whenever they called, always had to be short . . . , 'Yeah, mom, I am ok, but I am at work, I call you back when I have time.'" But now, she said, "They put everything over there what they do. . . . I do think . . . the loneliness . . . and the anxiety is lower in my case. . . . I am more in communication with them than before."[143]

Social media might also bridge greater distances. Eulogio Alejandre, a charter school principal in Utah, celebrated the fact that it helped him stay in touch with family in Mexico. When he and his family had arrived in Utah in the 1970s, it had been "extremely expensive" and difficult to call his aunts who remained in Michoacán. The town in which his aunts lived had no telephone, so Eulogio's mother would have to call a phone in the nearest city and then have someone send a taxi out to his aunt to arrange a time for a call. On the appointed day, his aunt would take a taxi to the city to receive the call. Each call cost at least a hundred dollars, plus the price of two taxi rides. The contrast with the present was stark: "Now, I'm sitting at home and then I get a little message from Facetime: 'Cousin, can you get online? We got something to show you.' OK. So I get online, on Facetime, and they're showing me the new song, or the new stuff they're selling" in their store in Michoacán.[144]

Diana Lopez, the community outreach coordinator, noted similar contrasts between past and present. She had emigrated from Michoacán as well, from the small rancho adjacent to Eulogio's town. Although she came more than a decade after Eulogio, she too had faced difficulties when trying to stay in touch with her relatives in Mexico before the spread of cell phones. "I remember having to call our . . . town phone and I would say, 'Hey, so and so, can you go get my grandma?'" Then Diana would wait and, if she was lucky, a little while later her grandmother would call her back. In contrast, she reflected "now you just call a cell phone, you can text back and forth, you're on Messenger. I Facebook with cousins that are in Morelia all the time. . . . It's so much simpler to . . . have that . . . instead of . . . one town phone. . . . Now it's easier communication."[145]

Social media and smartphones meant even more to the undocumented in the community. Eulogio had a friend whose mother had recently died in Mexico. Returning there was prohibitively expensive and risky, "so he attended mom's funeral on a phone that somebody there was just showing everybody. . . . I think that has brought people together."[146]

There were those who reported that connecting through social media or texting had helped them gain new friends within the United States as well. Ana, for instance, recalled how it helped her navigate the rocky terrain of adolescent friendship. She had come to the States when she was eight. She got a cell phone at age fifteen, which helped her make friends and become less isolated. Because of the "language barrier," communicating in English had at first been a social impediment that made her "nervous," because "friends . . . would be judging how I speak." But when she texted, she could "look up a word," if she did not know it, which made it "easier for me to interact with classmates and easier to make friends."[147]

As we sat in her dorm lounge at Grinnell College, H., a twenty-one-year-old from Minnesota, pointed out other ways that smartphones and social media could help strengthen social bonds. H. had been homeschooled, but she took some high school classes in the Twin Cities, where she made friends. She explained: "For the longest time I didn't have a cell phone. My parents are very antitechnology and I lived about forty-five minutes away from most of my friends, so Facebook was a really great way to chat and keep up with people. . . . I think it definitely helped my social life and I would've been a lot more lonely without it, just because there wasn't a lot of people my age around my house. (I lived out in the country.)"[148]

While some appreciated digital technology's ability to foster friendship, others credited it with helping them find romance. Anna noted, "I actually met my boyfriend on Match.com." It was easier for her to meet someone that way because "in bar situations, it's hard for me to talk . . . and I get awkward . . . so it's hard for me to meet guys at bars." The online matching service had eliminated that initial face-to-face awkwardness.[149] Likewise, the Winwards, Kent and JulieAnn, believed the internet had made them, at least, less lonely. The couple had met online—they both commented on the same blog and began to exchange messages. As a result, when asked whether the internet made them lonely, Kent, a lawyer and writer, declared, "Obviously not, since we're together." JulieAnn added, "There are a lot of people out there who are incredibly lonely and they reach out on Facebook . . . the only way they know how to reach out, . . . and it's kind of this pseudo-connection that comes in this very small little space of words, and it feels for a moment like they're actually connected because they get a response to their loneliness and their misery."[150] Some admitted that this indeed was why they turned to social media and other online gathering spots. As an online dater confided to a *New York Times* reporter, "When I'm lonely, it helps to know there is someone out there who is looking for me."[151]

But such raw emotional need for connection could also make people vulnerable to commercial forces on the web, for loneliness could be easily exploited. Christian described an acquaintance who had gone on the dating application Tinder, not to find a date, but instead to find new followers for his YouTube channel. If he reached a sufficiently high number, he could reap profits from advertisers. Christian recalled, "He created a bot . . . that would just auto-like people . . . and was auto-liking people in France, Tokyo, Australia, South Africa, all across the world while he was sleeping, . . . swiping right, getting 12,000 matches . . . and then directing this traffic to his YouTube channel to create a following."[152] Other people's interest in him, as well as their own longing for relationships, could all too easily be commoditized. While people might turn to social media to escape loneliness, there was no guarantee that they would find true social connection.

Those who were older had the greatest reservations about social media. Unlike the younger generation, many were uninterested in making new contacts or corresponding with old friends in an online setting, and they worried somewhat less about missing out on what was being posted online. As N., a Honolulu attorney in his fifties, told us, "I don't feel that I'm out of it because I'm not in it."[153] Eli had for a time believed Facebook might offer her connection, but ultimately came to doubt this. She observed, "In . . . Facebook . . . there's a lot about believing we can access something . . . else, and you know I think it really feeds the FOMO [fear of missing out] thing." Because she no longer wanted to worry about parties she had missed or occasions she had not been invited to, she deactivated her account.[154]

And yet if this cohort was not subject to the direct effects of the digital loneliness industry as much as the younger cohort, it was still something that they were mindful of. William and Debbie Speigle, a retired couple living in Ogden, Utah, had recently unsubscribed from a social media site called Nextdoor.com, which places people into online communities that correspond to physical geographic neighborhoods. William explained why: "As far as being a member on something like that . . . we're face-to-face people. We love to talk, visit, socialize. Just not on a computer." When someone posts on

Nextdoor.com, Debbie explained, "You see a name and you don't know them. Whereas when we are out walking and they say 'hi,' . . . you stop and visit 'Where do you live? Oh . . . up here.' . . . You actually put a face and location and . . . a name . . . and at least you had some face-to-face contact instead of an anonymous sentence on a computer."[155]

Some thought that online contacts were also insufficient for sustaining close familial bonds. Mary Kaplan, an eighty-seven-year-old retiree in Bettendorf, Iowa, occasionally used email to stay in touch, but found it largely inadequate to her needs. She reported, "When you really come to think of it, emails are not discussions and you don't get the nuances . . . and it's like I can't be bothered. I want to hear my kids' voices," so she ended up giving her laptop away.[156]

Many people in their forties or older believed online communication gave the illusion of connection, but not the reality. Eli noted, "I do believe that Facebook makes people feel connected. But . . . in some ways it's just connection in your own imagination. Because you are in a room by yourself. With a computer."[157] Allen Riedy, a retired librarian in Honolulu, Hawaii, said of Facebook, "I don't find any satisfaction [in it]." And while email performed some vital services, it was not "an adequate substitute for actually seeing people."

Beyond articulating a dissatisfaction with online media, Allen felt that they were impoverishing the public spaces where he was accustomed to meeting friends and acquaintances. In the gym "everybody . . . has ear plugs on. They don't tend to pay attention to other people so much." And more generally, Allen observed, "My experience is I've found that . . . gathering places have kind of diminished in a sense. There's fewer of them. I think more people are kind of contacting each other via cell phones, email. . . . There aren't the common gathering places. . . . There aren't so many of them. . . . They're not as frequented as . . . they were in the past. . . . I think . . . [people] tend to do a lot of kind of one-to-one kind of contact . . . or maybe small group contact rather than going someplace and meeting people."[158]

Although sociologist Zeynep Tufekci argues that cell phones have counteracted some of the isolating effects that cars have had on

American society, Allen had not registered that in his own experience.[159] He believed that "the automobile is so isolating," keeping individuals "inside this metal machine." The effect was that drivers were "not interacting with anybody. . . . You're just kind of isolating yourself." But cars were not the only technology to foster loneliness, he confided: "A lot of electronic stuff is isolating in that way also."

Allen was not too forthcoming about how these developments were affecting him personally but there were occasional glimmers of the loneliness he might have been feeling: "I think they're replacing . . . human interaction with this electronic kind of stuff. It just. . . . diminishes things." In the next year Allen acknowledged he was going to leave Honolulu and return to the mainland to be closer to his family.[160]

That sense—that the public sphere was being diminished by digital devices—was common, and many feared that Americans were becoming increasingly alienated from one another, uninterested in a shared social life. Sitting in his living room, Jeff Hellrung, a sixty-year-old trucker living in South Ogden, Utah, told us he believed that an array of technologies, including but not limited to the phone, were separating people rather than connecting them.

> We're becoming less personal with people that we feel that we don't have to be personal with, like, "What's the point?—Why should I be personal with the bank teller? Why do I need to go in there and see her face-to-face . . . and interact with her? You know, I don't need to do that. They're not a relative of mine, you know. There's nothing that they're gonna do for me; I'm not gonna do anything for them." I think that's the attitude. . . . "Why interact with somebody on a personal basis that's really not part of my life?"

Jeff thought this was foolhardy. "I think there might be missed opportunities. . . . Maybe it's hurting us, you know, that we're starting to . . . get into our . . . own little bubble, not being concerned with anyone else."[161]

If technology is commonly represented as a tool that can be used for good or ill, in the eyes of many people middle-aged or older, it had clear effects. It was desiccating more traditional public spaces and limiting the number of face-to-face interactions that they had. They remembered other ways of relating that the rising generation no longer knew or placed much less truck in.

Yet even younger people, who used their phones in just the way the older generations decried, realized that this might be affecting the public sphere. Xavier Stilson, a twenty-one-year -old Utahan, reported, "I think that in my life and in the life of lots of people . . . you have Facebook on your phone and . . . you'll be at a party or you'll be sitting down with someone and . . . most people, they've got their phones out instead of interacting with people. I think that it hinders it more than it helps it."[162] L. reported that looking at one's phones as a way to avoid face-to-face interactions was a self-conscious and widely shared strategy: "We use it for, like, kind of an escape. . . . If you're anxious or sitting next to somebody on the bus you aren't familiar with, you'll pretend to be looking at your phone to appear busy or distracted, just kind of as an escape."[163]

Skyler often walked across campus with his head down, staring at his phone. He did it because "it makes the distance seem shorter," but also because "I see everyone else . . . is doing it too, and everyone else is trying . . . maybe not to talk to each other or something like that, so it's kind of an excuse when you don't want to talk to anyone too. So when they see you on your phone they . . . [think] 'oh, he's busy.'"[164] To be thought of as busy, as independent and yet still connected, as not isolated or lonely, was often a goal of those using these technologies.

The phone propped up the myth of individualism. It allowed one to appear independent, but simultaneously kept one from seeming too alone, for onlookers might assume that Skyler, standing by himself but staring at his phone, was connected to online friends, that he was popular on social media, if not on the quad. Yet the phone's real-life social effects were often the opposite. The inability to appear alone might indeed render one even more so, isolating people from those around them.

Phones also gave the illusion that one might be in two places at once—in one's actual location and somewhere else, in the ethereal world of Facebook or Instagram. Phones might distract from the dissatisfactions of the here and now, and promise online relief. But in taking people's minds away from where they really physically were, it might mean the actual was shortchanged for the virtual, for phones promote the idea that where one was did not actually matter, that one could be anywhere and everywhere. In so doing, phones suggested that one could find friends and connections easily anywhere, but the actual physical landscapes they created—of people standing next to each other, but looking down at their devices rather than up at their companions—were often lonely ones.

The Eroding Tolerance for Solitude

In an appearance on *Conan*, the now-disgraced comedian Louis C.K. described how addicted people are to texting (and by extension to each other), even when they risk mortal injury to themselves or to others:

> You need to build an ability to just be yourself and not be doing something. That's what the phones are taking away, is the ability to just sit there. . . .
>
> That's why we text and drive. I look around, pretty much 100 percent of the people driving are texting. And they're killing, everybody's murdering each other with their cars. But people are willing to risk taking a life and ruining their own because they don't want to be alone for a second, because it's so hard.[165]

Louis C.K.'s riffs have circulated widely and resonated with a generation that can still appreciate the fact that they are using communication devices that offer unprecedented levels of connection. Yet even while they have access to heretofore unimagined technological powers, they still struggle to use them in ways that balance the need for connection with the need to commune with their inner selves.

These needs seem hard to balance in the twenty-first century, for technology has accustomed us to constant, albeit virtual, connection. Literary critic William Deresiewicz observes that "a hundred text messages a day creates the aptitude for loneliness, the inability to be by yourself."[166] Emily Meredith, the nurse in Oakland, California, had seen such effects, particularly on young people around her. In her sun-filled living room, she reflected, "Younger people have lost their capacity for being alone, you know, to actually spend time alone and enjoy it. . . . If they're awake, they're connected. . . . If they're awake, they're in touch with somebody all the time. . . . They always know where so and so is. . . . Even if they're not on their device . . . they're always aware of where all their friends are, what they're doing. . . . They can't really experience solitude or tolerate solitude I think, and that seems like a loss."[167]

This flight from solitude represents a significant shift from earlier expectations about loneliness and social connection. In the eighteenth and nineteenth centuries, men and women often felt lonely, but they endured the feeling in order to live up to notions of piety and self-sufficiency. They believed there might be a redemptive value in solitary existence. Rebecca Dickinson, the Massachusetts spinster, saw God's hand in her lonesomeness, believing her circumstances taught her virtues. Nineteenth-century Americans likewise found loneliness difficult, but believed it might enrich faith, give birth to revelation, or bolster a sense of independence. They also sought relief by frequently visiting with neighbors and using the post, or more occasionally sending a telegram or placing a phone call. They did not expect a massive number of friends stretched across the continent, much less the globe.

By the twentieth century, cities with bulging populations and telecommunications companies with affordable phone and telegram rates began to raise expectations about sociability, suggesting that endless contacts across the nation, and eventually across the world, were available and desirable. If one did not have such a large network, was not sociable, one was a loner, a failure, a misfit. Yet as Riesman demonstrated, while the outgoing men and women of the twentieth century often garnered many social connections, these did not inure them to loneliness or sadness in an age when those emotions were increasingly pathologized. Today, Americans, eager for gregarious sociability, use a whole new set of technologies to connect with distant people, yet these tools have only compounded rather than helped transcend the predicaments of the other-directed type. In effect, "the lonely crowd" has become the lonely cloud.

As Americans have come to worry about, measure, and dissect loneliness all the more, their tolerance for aloneness has diminished. Loneliness signals want and need and failure, and is therefore a condition few wish to admit to. Yet while most try to avoid the stigmatized condition, they also fret more about it, because new technologies as well as changing views of human psychology have undermined the capacity to experience it as something positive—that is, as solitude. One telling indicator is a Google Ngram that reveals the declining use of the word "solitude," just as use of the term "loneliness" has increased.[168] Because Americans employ the concept of solitude less in their language than they once did, they are increasingly incapable of distinguishing the feeling from loneliness. As they have relabeled and reinterpreted the experience of being alone, and changed it from a common and possibly positive state to an inherently negative one, they have also changed the actual feelings that go with aloneness.[169]

Today, even though the national discourse is full of celebrations of individualists, Americans' daily behaviors indicate that they are anything but. American literature is brimming with tales of solitary individuals, from *Walden* to *The Martian*.[170] Yet while books and movies may celebrate the independent person as a hero, they do not accurately reflect Americans' feelings about being on their own. People stare at their phones, text, and post on social media precisely because they want to avoid loneliness and even the appearance of being solitary.

What then of the debate about loneliness and technology that is engaging contemporary pundits and academics? Some suggest that technology is shallowing out what in other contexts might be deeper, more meaningful contact. Deresiewicz calls this state "faux friendship," while Sherry Turkle, the renowned MIT social scientist, despairingly

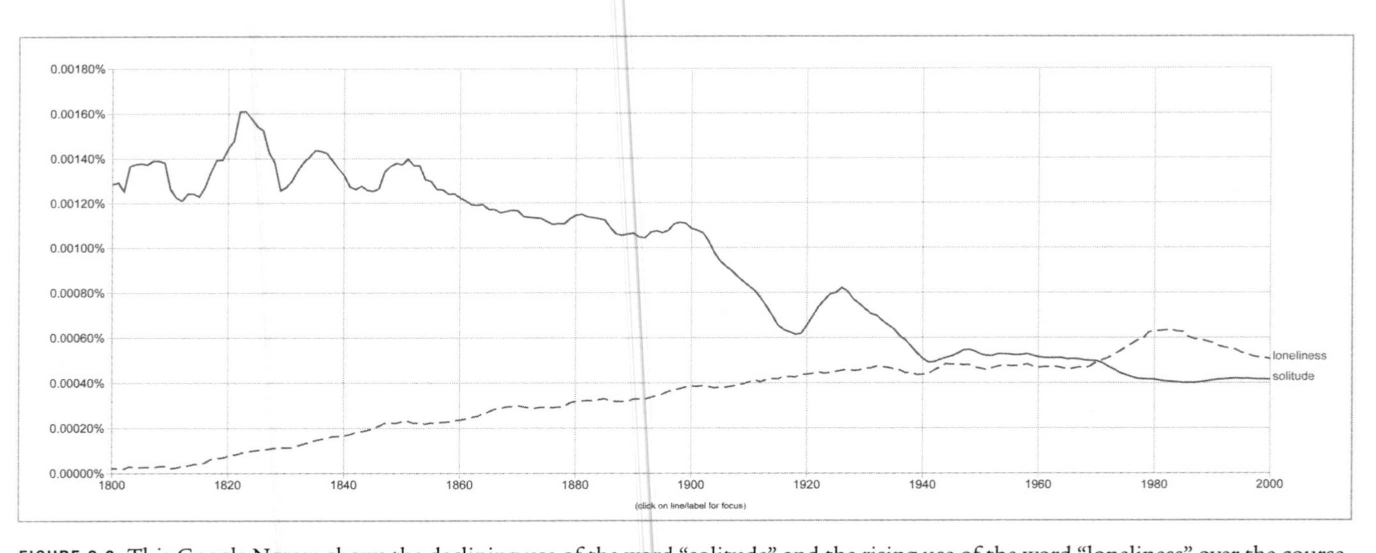

FIGURE 2.6 This Google Ngram shows the declining use of the word "solitude" and the rising use of the word "loneliness" over the course of the nineteenth and twentieth centuries. *Google Ngrams: http://books.google.com/ngrams*

refers to it as being "alone together." And in the viral video "Look Up" (which has been viewed over fifty-seven million times), the narrator begins with this lament: "I have 422 friends, yet I'm lonely. I speak to all of them every day. Yet none of them really know me."[171] On the other hand, many sociologists have discounted these worries, maintaining that such attempts to indict the internet as the perpetrator of loneliness smack of so much moral panic.[172]

Digital technology has undoubtedly changed the experience of loneliness, ramping up expectations for easy communication and connection, but often leaving users with unfulfilled hopes. However, it is impossible to say whether the incidence of loneliness has increased, for the emotion has not meant the same thing to different generations. The feeling nineteenth-century pioneers felt is not the same one twenty-first-century Facebook users experience. In reality, Americans feel loneliness differently than they did in the past, for the feeling has changed profoundly.

The experience of loneliness has changed in three major ways. Partly as a result of new technologies, from the radio to the smartphone, people are losing the capacity to appreciate solitude. Likewise, individuals' expectations about the number of social connections they should have and the geographical regions in which they reside are expanding rapidly and often unrealistically. And a growing loneliness industry encourages people to think of loneliness as a feeling that deserves redress, and it fuels the expectation that happy, sociable fulfillment is easy, always possible, and the norm.

Americans have walked lonesome valleys and endured lonely crowds. They founded Less Lonely Leagues and sought to stamp out loneliness in the digital cloud. They celebrated solitude on the side of Walden Pond and then, in a new century, turned their backs on loners and attempted to change the study of loneliness into a science. Where they once thought of loneliness as inevitable and unremarkable if also uncomfortable, today they see it as a dangerous emotion and a sign of embarrassing neediness. Modern technologies have made friendship, romance, and social connection look easier than ever, and therefore the absence of such relationships has become all the harder to bear.

Unlike earlier generations who believed the feeling was inevitable, many today believe it to be curable, a condition that can be mitigated by phones and computers. However, when those tools fail to bring the happiness, warmth, and sociability they promise—and to some degree, that failure is inevitable, given the outsized expectations—Americans have come to blame themselves.

CHAPTER 3

The Flight from Boredom

In May 1894, Edward Rich, a doctor in the Utah Territory, wrote in his diary, "A pleasant day. Dull! dull!! dull!!!" Such entries were typical in Rich's journal. Dullness seemed a part of life to him, as he noted the next month: "Nothing special Times awfully dull."[1] While he did not enjoy the monotony of his days, Rich seemed little motivated to alter them, and there appeared to be little he could do to change their texture and meaning. He was reconciled to the fact that some moments were unexciting. Such attitudes were common among nineteenth-century Americans. Life was often dull, time hung heavily, and, while not pleasant, few found that remarkable.

One hundred and thirty years later, *The Onion* published a satirical piece titled "Americans Demand New Form of Media to Bridge Entertainment Gap while Looking from Laptop to Phone." As *The Onion* noted:

> According to reports from across the country, citizens are loudly calling for a device or program capable of keeping them captivated as they move their eyes from a computer screen to a smartphone screen, arguing that a new source of video and audio stimulation is vital to alleviating the excruciating boredom that currently accompanies this prolonged transition.[2]

As the article suggests, Americans have become far less tolerant of boredom and dullness over the last century, particularly so over the last decade. Whereas in the nineteenth century, people were resigned to experiencing tedium and monotony in their lives, contemporary Americans are primed for constant change, novelty, and excitement. In fact, in the digital age, they have become so accustomed to continual stimulation that in a recent study, psychologists found that individuals would rather administer electrical shocks to themselves than sit bored and "alone with their thoughts."[3]

Despite this modern intolerance for boredom, the emotion is a feeling widely recognized in the United States today, and one that a majority of people have experienced. However it would have been an unrecognizable complaint in earlier centuries, for until very recently, the word "boredom" did not even exist. Only as people came to see solitude and empty moments as a problem, and came to see novelty and change as their due, did boredom become a widely recognized problem. The terms "bore" and "boring" only appeared in the English language in the eighteenth and early nineteenth century, respectively, while "boredom" made its debut in the mid-nineteenth century.[4]

Admittedly, there were feelings and words prior to 1750 that were in some ways similar to modern boredom. Historians generally trace the genealogy of boredom back to "acedia," a word coined by the ancient Greeks to describe listlessness.[5] Early Christians used the term to refer to the feeling that monks experienced as they retreated to the desert. They sometimes also called the condition "sloth" or the "noonday demon." Although monks were in the desert to keep their minds focused on God, the unchanging nature of their days often distracted them. As Evagrius, a fourth-century monk, explained,

> The demon of . . . [acedia], also called "noonday demon," is the most oppressive of all demons. He attacks the monk about the fourth hour and besieges his soul until the eighth hour. First he makes the sun appear sluggish and immobile, as if the day had fifty hours. . . . He stirs the monk also to long for . . . a much less toilsome and more expedient profession. . . . To these thoughts the demon adds the memory of the monk's family and of his former way of life.[6]

Acedia was somewhat different from modern-day boredom. As historian Ute Frevert noted, acedia was induced by a demon, who attacked those lacking moral strength and resolve. Church fathers perceived the feeling as caused by external powers that preyed on the weak, unlike modern perceptions of boredom as an inner state.[7]

Another distinctive feature of acedia was its victims. According to medievalist Siegfried Wenzel, Christian acedia was initially connected to one place and one type of person: the desert-dwelling monk who was losing interest in his calling and his relationship to God. When monks moved from the desert to newly created monasteries, acedia moved with them.[8] In the twelfth century, acedia spread beyond the cloisters and began to be identified as a sin that laypeople could commit as well. Unlike other deadly sins such as envy or avarice, acedia continued to be explicitly linked to religious duties. If individuals failed to contemplate God and were distracted from prayer, they might be guilty of it. Yet it did not have import outside of this religious world, meaning that as secular contexts began to emerge during the Renaissance and Enlightenment, acedia gradually lost relevance and currency.[9]

It was in these new circumstances that Europeans developed new concepts of—and words for—mental listlessness and lassitude. In the twelfth or thirteenth centuries, the French developed the term "ennui," which came from the Latin word "odium," but which gradually evolved to mean "to annoy." That definition continued to change, and the French began to use ennui to signify melancholy weariness and mental languor. The term entered the English language in the eighteenth century and was used to describe the condition of those with too much leisure and too little to fill it.[10] It was therefore generally a condition linked to wealth and privilege.

Other terms describing mental torpor emerged as well—dull, tedious, monotonous, drudgery—words that were often employed to describe days, labors, and landscapes. In the eighteenth century, "bore" began to be used to describe a person; in the mid-nineteenth century "boredom" emerged as a term to describe a state of mind.[11]

Perhaps boredom caught on and gradually replaced acedia and sloth because the latter were born in religious contexts, and a new, secular

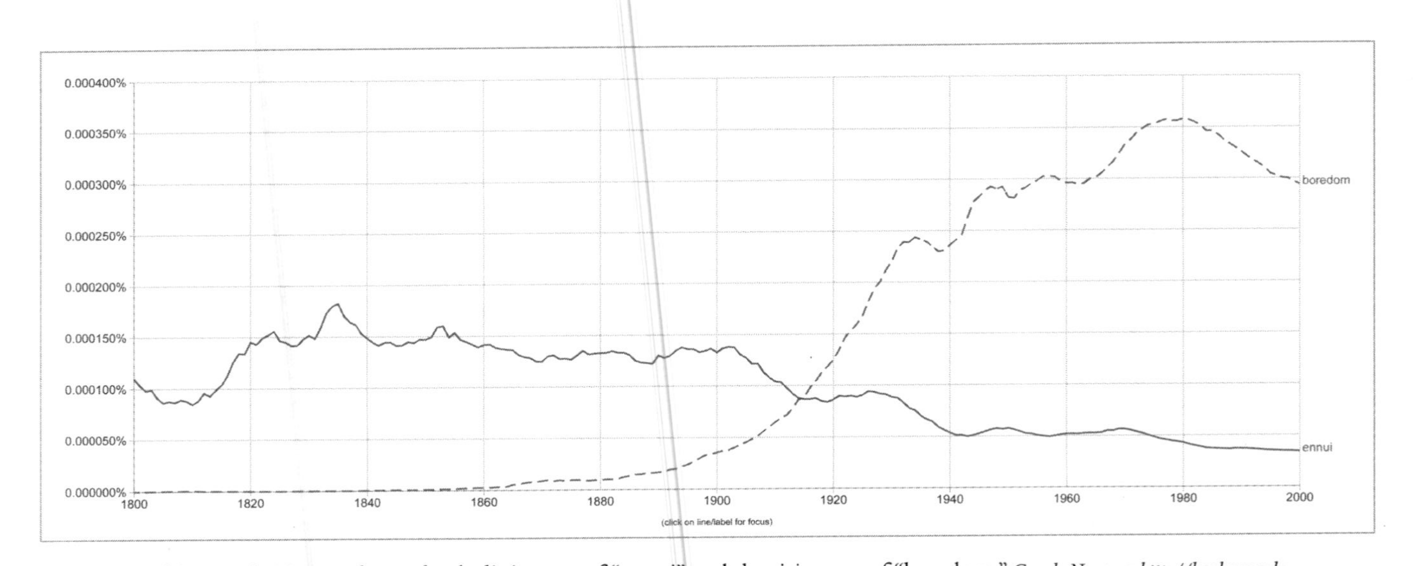

FIGURE 3.1 This Google Ngram charts the declining use of "ennui" and the rising use of "boredom." *Google Ngrams: http://books.google .com/ngrams*

word was needed to describe a feeling that was not always wrapped up with spiritual import. After all, acedia referred to a deep form of spiritual malaise while simple boredom did not. Boredom, at least in its initial usage, could describe an unpleasant and transient state in which a person perceived time as passing slowly and was unable to focus on an engaging activity.[12] And because it had more democratic connotations, boredom also gradually supplanted ennui, which was generally regarded as an aristocratic complaint.

Elite Ennui in America

Before "boredom" was coined, however, colonial elites used the term "ennui" and did their best to avoid the state. Thomas Jefferson offered his family instructions about what to do to stave off the feeling. When his daughter Martha was only eleven, he sent her a schedule so that she might stay busy and avoid the problem of empty time. He suggested the following:

> With respect to the distribution of your time, the following is what I should approve:
> From 8 to 10, practice music.
> From 10 to 1, dance one day and draw another.
> From 1 to 2, draw on the day you dance, and write a letter next day.
> From 3 to 4, read French.
> From 4 to 5, exercise yourself in music.
> From 5 till bedtime, read English, write, etc.[13]

Throughout her adolescence, he continued to worry that she might fall prey to ennui. In March 1787, he wrote to her, extolling the virtues of industry and reminding her to avoid indolence: "Of all the cankers of human happiness none corrodes it with so silent, yet so baneful an influence, as indolence. . . . Idleness begets ennui, ennui the hypochondria, and that a diseased body." To prevent such a condition, he recommended "Exercise and application," which "produce order in our affairs, health of body, cheerfulness of mind, and these make us precious to our friends." He warned Martha, "If at any moment, my dear, you catch yourself in idleness, start from it as you would from the precipice of a gulf."[14] Jefferson continued to worry about his daughter's daily habits, for two months later, in May, he wrote her again, encouraging her to develop virtues that "will guard you at all times against ennui, the most dangerous poison of life."[15]

Other colonists admitted that they too had to work hard to ward off ennui. John Adams complained of it in 1794, while he served as vice president, writing his wife that he wished he could join political clubs (which were discouraged in the new republic), for they "would save me from ennui of an evening, which now torments me."[16] When Jefferson succeeded him as president, and Adams was forced to retire from politics in 1801, Adams worried that he would miss the work of governing: "The only consolation I shall want will be that of employment. Ennui, when it rains on a man in large drops, is worse than one of our north-east storms; but the labors of agriculture and amusement of letters will shelter me."[17]

Jefferson and Adams were privileged members of the early republic. Their problems with ennui came from too much empty time. Those who enjoyed—or suffered from—similar leisure often shared their concerns about the feeling, for there was a widespread belief that ennui could cause pain, discomfort, and a host of other problems.

This idea was supported by medical literature. Up through the mid- to late nineteenth century, such texts used the word "ennui" to describe the troubling condition; thereafter, they also sometimes used boredom as a synonym. The French physician F. J. V. Broussais wrote of ennui in *A Treatise on Physiology Applied to Pathology,* which was translated and printed in the United States in 1826. In it he suggested that ennui might cause coldness, a feeling of "emptiness" and "uneasiness."[18] The Scottish physician Andrew Combe, whose works were reprinted in the United States, likewise described ennui's physical consequences: "*non-exercise* of the brain and nervous system, . . . inactivity of intellect and of feeling, is a very frequent predisposing cause of every form of nervous disease."[19] In his 1851 study of mental diseases and asylums in

Ohio, D. S. Welling traced how ennui might lead to insanity and even death: "Ennui is that tiredness, that disgusted and self-consuming state of mind, which marks those who live without occupation, or innocent engagement." This "painful and dangerous state of mind" could strike "the idle and profligate, as well as the luxuriant and wealthy pleasure taker." He also worried about those who engaged in "revery." He laid much at the feet of these twinned disorders, noting, "Nine-tenths of the miseries and vices of mankind proceed from indolence."[20]

As Welling noted, ennui was linked to both vice and disease. Some believed vice caused the feeling, while others maintained the feeling caused vice. Yet inevitably, whether ennui caused moral or physical debility or was caused by it, the consequences were serious. A journalist using the newly popular term "boredom" noted in 1875 that "boredom makes more patients than fever," while an 1894 article proclaimed "Ennui is one of the most powerful causes of ill health."[21]

Given its serious consequences, commentators and physicians tried to isolate its causes. Some blamed poor habits. Welling pointed to "card playing," which "dissipates the mind, leads to irregular hours," and which in turn produced "ennui, uneasiness, and feverish discontent."[22] Louis Kahn, in his 1870 work, *Nervous Exhaustion,* believed masturbation led to ennui, which in turn might lead to suicide.[23]

While some believed bad habits caused ennui, others claimed ennui led to bad behavior. Norman Shanks Kerr, a British doctor and temperance advocate, concluded that ennui led to "inebriety."[24] The *New York Tribune* agreed, running the headline "Ennui the Chief Incentive to Drink."[25] Ennui might also lead individuals to seek diversion in other vices as well. An 1888 article suggested that ennui motivated "indiscriminate novel-reading, flirtation, and scandal."[26]

Many doctors and political observers took comfort from the fact that ennui and, later, boredom were relatively rare in the United States. They believed such mental torpor was a condition that affected those with too much leisure. America, lacking an aristocracy and supposedly populated only by hardworking individuals, therefore saw little of this problem. In 1893, the *New York Tribune* termed ennui "A Misunderstood Malady from which Americans are Exempt, and Royalty the

Greatest Sufferers." It explained, "The majority of Americans are far too busy to experience the sentiment, while the leisure class, still in an embryo state are as a rule too new to their conditions to have the feeling." The *Tribune* therefore claimed that "Americans are absolutely exempt" from ennui, because "everybody is striving and rightly so, for something better. Now, as long as there is this feeling, there can be no satiety, and without satiety there is no ennui."[27]

In contrast to America, the condition was rampant in nations with entrenched aristocracies, and various well-known nobles were diagnosed with the condition. The *Tribune* suggested that the Prince of Wales was suffering from it, while throughout the nineteenth and early twentieth centuries, a host of other aristocrats were presumed to be wrestling with it as well.[28] "Just Tired of Life So He Dies; 'Sheer Boredom' Causes Parisian to Shoot Himself before His Guests," announced the *Daily Missoulian,* detailing the suicide of Fernaud Ravenez, a young society man, who killed himself in Monte Carlo. "Crown Prince Dying? Says It Is Boredom," read a 1921 headline describing the plight of the former Crown Prince Frederick William of Germany.[29]

Occasionally some accounts explored rare American cases. Two groups were singled out as vulnerable to it: social elites and privileged women, both of whom were presumed to have too much leisure in their lives. The *Arizona Republican,* for instance, called ennui "the great bugbear of society people."[30] Likewise, women who lived lives of comfort were naturally prey to feelings of uselessness and boredom, for they supposedly had little to do during the day. The *Virginia Enterprise* warned that young wives might feel their lives were "dull and lonesome." The paper counseled, "Avoid idleness, is the best advice that can be given to or taken by a young wife, for work is the only real talisman against boredom, ennui and hysteria."[31]

Sometimes commentators portrayed the condition as the burden of or perhaps even punishment for a life of wealth and ease. Framing it this way was an attempt to console those who did not have the luxury of riches, leisure, and ennui. The *Palatka News and Advertiser* described "the terrors of ennui," reminding readers that while they might believe that "wonderful possessions make for happiness," the reverse was true.

"These things never bring happiness. Never, never, never! . . . They bring, on the contrary, sadness, weariness, and ennui. Oh, is there anything on earth as awful as ennui? It bites into the very soul and saps life of all that is worthwhile. It is worse than sorrow, worse than sickness, worse than death."[32] Better to be poor and hardworking with full days, than to be wealthy, leisured, and bored.

Such attitudes were an outgrowth of the Protestant work ethic. As Max Weber argued, prior to the Protestant Reformation, work was not vested with spiritual import. Laborers in medieval Europe treated work as something that put food on the table and ensured their survival, but it was not an activity that held intrinsic religious or moral meaning. But with the rise of Protestant theology, perceptions of labor changed. Many began to consider work an activity that not only produced a living but as something that signaled personal worth and one's prospects for salvation. Work took on new meaning.[33] In America (and particularly in the North, where free labor was most celebrated), from the colonial era into the nineteenth century, industry, in all its tedium, was a sign of virtue. In contrast, those who did not work, those who suffered ennui, seemed suspect, different, perhaps deserving of the torture of mental lassitude and meaninglessness that came with a life of ease.

Monotonous, Tedious Work

Consequently, while some regarded the rich people who suffered ennui with a mixture of contempt and pity, in the first half of the nineteenth century, relatively little sympathy or concern was afforded workers who encountered tedium in their lives and labors. And such people were legion. Many Americans indeed found their days too full of work, not too empty, as the rich did. And they often found this labor grinding and dull. Yet their mental condition provoked relatively little discussion, either among commentators or among the working population itself.

The dullness that working people experienced differed qualitatively and linguistically from that which wealthy elites felt. The former had to cope with the tedium of their work, while the latter grappled with the feelings that came with too little work. Laboring people in the first half of the nineteenth century rarely used the word "ennui"; instead they employed "dull," "tedious," "wearisome," "monotonous," or "drudgery" to describe their labors, but, significantly, not themselves. Work or days or tasks were dull and tedious; an individual was not. Gradually, however, the new word "boredom" crept into their vocabularies and by the late nineteenth and early twentieth centuries, the word served as a way to describe an individual's inner state.[34]

Farm life was famously dull. Daniel Drake, who recalled his childhood in frontier Kentucky in the late eighteenth century, wrote of the repetitive tedium of his manual labor. Harvesting Indian corn and then transforming it into hominy was a memorably monotonous process. When he had to use a pestle to crush the corn, his mind "went up and down with the pestle. . . . [T]he pestle . . . still continued to move up and down, forever up & down!" He complained that "its power in developing the mind is not equal to its efficacy in developing the muscle of the arms."[35] James R. Stewart, a settler in Kansas, wrote of the long hours he spent establishing a farm. He wrote in May 1855, "Worked through the day, fidled [*sic*] read & wrote some, Saw, heard, nor experienced nothing uncomon [*sic*]." A month later, he recorded in June 1855, "Made rails, read, & built air castles, saw no unusual sights, heard no unusual sounds, did no unusual feats."[36]

If free people found their days and work tedious and repetitive, those who had to do forced labor found it even more so, for there was little respite from their efforts and when there was variation, it was often for the worse. Charles Ball recalled the labor he did while enslaved in Georgia. He recounted that except for the occasional trip to Savannah for his master, "We passed this winter in clearing land, after we had secured the crops of cotton and corn, and nothing happened on our plantation, to disturb the usual monotony of the life of a slave."[37] James Williams, a slave born in 1805, recalled, "The history of one day was that of all. The gloomy monotony of our slavery was only broken by the overseer's periodical fits of drunkenness, at which times neither life nor limb on the estate were secure from his caprice or violence."[38] Yet slaveholders maintained that their slaves were contented with their situation because they had all they could long for. As one writer noted in 1850, "We have always believed that there is no laboring class in the world as happy, contented and comfortable as the negroes of the South."[39] After the Civil War, a former slave owner claimed that African Americans were both physically and mentally "better adapted to the drudgery of farm work than any other race of people."[40] Believing that slaves were mentally suited for repetitive and backbreaking work, slave owners felt little concern about the way their bondsmen and women experienced the oppressive tedium of slave labor.

Although they had far more control over their bodies, their actions, and their time than enslaved people, the earliest industrial workers also encountered monotony in their daily routines. Young women who began to labor in the factories at the Lowell Mills in Massachusetts in the 1820s, 1830s, and 1840s were surprised by the unrelenting rhythm of their days.[41] Lucy Larcom, perhaps the most famous of the mill girls, recorded in her autobiography that "sometimes the confinement of the mill became very wearisome to me."[42] In the *Lowell Offering*, the literary magazine the young women produced, another worker, who signed her name "M.," described herself as "wearied with the dull monotony of a factory life," and eager for "relief from the flying machinery." She felt like "a long-caged bird."[43] In a less romantic mood, Malenda M. Edwards wrote to her niece, Sabrina Bennett, in 1839, describing her experience in New England factories. She confided, "I would not advise any one" to quit a position in order to take a job in the factories, "for I was so sick of it at first I wished a factory had never been thought of."[44]

The workers at Lowell who found their jobs tedious and dull often discussed their heightened awareness of time and how it passed while they labored. They commented on the bells that divided their days, regulated their work, enforced their schedules, and increased their boredom. For instance, a Lowell girl who called herself "F. G. A." described the monotony of the work in a short story published in *The Offering*. The heroine, Susan Miller, was just starting at the factory: "at first, the sight of so many bands, and wheels, and springs, in constant motion, was very frightful. . . . [T]he day appeared as long as a month had been at home." She eventually became accustomed to the work: "Though the days seemed shorter than at first, yet there was a tiresome monotony about them. Every morning the bells pealed forth the same clangor, and every night brought the same feeling of fatigue."[45] This generation of workers was the first to feel industrial time discipline and it was not always easy or pleasant to adjust to it. Some claimed it worsened their feelings of tedium.

Boredom and the more precise measurement of time did not always follow a steady and proportioned march. Sometimes, in fact, the relationship seemed reversed. For example, a short piece in *The Offering* described girls who had moved from the country to the factories. A worker named Ellen declared: "I am going home, where I shall not be obliged to rise so early in the morning, nor be dragged about by the ringing of a bell, nor confined in a close, noisy room from morning till night. I will not stay here." She lamented, "Up before day, at the clang of the bell—and out of the mill by the clang of the bell—into the mill, and at work, in obedience to that ding-dong of a bell—just as though we were so many living machines. I will give my notice to-morrow: go, I will—I won't stay here and be a white slave."[46]

While sympathetic to Ellen's complaints, the narrator reminded her that the life they had left behind in the country was hardly less dreary than what they had in the mills. She observed that in the country "people have often to go a distance to meeting of any kind—that books cannot be so easily obtained as they can here—that you cannot always have just such society as you wish. . . . [In the country] they have what is worse—and that is, a dull, lifeless silence all around them. The hens may cackle sometimes, and the geese gabble, and the pigs squeal"—but that was all there was for diversion in the countryside.[47] Subject to the factory bells, Ellen longed for the less regimented rhythms of the country. Yet while she found industrial work tedious and the time discipline wearying, she also realized that days in the country, while ostensibly off the clock, would actually move more slowly and be more dull.

It was not only the bells that bothered nineteenth-century Americans, it was also being subject to the schedule of others that led to complaints about monotony and feelings of despair. Tally Simpson, a South Carolina soldier fighting for the Confederacy during the Civil War, complained of this. He wrote home to family members, again and again lamenting the tedium he endured. To his sister Mary he confided, "The dull routine of camp life continues daily, and I am becoming entirely disgusted with anything that pertains to this form of life. Drill, drill, drill; work, work, work; and guard, guard, guard. Eat, e-a-t."[48]

Although they did not enjoy dull times, nineteenth-century Americans also seemed somewhat resigned to them and did not panic when they occurred. Instead, they devised solutions to their boredom, often turning to their own inner resources. To enliven her day, Larcom posted poems and newspapers next to her machine, and she asked to work near windows overlooking the river, so that she could gaze outside.[49] An operative who identified herself as Fiducia put her thoughts in verse:

> O swiftly flies the shuttle now,
> Swift as an arrow from the bow;
> But swifter than the thread is wrought,
> Is soon the flight of busy thought;
> For Fancy leaves the mill behind,
> And seeks some novel scenes to find. . . .

Fiducia's mental wanderings took her

> . . . to lands before unknown;
> I go in scenes of bliss to dwell,
> Where ne'er is heard a factory bell.

Her imagination took her on a grand tour, but finally led her to dark images of death and decay, making her realize that while mental wandering might bring some relief from the monotony of work, it brought frightful images as well. She concluded, therefore, that despite the regulated work day, she would resist mental escapes:

> O Fancy! now remain at home,
> And be content no more to roam;
> For visions such as thine are vain,
> And bring but discontent and pain.
> Remember, in thy giddy whirl,
> That *I* am but a factory girl;
> And be content at home to dwell,
> Though governed by a "factory bell."[50]

In short, she decided to endure the monotony of her daily work rather than distract herself with dark fantasies.

Not all reveries lead in such a doleful direction, and others working in early factories found that their mental escapes from tedious work could be enlightening. A mill girl named Sarah Bagley expostulated on the "pleasures of factory life." She explained that because the work was repetitious, workers could concentrate on other things. She asked, "Where can you find a more pleasant place for contemplation? . . . [H]aving but one kind of labor to perform, we need not give all our thoughts to that, but leave them measurably free for reflection on other matters."[51]

The mill girls' methods for dealing with dullness were common across the nation. Out in the countryside, James Stewart, the Kansas settler, coped with dull days by turning to his own thoughts. He recorded in July 1855, "I took a walk over to Wills house, found no one there, came home, fiddled built air castles."[52] Tally Simpson, the bored Confederate soldier, described some of his own tactics in an 1862 letter to his aunt. He wrote, "The times are dull and I amuse myself thinking of the gals and reading. I have thought of several gals and read several books lately."[53]

These nineteenth-century Americans seemed resigned to monotony. Dull, repetitious work was inevitable, and one did one's best in the face of it. Turning inward, to daydreams, to mental wanderings, to air castles, was a common response. Unlike the wealthy who struggled with ennui, workers harbored fewer fears about how to cope with the feeling or what it might do to them morally or physiologically. Perhaps because it was the result of virtuous labor, rather than viceful, luxurious leisure, their mental torpor did not constitute the moral threat that ennui did. Indeed, some claimed dull work offered moral benefits. The Unitarian minister William Channing Gannett composed a paean to monotony in his popular 1890 pamphlet, *Blessed Be Drudgery.* There he claimed "without much matter what our work be . . . it is because and only because of the rut, plod, grind, hum-drum in the work, that we at last get . . . attention, promptness, accuracy, firmness, patience, self-denial."[54] Like Gannett, many accepted dullness as a natural and even virtuous part of life and work; it was an experience that individuals could and should cope with, using their own resources.

Industrial Monotony, Industrial Entertainment

Yet this spirit of resignation was gradually challenged. As industrialization increased in the late nineteenth and early twentieth centuries, manual work became more regimented and more taxing, both physically and emotionally. More Americans also began to work for others rather than for themselves, so their work and leisure, their time on the clock and off the clock, became separate categories. At the end of the nineteenth century, a popular demand of industrial workers was "eight hours for work, eight hours for rest, eight hours for what we will," evidence that in the minds of many, labor and free time were becoming increasingly distinct. Indeed, the Eight Hour movement's song declared, "We mean to make things over; We're tired of toil for naught; We may have enough to live on, But never an hour for thought."[55]

Few who knew the reality of industrial labor could deny that it was physically taxing and often mentally deadening, and it seemed to be growing more so as the nation's factories adopted new technologies. Journalists recorded garment workers' discontent about increased monotony as their work was divided into piece work. Female glove makers disliked the idea of doing just one "kind of work by itself, nine hours a day." They tried to hang on to "the variety . . . change, contrast, interest," the "mental luxury" of diversified work, but to no avail.[56] A few years later, in 1913, when Henry Ford instituted the automobile assembly line, his workers protested loudly. Ford, like many other manufacturers, had consulted with efficiency expert Frederick Winslow Taylor and then used Taylor's time and motion studies to codify the minutes and seconds that workers should require for set tasks. At first, many rebelled at the assembly line and this routinization of labor. As one of Ford's biographers recalled, "So great was labor's distaste for the new machine system that toward the close of 1913 every time the company wanted to add 100 men to its factory personnel, it was necessary to hire 963."[57]

Ford himself, however, defended the assembly line, arguing that the perception that mass production increased workers' boredom was inaccurate. He claimed that while there was much discussion of "the monotony of repetitive work," in reality "this monotony does not exist as much in the shops as in the minds of theorists and bookish reformers." He admitted that the assembly line increased the "repetitive quality" of work, but he argued that this did not mean workers were more bored than they had been before, for he suggested that in his factories, workers encountered "frequent changes of task." Further, the net benefits surely outweighed the disadvantages, for he claimed that factories like his led to the "increasing supply of human needs and the development of new standards of living." He boasted that his systems, and others like them, made possible the "enlargement of leisure, the increase of human contacts, [and] the extension of individual range."[58]

Autoworkers subject to the discipline of the assembly line did not necessarily agree. Clayton Fountain migrated to Detroit, lured by "visions of hundred-dollar paychecks," and "daydreams full of new suits and Saturday nights with a pocketful of dough to spend in the speakeasies." His first job was as a "helper of a square shear," which had "no other qualifications than the traditional 'strong back and weak mind' necessary for most unskilled occupations in industry." The glamour of his dreams wore off the first night, as he worked piling up sheets of steel. "This was my introduction to the monotony of most mass-production operations in a big factory." As he worked, he was aware of the slow march of time: "I thought six o'clock would never come—and before it did, at five-thirty, the straw boss came by and told us to work overtime until seven-thirty." Tired and groggy when his work was finished, "I no longer dreamed of new suits and excursions to speakeasies; all I could think of was that wonderful soft bed waiting for me." As a result of such days, "the romance of working in an auto plant soon began to wear off. I began to think of getting into another business, one that could provide excitement and adventure in addition to substantial paychecks." He quit his job, with hopes of becoming an aviator, but eventually returned to the auto factories.[59]

Like Fountain, workers employed in factories of all types found the work increasingly boring as it became more and more regulated by efficiency algorithms and the tyranny of the clock. Labor reformer Florence Sanvill documented how widespread this perception was when she worked undercover in factories in the 1920s. She found the work dreary. Moreover, it was made drearier the more closely that time was watched:

> The work which fills these crawling minutes is not absorbing—it is not even interesting. . . . In fact, I found that the passage of the time becomes the most absorbing question of the day about an hour after work has commenced. In mills where the employer failed to provide a clock I quickly found that, as the discovered possessor of a watch, my life became a burden. . . . I ran the gauntlet of a continuous volley of questions—"What's the time please?" "Let's see your watch." . . . The second day I found it hard not to be impatient, despite the wistful questions. . . . The third day in self-defence I left my watch at home—although my penalty was to share the prevailing ignorance of how the day was passing.[60]

Some experts tried to allay the trial of monotonous work. Industrial psychologist Elton Mayo documented the problems of workplace boredom in a series of articles, speeches, and papers. He suggested that Taylor and other efficiency experts' obliviousness to fatigue and monotony in industrial work was "the blind spot in scientific management."[61] He was willing to share this critique with leading advocates of scientific management. In a 1924 address to the Taylor Society, a group dedicated to the principles of efficiency in business, Mayo waxed eloquent on what he termed "the vexing question of monotony and boredom" in industrial labor, noting "modern methods of industrial organization tend to impose on the average individual long periods of revery thinking. Machine operation, once the worker is habituated to it, does not demand a high degree of concentrated thought. On the other hand, it is impossible [for] him to concentrate his mind upon anything else. One finds in actual practice, therefore, that the mental mood which accompanies work is very frequently a low-grade revery of a pessimistic order." The effects of such mental states were not inconsequential: "All the authorities agree that an adult nervous breakdown originates in earlier pessimistic reveries."[62]

Mayo eventually came to suggest some ways to make work less boring—advocating rest pauses throughout the day, for instance. He also endorsed the idea that workers should offset the tedium of their work with recreation in their time off. In a radio broadcast he drafted entitled "What Do Workers Want?" he explained, "If our work has become in some degree routine, or even monotonous, we are compelled to develop, as a balance, the practical and the social aspect of things. The home and garden, the children, participation in neighborhood activities, all take on an added importance."[63]

That was how many workers, in the end, did make peace with the time clock and the assembly line—by seeing their work as a means to leisure.[64] They submitted themselves to monotonous work conditions because it allowed them to spend their off hours and their wages on entertainment and consumer goods that they would not otherwise have been able to afford. Clayton Fountain, the autoworker, recalled, "After working hard for a long shift in the factory, I sought relaxation in movies, dances, and bouts with bottles of prohibition liquor."[65]

Factories had not alleviated boredom—rather, laborers were willing to overlook (or had little choice but to overlook) the boring nature of their work, because of the leisure and consumer benefits it afforded them.[66] Workers adapted to these new regimented labor conditions but not because their work had suddenly become interesting. Just as in the 1840s Ellen, the Lowell mill girl, had ultimately come to terms with the tyranny of the factory bell because of the social activities available after work, so twentieth-century industrial workers made the same trade, in much greater numbers.[67]

While many felt there was little they could do to change the nature of their work, there was a growing sense that all people were entitled to pleasure and diversion, not just duty and drudgery. Leisure and entertainment seemed to be new and fundamental rights.[68] Many turned to the booming consumer culture for relief from boredom. Americans' embrace of consumerism depended on them accepting pleasure as legitimate and rejecting suffering, deprivation, and delayed gratification as divinely ordained.[69] This was a revolution in how many conceived of life's meaning; it altered their expectations for what they were entitled to. Rather than sadness and passive acceptance of routine drudgery, many came to believe that pleasure, happiness, excitement, and novelty were their birthright.

This was evident across the nation. Factory workers in cities and towns turned to the commercial pleasures of urban life to bring verve into their lives. Dance halls, theaters, amusement parks, and movies, as well as the cornucopia of goods offered at department stores and five-and-dimes, offered fun and diversion.[70]

For those living too far from cities to partake of such amusements, turn-of-the-century reformers, merchants, as well as farmers, cowboys, and other agricultural workers held out hope that the new technologies of the era might finally alleviate the empty hours and mental lassitude that rural dwellers endured. Some believed telephones would work this cure; others counted on the automobile or the radio. Many succeeded in using the inventions of the era to reduce boredom.

H. E. Wilkinson, born in 1892, recalled that as a child, the introduction of the telephone changed the nature of long winter nights in the Iowa countryside. He and his family had "no radio, no television, no record player, no car to dash to town in, and no movies to see if we did get to town by team and surrey," so they relied on simpler entertainments. One favorite was Pegleg Roach, "a one-legged peddler who played the jew's harp" and was "the sensation of the neighborhood." During the winter, he often stayed with a neighboring family, the

O'Donnells. Wilkinson recalled, "When they could, the O'Donnells cajoled Pegleg into performing for the whole neighborhood." To broadcast his performance, they relied on the telephone. They would call on the party line, using the "general ring," and everyone would pick up. Because the whole Wilkinson family was huddled around the receiver, "reception was somewhat difficult to say the least." Nevertheless, he wrote, "We thought the music was wonderful, we who were so hungry for diversion of any kind, and when Pegleg came to the end of the piece we applauded loudly and happily into the transmitter. It was entertainment, and it was coming mysteriously over that slender thread of wire strung on poles along our fence lines. The angry northwest wind might be whipping the snow into six-foot drifts outside . . . , but Pegleg Roach and his jew's harp were coming in with the faint but twangy notes of 'My Old Kentucky Home.' What more could you want on a wintry night in Lime Creek Township?"[71]

While the telephone surely brought enhanced sociability to the countryside through such impromptu concerts as well as more mundane conversations, some turned to other technologies to relieve boredom.[72] Ads for the phonograph suggested that families, both rural and urban, might find hours of amusement with their new music machines. One advertising flier read, "The happiest hours of life are those spent in the home, in easy enjoyment of pleasing melodies. No need to go to places of amusement when home is made bright and attractive by THE EDISON PHONOGRAPH." Another likened the purchase of the phonograph to the excitement of the circus. Families no longer had to travel in search of fun; it was now available in their own parlors. "Right inside Ladies and Gentlemen for the Greatest Show of the Age." It noted, "The Greatest of All Entertainment is the Edison Phonograph," and then promised buyers that with it, "You may hear the songs of great singers, the music of orchestras, the speeches of great speakers. You can reproduce the latest vaudeville hit, the popular songs that everyone is whistling, or the star part of a star opera singer, in your own home, to a circle of your own friends." And significantly, the ad noted, "An Edison Phonograph is especially needed in homes away from the large cities. It is the ideal amusement for the farm."[73]

Others turned to the new radio sets that began to grace many a living room by the 1920s. In a piece entitled "Static Days and Nights: The Boredom of Ranch Life Is Now Broken by Radio," Remington Schuyler lauded the diversion that broadcasts might offer to lonesome cowboys far away from the excitement of urban life. "'Static' describes perfectly the evening on the old ranch in South Dakota. So static were our evenings that in desperation we turned in along about nine o'clock of a winter's evening, bored to death with each other." The problem was that the cowboys saw the "same old faces, stories, and magazines," day after day. "We knew the magazines from cover to cover. We knew the advertisements with the same close intimacy. We knew every yarn of the other fellow's and every 'funny story.'" Such familiarity and repetition was bad for the spirit of the cowboys. "Living the same life, doing the same things, day after day atrophied our brains. Our conversation moved sluggishly in deeply worn channels, all too familiar and threadbare." Relief finally came with the radio, however; when the narrator returned after some time's absence, he realized that it had conquered boredom. "Each evening while I was there we had a radio banquet. Gone was the old dismal gloom of snow-bound isolation. A wider world had stalked across the frozen prairies and opened up their lives. They were *living* nowadays and happy."[74] This idea, that contact with the larger world, piped-in entertainment, and an end to routine were necessary for a good life, represented a new view of what it meant to be a fulfilled person, and this vision continues to hold sway today.

While few denied that there were still dull jobs, many, like the cowboys and assembly-line workers, seemed to be making a trade. They would endure boredom on isolated ranches or on fast-moving assembly lines in exchange for amusement in their leisure hours. This represented a shift in attitude and expectation from earlier generations, who had often regarded tedium and monotony as their lot. Now, as commercial entertainments proliferated and as work and leisure became more distinct—and as "boring" became a new category of experience—individuals made new calculations about how much boredom at work they were willing to trade for diversion later in the day.

And this trend extended beyond farmers, cowboys, or Ford factory workers. Increasingly, individuals across the country, in all walks of life, looked for ways to fill their free time, often turning toward commercially produced entertainment. They took great delight in the amusements that consumer culture offered, believing that they enriched the day, freed them from drudgery, and added vitality to life. For instance, Harry Bean, of Somerville, Massachusetts, was an early radio listener to a Medford, Massachusetts, radio station. He wrote the announcers a fan letter in 1923, offering "just a few lines of appreciation from a family of radio fans. We all appreciate your concerts very much and always listen in a few moments ahead of the announced time of starting if possible, so as not to miss even one word. Mrs. Bean who is an invalid gets a great deal of pleasure from your programs and wishes to add her thanks for what you are doing for the shut-ins."[75] E. E. Ricker, a listener from Lynn, Massachusetts, joined the chorus of praise, writing, "I am a shut-in and have to sit all day 'day in and day out' away from the busy activities of life which I used so much to enjoy. But I find in my Radio my constant companion, affording me an almost constant source of entertainment, and instruction, and as I eat my dinner and at the same time listen in to 'WGI.'"[76] In radio broadcasts the "shut-ins" found that they no longer needed to endure monotony; instead, relief was at hand. The intentional flight from boredom was becoming normal and laudatory.

Not all were comfortable with this change. Some worried about the wisdom of looking for a cure for boredom in commercial amusements. In particular, social commentators worried about city dwellers and exhausted factory workers who sought relief from the dullness of labor in mass culture. Many observers believed that this widespread search for stimulation was a result of a constant engagement with the novelty of urban life; they feared that that novelty was rewiring the human psyche, making Americans addicted to nonstop diversion and excitement.

Psychiatrist Abraham Meyerson wrote that city conditions simultaneously brought both greater tedium and greater excitement. In 1921 he described "the growing monotony of most labor." He explained,

"The factory, with its specialized production, reduces the worker to a cog in the machinery. In some factories, in the name of efficiency, the windows are whitewashed so that the outside world is shut out and talking is prohibited; the worker passes his day performing his unvaried task from morning to night." When workers did leave the factories behind, however, they sought out amusements designed to assuage their boredom. "Restlessness, eager searching for change, intense dissatisfaction are the natural fruit of monotony." He believed there was much stimulation to find in America's metropolises, for "cities are exciting. The multiplicity and variety of the stimuli of a city—social, sexual, its stir and bustle—make it difficult for those once habituated ever to tolerate the quiet of the country." This was a problem because individuals' personalities were permanently changed by exposure to such attractions. "Excitement follows the great law of stimulation; the same internal effect, the same feeling, requires a greater and greater stimulus as well as new stimuli. So, the cities grow larger, increase their modes of excitement, and the dweller in the city, unless fortified by a steady purpose, becomes a seeker of excitement."[77]

Sociologist Robert Park agreed, writing in 1925 that workers increasingly had to focus on "some minute detail" in their labors, which narrowed their range of vision. Yet while work might be monotonous, "Leisure is now mainly a restless search for excitement. It is the romantic impulse, the desire to escape the dull routine of life at home and in the local community that drives us abroad in search of adventure. This romantic quest which finds its most outrageous expression in the dance halls and jazz parlors, is characteristic of almost every other expression of modern life. . . . We are everywhere hunting the bluebird of romance, and we are hunting it with automobiles and flying machines." Modern Americans' ceaseless pursuit of novelty and excitement was a "reflection of a corresponding mental instability." And, he suggested, the quest for constant entertainment, the flight from dullness, ultimately led nowhere. "This restlessness and thirst for adventure is, for the most part, barren and illusory, because it is uncreative. We are seeking to escape from a dull world instead of turning back upon it to transform it."[78]

What concerned observers like Park was that not only was work boring people, deadening their instincts, rendering them less human, but so too were the respites from work. A journalist reported in 1930, "Most of us lead the monotonous life of robots, and the cumulative effects of monotony are deadly. They slowly suffocate man's spirit. In this toxic atmosphere only stimulants can rouse us, and thus we turn to the over-emphasized beat of jazz, the melodrama of the screen, the dazzling gyrations of lights. Only through artificial respiration does the jaded spirit revive."[79]

Because work in industrial society was dull, workers sought entertainment, and the more they became habituated to it, the more they required. Psychologists, sociologists, educators, and laypeople alike recognized that modern Americans required a rapidly increasing amount of stimulation to feel truly alive. Social commentators did not always like the changing personality traits they saw about them, but they could not deny them. And as they discussed them, they sketched out a new type of American—one who differed in behavior and emotion from those who had populated the nation in earlier years. In the nineteenth century, there had been little discussion of boredom as soul deadening and stimulation as essential; however, by the early twentieth century, it seemed that boredom was becoming more common and less tolerable, and that diversion was all the more necessary to lead a truly fulfilling life. Indeed by 1938, psychoanalyst Otto Fenichel declared that individuals had "the right to expect" some diversion from the world at large.[80]

While this new view of the modern individual's psychology would become commonplace by the late twentieth century, concern about boredom and stimulation abated temporarily during the Depression and World War II, as work and leisure's meanings were upended. During the Depression, Americans found the prospect of industrial work—boring though it might be—far more attractive than empty and impoverished days. For instance, Clayton Fountain, the assembly-line worker who had labored at auto plants in Detroit, eventually quit because he disliked the work. But when the Depression hit in earnest and he found himself facing long-term unemployment, he jumped at the chance to work the third shift as a punch-press operator at a General

Motors plant in 1933.[81] Monotonous work was better than none. As a result, many worried far more about the mental state of the unemployed than about that of those gainfully employed doing repetitive factory labor. For instance, in 1932, YMCA leaders in Waterloo, Iowa, argued that their new mission was to help fight "the supreme evil of unemployment," which led to demoralization, "despair," and "the hideous boredom of having nothing to do."[82]

World War II amplified the need for assembly-line labor and workers who could withstand its repetitive nature. Some observers worried that these new industrial jobs might cause widespread fatigue, which a 1942 article labeled "the enemy of war production." War workers might feel "boredom in some types of work, where the same thing happens over and over again, as in feeding a punch press or inspecting an endless line of finished products."[83] As a result, both the government and businesses worked to change the image of industrial assembly-line labor. They idealized workers, such as Rosie the Riveter, as paragons of patriotism:

> All the day long, whether rain or shine
> She's a part of the assembly line
> She's making history, working for victory
> Rosie, brrrrrrrrrr, the riveter.[84]

Glenn Miller recorded a similarly themed song in 1942, "On the Old Assembly Line," which made factory work sound positively redemptive:

> On the old assembly line
> On the old assembly line
> Everything is hum-hum-hummin?
> On the old assembly line
> When the overalls combine
> With the mighty dollar sign
> there'll be miles and miles
> Of American smiles.[85]

Perhaps such campaigns had the desired effects, for a young African American woman, described as a "champion arc welder," explained her attitude toward her work in this way: "When you work in a war plant . . . you talk very little while you work but you do a lot of thinking. And with the roar of machinery you sort of get a message which seems to say to you that this job must be well done, because the stake in getting it out is perhaps the life of some boy fighting on the beachheads. That's what I keep thinking."[86] Newsreels and propaganda campaigns likewise glamorized assembly-line work, offering up images of contented and highly motivated workers who derived great satisfaction from their jobs. As a result, by the closing months of World War II, many Americans seemed to at least grudgingly accept the demands of the mass production economy, ruled by time-bound efficiency standards. Whether they liked it or not, assembly-line work, and the boredom it entailed, was here to stay.

Leisure and Boredom

Some argued that it was not work but leisure—that is, shorter working days and too much free time—that was the true cause of boredom. The *New York Times* reported in 1949 that after a trip to the United States, the French playwright Jean Cocteau thought Americans were noticeably bored. The reporter summarized Cocteau's impressions that "boredom is the great terror of New Yorkers. They never dare to mention it, but 'they drink to escape it, they go to the movies, to their psychiatrist, they sit in front of television, there is no conversation.'" While the *Times* reporter took issue with some of Cocteau's characterizations, the journalist did believe that boredom was increasingly widespread in the postwar world.

> There is plenty of boredom in New York, . . . there are, indeed, far too many people at loose ends, far too many people who do not know yet how to use the leisure that our wonderfully productive economy has given them. A vacuum has been created. Promoters of television, radio, movies and spectator sports have rushed in

> to fill it. . . . Great new schemes are under way to fill the void that has been created by shorter hours of work in the shop and in the home. Television is to complete the job begun by radio. . . . 'Entertainment' is to be about as all-pervasive as the air we breathe.

But, the writer continued, despite the array of amusements, "the worst features of boredom are likely to persist and to become more disturbing." He held little faith that television or radio or other technologies ultimately could solve boredom; rather, individuals in the new leisure society would have to find meaningful ways to spend their time.[87]

This was easier said than done, for many who observed the nation's leisure patterns during the 1950s believed that Americans had not found truly rewarding ways to spend their off hours. Nothing seemed to make them happy or fulfilled. In an article entitled "TV Boredom," reporter Hal Boyle noted that few in the mid-twentieth century were satisfied or stimulated by the devices of modern life. "One of the biggest problems of this wonder-filled modern world is to keep from yawning while you are being entertained. This is the pitiful plight of a pampered people in a time of plenty—too many are being entertained into boredom. They become depressed while trying to enjoy themselves." This, he claimed, was a problem unique to the era: "No previous age in history ever surrounded itself with more gadgets designed to give fun, pleasure or relief from care." Americans were "fatigued with the dull parade of nonsensical murder mysteries and repetitious comedians on his television screen." Mass entertainment was passive. To escape its trap, people needed to learn to "have fun and entertainment within" themselves, or they would "never escape . . . boredom."[88] Journalist Harriet Van Horne went even further in 1961. She noted that there were 47.9 million TVs in the nation's homes. The common feature of these households was "boredom." "You have to be bored to near-frenzy to turn on the television set these days. Once you've turned it on, you merely exchange passive boredom for active boredom." The emotion was so widespread that "boredom begins to look like a national malaise. Where, one wonders, is the antidote? What will be the next national enthusiasm, giving us a sense of expectation and awe?" She suggested that whatever astonished Americans next would not do so for long, for they had become all too accustomed to novelty and disappointment. "It seems to be a national ailment right now, this heavy, soul-draining boredom. Our eyes are jaded, our span of attention short. Few things delight, and never for very long."[89]

Beyond Boredom

These Americans—bored both at work and at play—generated considerable interest among psychologists and journalists, many of whom began to conceive of boredom as a new social and mental problem. A landmark 1957 study of boredom, which termed it a "pathology," appeared in *Scientific American.* The author, Woodburn Heron, laid out this concern: "Monotony is an important and enduring human problem. Persons who have to work for long periods at repetitive tasks often complain of being bored and dissatisfied with their job, and frequently their performance declines." This problem was worrisome in the present day, because of modern technologies: "In this age of semi-automation when not only military personnel but also many industrial workers have little to do but keep a constant watch on instruments, the problem of human behavior in monotonous situations is becoming acute."[90] The expanding economy and the military-industrial complex required regimented work and workers, and it was therefore necessary to find ways by which people could cope with such boredom.

But Heron did more than just point up the need to cure boredom in the workplace; he also presented a new view of human nature. He reported on the results of a study that required subjects to be in isolation for as long as they could. Participants found it difficult to do this for days on end, and the report, extrapolating from their experiences to humans more generally, concluded that what made men and women psychologically ill was sameness, monotony, dullness, repetition. What made them healthy was variety and constant stimulation. As Heron noted, "Prolonged exposure to a monotonous environment then, has definitely deleterious effects. The individual's thinking is impaired; he shows childish emotional responses; his visual perception becomes disturbed; he suffers from hallucinations; his brain-wave pattern changes. . . . The recent studies indicate that normal functioning of the brain depends on a continuing arousal reaction . . . which in turn depends on constant sensory bombardment." He continued, "A changing sensory environment seems essential for human beings. Without it, the brain ceases to function in an adequate way, and abnormalities of behavior develop. In fact, as Christopher Burney observed in his remarkable account of his stay in solitary confinement, 'Variety is not the spice of life; it is the very stuff of it.'"[91]

Heron presented this argument as a depiction of universal human nature, but in reality, it reflected a portrait of humans at a very particular time and place—accustomed to innovation, change, recreation. Constant stimulation was what modern individuals expected and needed in order to thrive. Monotony and sameness were mentally tiring—and damaging too. Whereas once individuals had endured dullness as their lot or even their religious calling, and later generations had worried about the problems of too much stimulation, the mid-twentieth-century psychologists studying boredom began to sketch out a new vision of human nature, one in which the psyche could thrive only when constantly engaged by an ever-changing array of diversions.

That mantra was picked up in pop psychology pieces as well. *McCall's Magazine* asked "Is Boredom Bad for You?" The answer was yes, and the magazine warned, "Don't underestimate the dangers of being bored. According to experts on the subject, it's one of the most destructive of your emotional states." Why was it dangerous? "It not only exhausts you through indecision, it frustrates you, causes nervous fatigue, feelings of insufficiency, and leads to introspection—the root of many neuroses. Boredom also takes up much more energy than you would think. In fact, one half-hour of acute boredom can often dissipate more nervous energy than a whole day's work."[92] Writer V. S. Pritchett summarized this new focus on boredom, noting that while it had once been a fashionable feeling among the wealthy who boasted of their ennui, in modern America "boredom (they say) is an illness.

All the same, it has not invariably been considered so. One hundred and sixty years ago—even less—it was the fashion all over Europe. Everyone who cared about the figure he cut in the world went in for boredom. It produced a remarkable literature: Byron's *Don Juan* in England; in Russia, Lermontov's *A Hero of Our Time,* in France, *Stendhal* and *Merimee.* But not today. If I am bored now, I am a case; I am a danger to society. It is the central sickness (they say) of industrial man, and psychiatrists are working on it."[93]

Some psychologists, science writers, and journalists suggested that boredom was an emotional state or illness symptomatic of modern times and industrial society. For instance, Peter Chew noted in *Science Digest* in 1972, "Boredom has always been with us. But behavioral scientists make a strong case that chronic boredom is epidemic in our industrial society. They call ours 'the land of the free and the home of the bored.'"[94] *Life Magazine's* cover article in September 1972 was entitled "Bored on the Job." It described the plight of autoworkers, suggesting that "their state of mind spells trouble" and declaring that "a new factor" affected factory workers—"job boredom."[95] Some suggested that the condition had long existed; what was new was that Americans found it more oppressive than they had in the past. In a 1974 *Harper's* piece, Estelle Ramey noted, "There's nothing new about boredom, just as there's nothing new about its emotional cognates—sloth, despair, hopelessness, alienation, passivity, and anomie. It's simply that we are less willing to endure the condition than our bored forebears."[96]

While boredom was now widespread in America, it had not always been. In the eighteenth century, it did not even exist yet as a feeling. In the nineteenth century, it was deemed a rarity in the United States, a feeling that was largely unknown to a nation of hard workers. However, in the twentieth century, with the spread of the word, and with a declining faith in the redemptive power of both industry and leisure, it had become a problem. Suffering through dull times no longer offered moral gifts; instead, it was a problem emotion in need of a cure.

As a result, psychologists began to study boredom more intensively. In 1975, Mihaly Csikszentmihalyi published *Beyond Boredom and Anxiety* in which he introduced the concept of "flow." As Csikszentmihalyi argued, when people's skills were not sufficiently challenged, they suffered from boredom; when their skills were not adequate to meet a particular task that they were trying to accomplish, they suffered from anxiety. In between these poles, when people felt challenged but could not quite meet those challenges, they experienced a state called "flow":

> Poised between boredom and worry, the . . . experience is one of complete involvement of the actor with his activity. The activity presents constant challenges. There is no time to get bored or to worry about what may or may not happen. A person in such a situation can make full use of whatever skills are required and receives clear feedback to his actions. . . . From here on, we shall refer to this peculiar dynamic state—the holistic sensation that people feel when they act with total involvement—as flow.[97]

Unlike more arcane concepts in psychology, flow was adopted by a larger public, eager to identify experiences that might eliminate their own boredom. In 1990, Csikszentmihalyi wrote a lay treatise, *Flow: The Psychology of Optimal Experience,* presenting his research, which had already been gaining popularity over the interim decade and a half. And with its publication, *Flow* went mainstream—President Clinton pronounced it to be one of his favorite books, it was translated into fifteen languages, and business schools started to use flow as an analytical category.[98]

By then, psychologists had declared boredom "America's no. 1 disease."[99] In 1986, psychologists Richard Farmer and Norman Sundberg developed a boredom proneness scale, explicitly modeled after the aforementioned loneliness scale.[100] The scale, which is now readily accessible as a questionnaire with automated scoring on the Web, comprises twenty-eight statements. Examples include "It seems that the same things are on television or the movies all the time; it's getting old"; "In situations where I have to wait, such as a line I get very restless"; and "Time always seems to be passing slowly."[101] The questionnaire, which helped to formalize and quantify the measurement of boredom, inspired psychologists to see whether the feeling correlated with other psychological conditions.[102] With the boredom proneness scale, the emotion became quantifiable and predictable. No longer an inevitable accompaniment to human life, it is now a malady to be avoided.

And those unlucky enough to fall prey to it must be treated. This has been evident in the growing numbers of youth diagnosed and treated for attention-deficit / hyperactivity disorder (ADHD) in the late twentieth and early twenty-first centuries. In the 1980s and 1990s, ADHD diagnoses escalated significantly, and prescriptions for ADHD drugs went up a whopping 700 percent.[103] According to Les Linet, a child psychiatrist who studies ADHD, the condition might better be described as a "search for stimulation" disorder.[104] In Linet's view, young ADHD sufferers have difficulty finding enough stimulation from their everyday environments and become easily bored. In order to focus, they require greater excitement and stimulation than other children.[105] As measured by the boredom proneness scale, those with ADHD are more likely to suffer from boredom than those who do not have ADHD.[106] While this feeling might have been ignored before ADHD was defined as a medical condition, with the wide-scale prescription of drugs like Adderall and Ritalin, the type of boredom that accompanies it is now regarded as a malady to be treated medically.

Boredom at Work and at Play in the Digital Age

More and more people are looking for ways to fight boredom today—at work and at play. While manual laborers long ago had to accept the monotonous nature of industrial labor, workers from a wider range of classes are having to do so today, as their jobs get automated. White-collar workers and college graduates who once were able to find work in cognitively demanding fields are increasingly less able to do so. The threat to mentally engaging work is now being felt by professionals, and the class divisions that historically influenced how boredom was experienced and expressed have become more amorphous.

The same processes that transformed manual labor at the beginning of the twentieth century are now reshaping white-collar professions. During the 1970s and 1980s, the nature of work began to change in office settings, and its effects were felt by white-collar workers. Through extensive ethnography and first-person interviews, author and social activist Barbara Garson documented how social workers, managers, and stock brokers found their jobs less interesting as more and more of their tasks were turned into procedures mediated and guided by computers. At McDonald's, not just fast-food workers but also their managers found their work tedious as a result of its regimentation and automation. A griddle cook reported: "You follow the beepers, you follow the buzzers. . . . to work at McDonalds' you don't need a face, you don't need a brain." Garson discovered that a McDonald's manager, Jon DeAngelo, faced similar conditions in his work:

> When I first came to McDonald's, I said, "How mechanical! These kids don't even know how to cook. . . . It's the same thing with management. . . . [Y]ou have to trust the computer to do a lot of the job. . . . [T]he manager . . . has to follow the procedures like the crew. And if he follows the procedures everything is going to come out more or less as it's supposed to. So basically the computer manages the store.[107]

As Garson observed, the McDonald's corporation had "broken the jobs of griddle man, waitress, cashier and even manager down into small, simple steps . . . and have siphoned the know-how from the employees into the programs." This happened to managers at McDonald's as well as to stock brokers. As Garson recounted after interviewing a stock broker,

> Len Deusenberg used the phrase "grid of information" to convey something that he felt closing in around the professional broker. Many data clerks, telephone operators, fast food managers and airline reservations agents grope for that same concept as they find themselves hemmed in by multiple numerical ratings. . . . [C]omputers can . . . measure performance on many scales at once. Eventually . . . there's only one possible way to move your arm or your head. Only one way to do the job.[108]

The changes in office work in the late 1970s and 1980s anticipated developments that have taken place since then. As Ben Sand of York University and Paul Beaudry and David Green of the University of British Columbia argue, the cognitive challenges that college graduates face in their jobs have declined significantly since the year 2000. Moreover, increasing numbers of "high-skilled workers have moved down the occupational ladder and have begun to perform jobs traditionally performed by lower-skilled workers. This de-skilling process, in turn, results in high-skilled workers pushing low-skilled workers even further down the occupational ladder."[109] More recently, Matthew Crawford in *Shop Class as Soulcraft* writes that "genuine knowledge work comes to be concentrated in an ever smaller elite," while the rest of white-collar workers become part of a "rising sea of clerkdom."[110] And in *The Glass Cage,* Nicholas Carr posits that pilots and other white-collar workers might also lose interest in their work as a result of their jobs being digitalized and automated.[111] The result is that boredom is spreading across the occupational world.

"Ernie" gave a name to such a phenomenon. In 2005, he confided on an attention-deficit disorder (ADD) forum that he took Ritalin every day to cope with the boredom he felt while doing computer-assisted drafting (CAD).

> I started a new drafting job two weeks ago. It's only for 4 hours a day. The work is very simple and well below my skill level. And I am bored to tears. If I could just get past the boredom issue it could be a really sweet deal. . . . I have a nice sized cubicle and can listen to music if I want. Right now I take 50mg of Ritalin before I leave for work. . . . I'm wondering what doses of Ritalin other people are taking. I'd really hate that my ADD could screw up a good thing.

When some on the forum suggested he work harder at conquering his boredom, he pointed the finger at his job, which consisted of turning drafting sketches into CAD. He explained:

> I feel like I have this spoiled brat inside me that demands that any and all work to be fun and changeling. [*sic*] I call him "the little brat." It's only 4 hours a day. I really, really hate ADD. I'm not in a position to come up with new ideas. Cad drafting is pretty straightforward. People give me a hand drawn sketch and I turn it into a Cad Drawing. . . . A Cad Drafter is a Cad Drafter.[112]

Just like the workers Garson documented in the 1980s, Ernie and other contemporary workers across the class spectrum are having their jobs—or parts of them—automated, and they find the resultant repetitive duties monotonous. In an era when such feelings of boredom have been pathologized, many worry about their boredom and sometimes try to cure it through medication.

Others are turning to digital amusements to offset the tedium of work. As Greg Bayles, a Las Vegas native who designed medical apps for a living, noted, "We have lots of people who are employed in jobs where they sit at their desks . . . and every once in a while they have things to do, but a lot of times they're spending their time on Facebook or YouTube or whatever. . . . [E]mployers are saying 'you're worthless,' . . . and employees are saying 'give me something meaningful; this technology is way more stimulating than whatever you're giving me.'"[113]

Indeed, with the emergence of smartphones, people can now access entertainment anywhere and anytime, offering a new way to deal with circumstances that would have conventionally triggered bouts of boredom, whether they be empty moments at work when there is nothing to do, delays in doctors' waiting rooms, or long commutes in traffic jams. In an attempt to capitalize off these moments, Motorola went so far as to coin the term "micro-boredom" to describe small moments of idleness that their phones might alleviate.[114] Software developers creating apps and social media sites also have been successful in promoting their programs as antidotes to the feeling. "An App a Day Keeps the Boredom Away" reads a tee shirt, while Tumblr advertises "Download Tumblr: Never Be Bored Again."[115] The tech companies' discussion of the feeling represents one more example of journalists, psychologists, medical experts, and digital businesses calling attention to the problem of boredom and looking for ways to commodify its cure.

Such efforts have been successful, for the ubiquity of apps, the Web, and social media makes Americans feel that they should not endure the emotion, and that when they do feel it, something is wrong. As a twenty-one-year-old Grinnell College student told us, "I think having technology at my fingertips makes boredom feel much more dangerous. . . . It's like if you're bored, fix it fast."[116]

Many of those whom we interviewed noted that they used their devices precisely for these purposes. Skyler Coombs, who had recently returned from an eighteen-month Church of Jesus Christ of Latter Day Saints mission, noted that the increasing portability of his devices allowed him to flee from the spectre of boredom. "[Before my mission] I had an iPod touch and a flip phone. So when I was at home I could just get on Facebook. . . . But now . . . it's anywhere. You can . . . do it whenever you are bored or just sitting there. So I think it has changed a little bit. . . . You can use it wherever you go."[117] For Hailee Money, the college student who had spent her early teenage years in Aurora, Colorado, but now lived in Utah, the sentiment was much the same: "Every time I'm bored, I pull up my phone."[118] Alta Martin, who had been waiting for a while before we ushered her in for an interview, registered similar sentiments. Only for Alta, the use of the phone to combat boredom was concretely illustrated by what she had been doing just prior to our interview: "Boredom is definitely a thing. . . . Anytime I have five minutes. I was just doing it in the hall. You know, I have five minutes, so I got on and checked real quick."[119]

These firsthand accounts corroborate observations made by the media as well. A *Huffington Post* reporter wrote in 2012 that Americans use Facebook "to cure a simple ailment: boredom." Citing research by communication scholars, the article reported, "During the study, Facebook was most often used when subjects were bored, [as] opposed to connecting with individuals to cultivate relationships." The author concluded that the near constant use of the application might be a sign of modern individuals' expectation of and need for "continuous" amusement.[120] Likewise, the *Atlantic* reported in 2013 that the second most commonly cited reason for why people use Facebook was to relieve boredom.[121]

Young adults might seem like avid and uncritical adopters of smartphone technology and some of the most willing to embrace corporate America's effort to banish boredom. (As one twenty-three-year-old we interviewed intoned, he updated his iPhone "every two years . . . almost religiously."[122]) Yet, they too recognized that the flight from boredom had costs. In spite of the fact that phones appealed to them as tools that could assuage boredom, they also noticed that using them in this way adversely affected their productivity and their conversations with peers. For example, Hailee explained, "It's really hard for me to focus on homework and not look at my phone. Because I don't want to do my homework and it's a way to distract myself so I don't have to [do] my homework." She wished that she could take only short pauses from her studies, but found that difficult.[123] This was a common late-night study anxiety that other students also expressed. For example, Alta observed, "I find when I'm doing homework I have to go put my phone in a different room and ignore it. Otherwise, if I get bored or I start mind wandering I'll grab it and start checking. So it wastes a lot of time. Not productive time."[124]

Many also registered concerns that their efforts to escape boredom affected their sociability. Digital devices were implicated in the rise of boredom and loneliness; often the emotions went hand in hand. To avoid these feelings, people would retreat to their phones, which ultimately might heighten the emotions, for they curtailed live social interactions. When asked if he turned to his phone because he found people boring, Skyler replied, "I don't think other people are boring. I think it's just like a time filler. . . .'Cause I don't think I get bored by other people. I enjoy being . . . [in] other people's company. It is a little weird when you are with a bunch of other people and everyone is on their phone. And you're like the only one who is not. . . . [Y]ou get on your phone because everyone else is. It would be weird if I started talking with someone and they are all doing stuff on their phone." Were these situations, then, kind of boring? "Yeah. Because everyone else is on their phone. Instead of having an interaction you kind of just do what they are doing."[125]

Alta too expressed concern about the way the phone was dampening conversation with her friends:

> It annoys me when I'm with my friends and instead of like just being at dinner they whip out their phones and I'm like "I never get to see you. Put your phone away. And let's talk instead of you showing me your phone." . . . I have friends where they literally can't put the thing down. And it's like "I came to hang out with you. Let's go do something and get off your phone," and they won't. And that's when I get annoyed.[126]

Our interviewees noted that face-to-face sociability was waning and that there was some association between this decline, their feelings of boredom, and the presence of smartphones. Some could trace it directly to digital media. There had been a sea change during their lives, and they realized they turned to their phones more than to those around them, which often led to fewer live conversations. As Anna Renkosiak explained, at her own family's gatherings, all of the young adults were at one table, but there were hardly the convivial conversations of old: "We're all down here but we're all on our cell phones." Eager for entertainment, fearful of awkward, dull moments, they turned away from each other.

"I think honestly . . . our toleration for boredom has gone down—a lot," Anna concluded. She thought Americans no longer had "tolerance for boredom. . . . [E]verybody constantly has to be stimulated." In the past, kids had had to endure boredom, whether it was long car rides or sitting through dinner, but today "you go into restaurants and . . . you see people who purposely bring iPads—not just one iPad for all their children—each kid has an iPad. . . . [T]hat's the way they pacify their children."[127] H., the twenty-one-year-old Grinnell student, agreed, noting that while today she had found it difficult to endure boredom, that was not how she had always regarded the feeling. Before she became immersed in the digital world, she felt she could cope with

FIGURE 3.2 A cartoon about the way the internet has reshaped boredom. *"Before the Internet" by xkcd.com licensed under CC BY 2.5*

the emotion. "I remember feeling bored as a kid and that was like a fact of life, and you just like kicked around a stick till you came up with something to do. There wasn't a lot of technology in my life. And now boredom has to immediately be filled in with something. I think that's a product of technology. Always having something at your fingertips." She said of her peers, "I think definitely our generation is always desperately in need of something to do and we're bad at entertaining ourselves as we have technology to do it for us." But despite these hesitations, even as they were aware of their phones' downsides, teenagers and young adults continued to use social media and apps to allay boredom.[128]

Older adults, however, were more willing to abandon their use. When we interviewed William and Debbie Speigle, we asked whether phones were a cure for boredom. William responded, "If [a phone] is a cure for boredom, it's a bad cure." Debbie noted, "To me as far as the boredom, I always liked to read. . . . The changes I see is people going to work used to always carry newspapers, magazines [or] books to read. . . . Now everybody is just staring in their hands at their phones. There's no interchange. . . . They're self-absorbed. . . . If you had a good book, someone would say 'Oh, that was really good' or . . . 'Can I read it after you?' There's just no interaction. . . . They're just totally isolating themselves onto one little three-by-five instrument and that's it. So I see a loss of socialization. . . . Nobody has any idea what you are reading or what you are looking at. . . . People use it as a form of retreating from social situations, social interactions."[129] One might retreat into one's phone because the world seemed inhospitable, uninteresting, boring; by doing so, one made it even more so for others.

Some of the middle-aged interview subjects volunteered that such uses of smartphones had other downsides as well, because boredom could actually be a virtue. Barry Gomberg, an attorney from Ogden, Utah, observed: "Boredom is probably a healthy thing to feel now and then." Emily Meredith, the nurse from Oakland, California, noted that boredom "can be . . . a fertile breeding ground for ideas, or new thoughts, or trying something new." For that reason, she said, "I hope boredom continues. . . . I think things come out of that."[130]

Many young people realized their elders held such opinions. H. noted that, to her, boredom felt "dangerous," but to her mother it did not. "I think my mom thought boredom was healthy. I always complained about boredom when I was younger and she was like, 'Good, go find something to do.'" H. said her mother believed "boredom breeds creativity."[131]

Some prominent media scholars share this view and have reacted against efforts to stamp out all forms of boredom with digital devices. As Sherry Turkle observes, with the advent of the smartphone, Americans have reached a point where they might actually come close to eliminating some of the more obvious sensations of physical disconnection and boredom. But Turkle does not perceive this as an unalloyed good. In her view, never being bored and never experiencing a moment of aloneness rendered people less human. To counter that development, she counsels readers to "look up" from their cell phones and start to "look at one another": "Most of all we need to remember—in between texts and e-mails and Facebook posts—to listen to one another, even to the boring bits, because it is often in unedited moments, moments in which we hesitate and stutter and go silent, that we reveal ourselves to one another."[132]

Like Turkle, William Powers, in his book *Hamlet's Blackberry,* promotes so-called Walden Zones, or physical spaces that people could retreat to when the din of digital entertainment threatened to overwhelm their ability to focus. Published in 2010, Powers's book was one of the first to suggest that something must be done to combat the digital deluge brought on by smartphones.[133] He has been followed by an array of others who have endorsed the idea of getting off the grid and embracing "digital Sabbaths."[134] In a seeming paradox, digital applications with names like Freedom and Pause attempt to monitor and curb digital addictions via digital means. And at the June 2018 Apple Developer Conference, the company promised similar functionality in their next iPhone software release. As Craig Federighi, Apple's senior vice president of software engineering, confided, Apple was building a "set of built-in features to help you limit distraction, focus and understand how you're spending your time." These features, with names like Screen Time and App Limits, would help users "make decisions about how much time you want to spend with your device each day."[135] Both Freedom and Pause, and Apple's soon-to-be-released features, allow users to monitor their digital habits and discourage overuse. These initiatives are complemented by the Center for Humane Technology, which was founded by a former Google design ethicist named Tristan Harris and other "former tech insiders." Their website claims they are dedicated to "reversing the digital attention crisis and realigning technology with humanity's best interests," and it urges acolytes to "Join the Time Well Spent Movement."[136] None of these initiatives endorse boredom per se. But they suggest that there might be more to fear from the entertainments that people are consuming to assuage the emotion than in boredom itself.

Along similar lines, Manoush Zomorodi, the host of a WNYC radio program called "New Tech City," sponsored an initiative called "Bored and Brilliant: The Art of Spacing Out," which challenged participants to take a recess from their smartphones. Zomorodi promoted her initiative on the premise that bringing back some degree of boredom could actually spur more creative thinking. As she stated in an NPR interview, "I kind of realized that I have not been bored since I got a smartphone seven years ago." But she did not think this was a good thing. "Our brains are doing some really important work when we think we're doing nothing. . . . [W]e get our most original

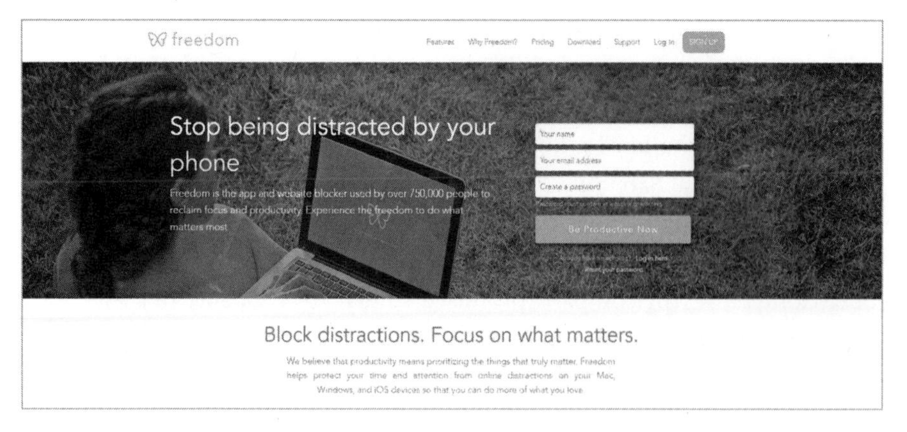

FIGURE 3.3 As advertised on Freedom's website, Freedom is an app that blocks "online distractions on your Mac, Windows and iOS devices." Freedom was once just a fringe technology. But in June 2018 Apple announced that it would include a spate of apps to manage distractions in iOS 12. *Courtesy of Freedom.*

ideas when we stop the constant stimulation and we let ourselves get bored."[137]

In a similar vein, in *Play Anything: The Pleasure of Limits, the Uses of Boredom, and the Secret of Games,* Ian Bogost, a professor of interactive computing at the Georgia Institute of Technology, reminds his readers that life can be more fun when limits are imposed. To illustrate his point, Bogost describes how his daughter turned a forced march through a mall into a fun activity by telling herself not to step on any of the grout lines on the tile floor. She played the game "step on a crack, break your mother's back," a seemingly banal activity. Bogost views it with greater existential import:

> Boredom is the secret to releasing pleasure. . . . My daughter encountered that principle in the mall. There, smothered by the incapacity of being dragged on a stupid errand, she became so disgusted as to move beyond disgust to find something—anything—that she could grasp onto as the bottomless pit of dark tedium cascaded over her. Otherwise how would she even have noticed the grout and the tiles, let alone fashioned something new out of them?

As Bogost sees it, his daughter initially regarded the walk as boring because it offered limited choices and opportunities; there were built-in constraints to what his daughter could see and do. Yet these very limits forced her to find ways to amuse herself; it was through boredom that she arrived at fun and novelty:

> She was having fun, but her fun emerged from misery. Fun isn't pleasure, it turns out. Fun is the feeling of finding something new in a familiar situation. Fun almost demands boredom: you need the sense that nothing good could possibly arise from an experience in order for the experience of finding something there to smolder with the hot pleasure of surprise.[138]

Bogost, Turkle, Zomorodi, and others are working to heighten opportunities for boredom; they see in it potential for creativity. Meanwhile, there are concerted efforts on the part of social media companies and smartphone manufacturers to redress boredom while making a profit, offering consumers unlimited entertainment, articles, status updates, and Candy Crush games.

"Boredom" is a word with a range of meanings for modern Americans. Most commonly it was regarded as an experience with negative connotations, particularly by the young. And when it was regarded that way, subjects often believed others suffered from it, rather than admitting that they themselves did.[139] Frequently in these circumstances, talk about boredom accompanied talk about generations, with elders indicting more youthful cohorts for being bored and for being incapable of coping with the emotion.[140] Alternatively, when boredom was regarded as a state that catalyzed creativity, older generations suspected younger generations of not being able to tolerate boredom. "Boredom" thus construed is not only a word that holds multiple meanings, it also can be used to buttress generational (and perhaps also class) divisions. People can admit to experiencing the feeling, as they sometimes do, or identify it in others. In either case, it shapes and defines what groups they identify with. The embrace of boredom as an opportunity, or the rejection of it as a waste, can reflect deeper sensibilities about how time is best spent and the mind best engaged. But it can also be used to fortify identity.

Moreover, at the same time that digital technology has provided the tools that purportedly conquer brief moments of boredom, some people are acknowledging that more existential types of boredom persist. "Ennui" may no longer be a word that people use to describe boredom today but it captures a certain mood in an age when people have become satiated with too many diversions and too much leisure. Ennui was once associated with the lives of the leisured rich, but something akin to it now can be found among the media-saturated population of all classes in the United States. The flight from boredom and toward easy entertainment might quickly become wearying. Diana Lopez confided, "I have a limited tolerance for being online all day, . . . even if it's social media. I can only look through so much Facebook before I can't anymore; it all kind of runs together."[141] L. had noticed this too. "If you are constantly using it to relieve your boredom you might become bored of that action." A younger relative of hers had been on Instagram and then posted a picture of herself with the caption "I'm bored." L. said, "I just thought that was kind of funny, how that can change so quickly. . . . Something that you once used to relieve your boredom is also what is making you bored."[142]

This was a recurring theme. William found that Facebook, rather than allaying moments of monotony, increased them, noting, "It became boring. . . . Facebook bored me." So he quit the site.[143] Cooper Ferrario, who was at home for the summer and had to rely on social media to replace the lively social scene on campus, noted that while the phone helped alleviate his boredom, it was not the same: "I would definitely say that there is not as much fulfillment from doing those kind of activities; . . . I still feel a little bit bored, you know. . . . You're not interacting with people and you're not doing things, you're not seeing things in real life, there's definitely a difference there."[144]

Some described a craving that sent them to social media but that ultimately could not be satisfied. K. described his unsuccessful attempts to assuage his boredom on Facebook. He would log in out of "sheer boredom." But the search for an end to boredom was fruitless, he found.

> I'd be checking . . . twice, three times a day. . . . It would be kind of dumb. There would be no new posts and you'd [keep checking]. It's kind of that moment when you look in the fridge and there's nothing and you . . . go back with that craving that there would be something good but still not there. It's like nothing changed but I'm still coming back. Why am I coming back? But you still do because you have that craving. I think . . . Facebook . . . was like that for me. Just like having that craving for some sort of like new . . . some gossip. And you open it up just to be disappointed. . . . And you just constantly check.[145]

Hailee agreed. "Every time I'm bored I pull up my phone and the first thing I do is Instagram and scroll through it until I catch up to where I was. Then I pull up Facebook. And it's like a pattern, it's like a habit." Yet, she confided, "I still feel bored."[146]

Emily suggested that these attempts to ward off any momentary boredom might result in a deeper type of the emotion, a true malaise. She did not think smartphones and the internet spelled "the end of existential boredom . . . but it could be the end of micro-boredom," which she defined as "the little moments in your day when you just have to sit there and there's nothing to do [but] stare at the wall." But, ultimately, filling those moments with online games might actually increase rather than decrease boredom. She observed: "It maybe exacerbates existential boredom in a way because you're just filling your time. . . . It accumulates . . . when you look back . . . at a few weeks and you're like 'what did I?. . . . What's going on in my life?' and it's just a hodgepodge of . . . all these online things you've done. And I think that might be depressing or contribute to that feeling of existential boredom."[147]

Perhaps this sense of dissatisfaction in the quest to cure boredom springs from heightened expectations that Americans have for their lives. Over the course of the twentieth century, psychologists, the media, and the consumer economy have told Americans that boredom was a problem and that it might be possible to solve it. As a result, many have hoped that at some point they would find sufficient stimulation in their TVs, radios, and phones to keep dullness at bay. Such expectations have been raised again and again over the last century, but our reach exceeds our grasp. Whereas earlier generations were somewhat more accustomed to dullness, modern Americans are not, and they relentlessly pursue new ways to alleviate the condition. Each new game or social media site temporarily banishes boredom, but only for a time, and it ultimately increases longings for something new and better.

This intolerance for boredom has grown over the last century, for it emerged out of the conditions of modernity. In *Essays on Boredom and Modernity,* Barbara Dalle Pezze and Carlo Salzani claim that "modernity and boredom are . . . inextricably connected and inseparable." They maintain that "a new social economical and political reality engendered a new psychological situation." The rebellion against dullness and the need for novelty were byproducts of the Industrial Revolution, which mechanized time and made it drag on in the textile mill and on the assembly line. As America industrialized, the meaning of work became secularized. What had once been seen as morally redemptive toil came to be understood as tedious drudgery. These feelings in the workplace were magnified by the development of leisure for the masses and a simultaneous growth in commercial entertainment and consumer culture. While these afterwork activities made factory labor seem more bearable, they also heightened Americans' expectations about what life could offer. Many began to feel that it was their right as humans to be constantly engaged by the world: an idea that psychology enshrined as well. Often they turned to new technologies of amusement—phones, radios, televisions, computers—in an effort to fulfill such hopes. When those expectations went unrealized, boredom was often the result.[148]

Boredom has not always been with us—either as a word or as an experience. Born of a new expectation for constant excitement and change, it reflects a new vision of human nature dependent on constant stimulation. Whereas earlier generations did not like dull, monotonous drudgery, they were not surprised by it. In contemporary America, however, individuals have come to regard change, excitement, and constant activity as their due, as the key to emotional fulfillment. They expect to have unlimited diversions and entertainments, and consequently have come to see boredom as dangerous. With phones in hand, they attempt to flee it.

CHAPTER 4

Pay Attention

Is the internet blunting our intellects? Many Americans have wondered this, among them Nicholas Carr, who posed the question in an *Atlantic* article titled "Is Google Making Us Stupid?" and in a subsequent book entitled *The Shallows.* Carr worried that the distractions of digital culture made it more difficult for him to concentrate. Instead of reading and thinking deeply, he found himself skimming through articles, skipping blithely from one link to the next. At first, Carr believed his mental laxness was a simple result of aging. But after a time he concluded that the internet was rewiring his brain and radically changing the way he read and thought. In a fit of nostalgia, Carr observed, "Once I was a scuba diver in the sea of words. Now I zip along the surface like a guy on a Jet Ski." He lamented, "I missed my old brain."[1] And apparently with good reason—for research indicates that even fish can pay attention longer than the average human in the digital age. A 2015 study underwritten by Microsoft found that goldfish had longer attention spans than people nowadays: the former could pay attention to something for nine seconds, while humans could do so for only eight seconds, a significant decrease from the average twelve-second human attention span measured in 2000.[2]

Yet despite such findings, other pundits have been more sanguine about the relationship between the internet and cognition. The most notable of these include Steven Johnson, Clay Shirky, and Clive Thompson, who believe that the internet facilitates intellectual collaboration and has expanded access to data by enabling people to outsource most memories to the cloud.[3]

But whether positive or negative, few of these assessments has confronted the fundamental fact that, like emotions, definitions of cognitive ability are, and have been, historically variable, differing across space and time. To ask whether individuals are becoming smarter or dumber is really to ask how well they have adjusted to the demands of their society and economy, to the particular tasks required of them.[4] Today, Carr and many other Americans regard the capacity to concentrate and pay attention as fundamental to cognitive ability.[5]

Yet this conception of intelligence and mental capacity is of relatively recent vintage. It gained prominence in the late nineteenth century. As America industrialized, it became increasingly more interconnected, and this in turn multiplied and complicated the web of human relationships. As life became more complex, Americans placed a higher premium on the cognitive capacities that could untangle this web. Many educators, psychologists, and success advisors began to suggest that in this new age, it was important to have the ability to pay attention. But while industrialization led business leaders and educators to promote focused concentration, not all agreed that it would increase human productivity or happiness. It was, after all, a new way of channeling intellectual effort, and some feared it was dangerous. Today, most have let go of many of these anxieties; however, new ones have emerged as Americans struggle to understand the power and limits of their brains in the digital age.

Intelligence in American Life

No matter how intelligence or cognitive ability has been defined, few cultures have made it so central to their modes of evaluating individuals as have modern Americans. Perhaps worries about the intellect are particularly acute in the United States, because celebrations of meritocracy and individualism emphasize the talents and achievements of individuals, rather than their lineage, group membership, or religious affiliations. As American political scientist Lewis Anthony Dexter observes, "In most societies, the stupid are not victims of the same overt discrimination as in our society. For in other societies, race, clan-membership, ancestry, religion, status, physical prowess, and probably appearance, play more of a part in determining what rewards one gets and what values one is deprived of than in ours. A stupid person with the right ancestry, for instance, can 'get by' better than with us. A society which increasingly focuses on 'excellence,' meaning thereby intellectual excellence, as does ours, tends more and more to discriminate against stupidity."[6]

Because Americans have grown up in a society that ranks people based on perceived cognitive differences, they are frequently blind to the fact that this is a culturally contingent category and means of judging people. In feudal Europe, people discussed intelligence or wit, as they called it, but status was more often defined by family and ancestry. Others focused on religious piety as a way of ranking individuals. One historian suggested that these two modes of evaluating people were based on the idea of honor (which was derived from family standing and nobility of person) and grace (which was a sign of God's favor); this system held sway through the Renaissance and Enlightenment.[7] Not only was status determined more by birth or piety than by mental ability, but many believed that true, super intelligence—genius, omniscience—was something that only God possessed.[8]

During the Enlightenment and the birth of capitalism, however, as achievement came to supplant family standing, mental ability began to assume new prominence in social rankings. Enlightenment philosophers, with their faith in the power of individual reason, accorded certain types of intellectual activity greater significance and suggested that intelligence was the possession of humans as well as divinity. An array of writers—from Helvetius to Thomas Jefferson, Adam Smith to the Marquis de Condorcet—tackled the subject, trying to understand if all people had the same mental capacities, or if some existed on a higher plane. Some believed that only select individuals had access to the insights and flashes of brilliance that defined genius or super intelligence. Others, however, suggested that all could be equally intelligent, given the right education and opportunities. In their view, individuals had the power to improve their minds and sharpen their intellects through purposeful effort. In *Divine Fury: A History of Genius,*

Darrin McMahon suggests that Europeans were more likely to seize upon the idea of genius, whereas Americans, schooled in a democratic culture, were more likely to believe that everyone might be equally intelligent and possess particular capacities for individual greatness.[9]

But on both sides of the Atlantic in the eighteenth and nineteenth centuries there were philosophers, theologians, and writers who contended that, regardless of one's place in society, it was possible to sharpen one's intelligence. One way to do this was to learn to "fix" one's attention. In 1798, Maria Edgeworth, an English writer, believed that "to fix the attention of children . . . must be our first objective in the early cultivation of understanding." To Edgeworth and many of her generation, control of one's attention was a sign of self-mastery—and a virtue to be cultivated.[10]

Some imbued it with religious meaning. The English Protestant clergyman Isaac Watts wrote in his book *The Improvement of the Mind* that those wanting to practice attentiveness should closet themselves away from distractions, choose subjects that they appreciated, and, finally, "fix and engage the mind in the pursuit of any study by a consideration of the divine pleasures of truth and knowledge—by a sense of our duty to God . . . by the hope of future service to our fellow-creatures, and glorious advantage to ourselves both in this world and that which is to come."[11] Piety would enable individuals to focus their minds—indeed, piety was the real purpose for doing so.

This vision of attention as a virtue was firmly ensconced in the United States by the early nineteenth century, for Americans had enthusiastically embraced the idea that mental and physical diligence was godly. Educator Samuel Read Hall suggested that in order to "stimulate the student to make vigorous application in his studies," teachers must tell them of the importance of learning. They must "present their obligations to study as a duty," which they owed themselves, their teachers, their families, and "Him who made them. He requires them to make a due improvement of their time."[12] Those who failed to work hard both in school and later in life were disobeying god and dishonoring themselves, for all were capable of attaining success if they only worked and resisted the temptation of idleness. As minister Henry

Ward Beecher lectured young men in the 1840s about the dangers of laziness and the virtues of industry, he connected those qualities and vices to mental states. He wrote, "God rewards with a peculiar satisfaction, a mind in the daily discharge of its duty," and claimed that "the consent of every faculty of the mind to one's occupation" would help people find virtuous happiness. In contrast, the inability to focus was, he believed, a form of sinful sloth: "indolence . . . scatters the attention; puts us to our tasks with wandering thoughts, with irresolute purposes, and with dreamy visions."[13] Concentration was godly; mental wandering was not. In the end, mental control, and the success that it might bring, all came down to a matter of will, virtuous industry, and a dedication to God.

And yet these ideals of attention and industry were rather relaxed by modern standards. Intellectual labor often might be interrupted. In fact, schools ensured that it was, for in the eighteenth and early nineteenth centuries, there was little silence in which to concentrate. In the 1840s, Daniel Drake recalled his school days a half century before on the Kentucky frontier as raucous and noisy, writing, "The fashion was for the whole school to learn & say their lessons aloud, and a noisier display of emulation has perhaps never since been made. This fashion, in those days, was common to all our schools." He claimed that his education had taught him how to do "'head work' in the midst of great noise."[14] Warren Burton remembered being a student in the early nineteenth century and described the "idle and turbulent habits" of his school. Even when teachers tried to enforce discipline, the scholars often became "uncommonly riotous."[15] A Pennsylvania writer recalled that when he was a boy in the 1830s, he and his classmates had been "accustomed to the pretty free use of our tongue during school hours" and were positively shocked when a new teacher "called us to order and silence."[16] Teachers who insisted on silence would become more common, but during the first half of the nineteenth century, discipline was still somewhat lax and opportunities for silent, focused study scarce.

And while the Protestant ethic extolled labor, hard work—whether manual or mental—was offset by long, often unregulated breaks, meals, and frequent drinking. As historian Daniel Rodgers notes, "Work moved in irregular, often leisurely rhythms in preindustrial America."[17] Not only were nineteenth-century Americans' sense of attention and work rhythms different from our own, so too were their expectations about workers' commitments to their jobs. Embedded in the definition of virtuous industry that Beecher and others promoted was the expectation that workers would devote time to endeavors other than work, and engage in activities that might benefit their larger religious and civic communities. In short, in antebellum America, individuals, moral leaders, and educators believed that people should be diligent, pious, and well rounded.

The New Need for Attention

In the closing years of the nineteenth century, however, many Americans began to worry that diligence and a well-rounded intelligence were insufficient to the conditions of the day. The world seemed increasingly complex, a place that called for more than merely the powers of simple comprehension and industriousness. Some scholars suggest that a crisis in intelligence and intellectual understanding occurred at the end of the nineteenth century. For instance, historian Thomas Haskell describes a "recession of causation" that accompanied transportation and communications revolutions. Whereas it once had seemed clear how things worked and related to each other, with the rise of transcontinental and later transoceanic connections—railroads, steamships, telegraphs, and telephones—it became difficult to figure out cause and effect.[18] Many Americans had previously been isolated from each other, living in what historian Robert Wiebe terms "island communities," and consequently they had based their knowledge and explanations on causes and forces close at hand.[19] Suddenly, they were presented with the reality that their welfare was affected by people living far distant from them. As they were interpolated into regional, national, and international markets and networks, causal connections seemed increasingly remote and they came to believe that they lived in a more complex world.

Some pointed to the telegraph, others to the railroad or the telephone as fundamentally changing the social universe of knowledge. In 1858, Reverend William Sprague proclaimed that, as a result of the telegraph, "the world itself will become a vast school, in which its brightest and most accomplished minds will, as if sitting together in one grand convocation, alternately impart and receive of their collected treasures."[20]

Newspaperman William Allen White believed the railroad had transformed the intellectual and social order of rural Kansas. "Then the railroad came," he recalled, "and everything had changed. I did not know that the smell of coal smoke, which first greeted my nostrils with the railroad engine, was to be the sign and signal of the decay of a town and indeed of pioneer times, when men made things where they used them—all the things necessary to a rather competent civilization."[21] And yet others claimed it was the telephone that had altered the intellectual landscape of the world, particularly once long-distance connections were available in the early twentieth century. As a journalist noted in 1919, "The telephone actually carries a man a thousand miles in an instant," for it connected minds across the nation: "The man of today, sitting at his desk in New York City, can work and he can be with his brain in Boston, in Chicago, in Baltimore, Philadelphia, a hundred cities, in the course of a single morning. Soon this condition may be extended to Europe."[22]

But whether they pointed to the telegraph, the railroad, or the telephone, many in the late nineteenth and early twentieth centuries were convinced that the world had been fundamentally altered. And sometimes they regarded these changes with trepidation. As physician Luther Gulick put it in 1908, "A great many people are afraid of the complexity of modern life. They long . . . for anything . . . which would enable them to flee from our many-sided and highly organized world of to-day and get back to simple habits and simple needs. . . . Such longings are wasted for the most part. In a day of apartment-houses and telephones and prepared foods and domestic science, complexity must be accepted."[23]

That complexity grew out of the fact that there was more to know. To thrive in this transformed culture, individuals needed to develop new intellectual capacities and new habits of thinking and acting. Yet this could be a daunting prospect, for with so much information to master, how could one assimilate it? Some suggested the answer was to focus. In an essay translated for American readers and published in *Popular Science Monthly,* French physiologist Charles Richet discussed the "mental strain" that the new, interconnected world imposed on individuals. He wrote, "Civilization has certainly enormously extended our knowledge of every kind. A well-informed man to-day must know some three times as much as he would have had to know two hundred years ago; and in another hundred years he will have to know as much more." The way to navigate the torrent of information was to narrow one's focus, for "there is a limit to our mental capacity. We must learn to restrain ourselves. Instead of being encyclopedists, we shall have to be specialists."[24]

To become a specialist and avoid drowning in the vast sea of information, it was necessary for individuals to hone their powers of concentration and attention. They must focus, close themselves off from distraction, and follow only a single pursuit. Art historian Jonathan Crary suggests that this idea took firm hold in Europe and the United States in the late nineteenth century. He observes that "in the late nineteenth century, . . . the problem of *attention* becomes a fundamental issue. . . . Inattention, especially within the context of new forms of large-scale industrialized production, began to be treated as a danger and a serious problem."[25]

In psychology laboratories in Europe and the United States, investigators tried to understand the mental processes that enabled attention. As one Swiss researcher put it in 1896, "In the series of questions which concern psychology, attention takes the first and foremost place."[26] A host of psychologists, philosophers, and educators on both sides of the Atlantic, including Wilhelm Wundt, the father of experimental psychology, and William James, tried to unlock the nature of attention.[27] Less famous researchers produced dozens of books and articles with titles such as *Attention, Lectures on the Elementary Psychology of Feeling and Attention,* "Attention and Distraction," and "The Overcoming of Distraction and Other Resistances."[28]

While the general public was not necessarily aware of this academic research, they encountered an increased emphasis on attention and concentration in popular culture and daily life. Across the nation, observers suggested that the ability to concentrate distinguished the winners from the losers. Distractedness and wandering thoughts were all signs of an undisciplined intellect, an individual inattentive to the task at hand. While Henry Ward Beecher had focused on mental industry and diligence as signs of morality, the new turn-of-the-century creed began to downplay the religious qualities of concentration. Though some continued to think in terms of virtue and vice, many came to reframe the need for focus as a psychological and physiological battle between concentration and distraction, one that ultimately would result in success or failure.

Joseph Baldwin, in *Elementary Psychology and Education,* brought this message to students. He began his 1888 volume with a chapter on attention, which he believed was of paramount importance, and he offered tips on how to cultivate it. He declared, "He who can concentrate his whole being, all his energies and his capabilities . . . is obviously more potent in behalf of that object of his endeavor than would be possible were his energies divided."[29] Similarly, success writer Orison Swett Marden interviewed some of the most prosperous and prominent men in America at the turn of the century and found that they agreed that the ability to focus attention was a source of power. Thomas Edison told Marden that the key to success was "the ability to apply your physical and mental energies to one problem incessantly without growing weary."[30]

Fellow inventor Alexander Graham Bell, who had created many of the devices that would be blamed for distracting and interrupting thought, recognized that his own success depended on avoiding such distractions. "I begin my work at about nine or ten o'clock in the evening, and continue until four or five in the morning. Night is a more quiet time to work. It aids thought." Department store magnate John

Wannamaker confided, "The chief reason . . . that everybody is not successful is the fact that they have not enough persistency. I always advise young men who write me on the subject to do one thing well, throwing all their energies into it." Of Wannamaker, Marden intoned, "A feature of his make-up that has contributed largely to his success is his ability to concentrate his thoughts. No matter how trivial the subject brought before him, he takes it up with the appearance of one who has nothing else on his mind."[31] Elsewhere, John D. Rockefeller claimed that failure occurred "because we lack concentration—the art of concentrating the mind on the thing to be done at the proper time and to the exclusion of all else."[32]

Again and again, success advisors told young men that they could make their mark in the world if they only fixed their attention and concentrated on a single subject or goal. In his book *Business Psychology: A System of Mental Training for Commercial Life,* T. Sharper Knowlson told readers that the difference between the "men who are leading in business, finance and the professions" and those "fellows who lag behind" was "*the intensity of their thought-forces.*" According to Knowlson, such intensity was not innate but must be acquired through practice: "The habit of concentration needs time for its formation. . . . Begin with brief periods at first and increase the time slowly. Half an hour may be long enough to start with—later you may find three hours no burden."[33] Charles Gerstenberg, the head of the finance department at New York University, advised aspiring businessmen, "Concentration will go far toward increasing efficiency, but concentration is impossible without attention."[34]

Business leaders also endorsed the idea that extremely narrow specialization was necessary for success. Entrepreneur and business educator Roger Babson presented readers of his book, *Making Good in Business,* with a flow chart designed to help them pick a career. He explained that, to flourish, "young people must specialize in something. It may be apples or bricks or clocks or pencils or clothing or shoes. By the way, if you specialize in shoes be content to take only a part of the shoe, such as the sole, the heel, . . . the shoestring, or the eyelet. It would be very difficult for any person to know all about shoes, but one could very easily become a specialist in . . . heels, soles, eyelets, or shoestrings."[35]

In suggesting this, business leaders like Babson were overturning older notions of what it meant to live a good, balanced, and productive life. Knowlson offered a more explicit repudiation of the well-rounded individual, warning that those who were not single-minded in their specialized pursuits risked failure. He described two brothers: "George Burton keeps a store say in Lincoln, Nebraska, and his brother Sam has one in Toledo, Ohio. Both are honest and highly respected," but George was more prosperous than Sam. The reason? "Analysis shows that George is a 'one thing only' man: he lives, and thinks, and dreams for that store." Alas, poor Sam. Knowlson described him as having "other objects in view besides his store. No one would accuse him of a lack of industry—at any rate during working hours, but when the store is closed, he turns his attention to church work and his favorite study of history." Knowlson continued that there was "no reason in the world why we should not commend him for this"; however, "if he fails to take competition seriously," then competing stores would drive him out of business.[36] Whereas earlier generations would have regarded Sam as an ideal citizen and a pious man who ably balanced vocational interests with civic and religious concerns, his well-roundedness was becoming a deficit, at least in the minds of success writers.

Universities Train the Mind

To produce narrowly focused specialists, American businesses and government organizations increasingly turned to universities, which, in the late nineteenth century, were taking on their modern form. Before their emergence, males seeking education beyond high school attended what historians term the "old-time college," where students followed set curricula dominated by the study of Greek, Latin, and moral philosophy. While fluency in these subjects took effort to acquire, the educators who promoted the ideal of the old-time college were mainly concerned with developing well-rounded graduates rather than experts in highly specialized fields. These views were epitomized by the Yale Corporation's 1828 curricular manifesto, which flatly rejected narrow professional education in favor of learning that was more general in character and that fostered a free-ranging intellect:

> The great object of a collegiate education, preparatory to the study of a profession, is to give that expansion and balance of the mental powers, those liberal and comprehensive views, and those fine proportions of character, which are not to be found in him whose ideas are always confined to one particular channel. He who is not only eminent in professional life, but has also a mind richly stored with general knowledge, has an elevation and dignity of character, which gives him a commanding influence in society, and a widely extended sphere of usefulness. His situation enables him to diffuse the light of science among all classes of the community.[37]

Such a broad education was necessary, for life should have as its purpose something beyond "obtain[ing] a living by professional pursuits." Each graduate had "duties to perform to his family, to his fellow citizens, to his country; duties which require various and extensive intellectual furniture."[38]

This educational philosophy dominated American higher education up through the Civil War. However, as America industrialized and began to confront the social and economic challenges that the machine age produced, educators started placing greater emphasis on the development of experts with the requisite knowledge to tackle new and complex problems. As a result, the curricula and the mission of the old-time college were eclipsed by a form of education that placed a high priority on research, advanced degrees, and intellectual specialization.

To be sure, Americans had not denigrated mental focus when the old-time college was in its heyday and then suddenly embraced it as the culture of professionalization grew up alongside universities. The old-time college, after all, was very much interested in disciplining students' mental faculties through the repetitive and laborious study of

Greek and Latin. But the new emphasis on professional knowledge, and the fact that the professions required potential practitioners to spend years of extra time studying, changed how this particular dimension of intelligence was conceived. At universities, students were to focus more intensely than before, and they were required to sustain that level of attention over many more years. An early proponent of specialization, Charles Eliot, captured the mental consequences of professionalization in a letter he wrote to a friend in 1854, when he was twenty years old:

> What a tremendous question it is—what shall I be? When a man answers that question he not only determines his sphere of usefulness in this world, he also determines in what direction his mind shall be developed. The different professions are not different roads converging to the same end; they are different roads, which starting from the same point diverge forever.[39]

Eliot decided on a career in chemistry and mathematics, starting down a road that ultimately led him to the presidency of Harvard. He presided over that institution as it came to offer the rising generation highly specialized courses of study, which were pathways to the professions. His own career, and the direction in which he turned Harvard, was emblematic of the nation's increasing interest in cultivating focus and attention. During his tenure as president, Harvard added a graduate school and business school, and bolstered its medical and law schools.[40] Across the nation, other institutions of higher education did the same, as they increasingly came to focus on professionalized training and specialized fields of knowledge.[41]

While the culture of professionalism grew hand in hand with industrialization, it did not succeed in making everyone equally enamored of specialization. In the university, educators who harbored nostalgia for a more well-rounded course of study raised objections against the direction of modern higher education. For example, Irving Babbitt, a professor at Harvard, wrote in the *Atlantic* in 1902,

> The work that leads to a doctor's degree is a constant temptation to sacrifice one's growth as a man to one's growth as a specialist. . . . The old humanism was keenly alive to the loss of mental balance that may come from knowing any one subject too well. The whole problem is a most difficult one: the very conditions of modern life require us nearly all to be experts and specialists, and this makes it the more necessary that we should be on our guard against that maiming and mutilation of the mind that come from over absorption in one subject.[42]

Similarly, John Bascom, the president of the University of Wisconsin, expressed distaste for an education that made the mind "microscopic in vision and minute in method, rather than truly comprehensive."[43]

And students too sometimes seemed discontented with the intellectual demands placed on them. Ruth Nelson, a student at Cornell University, wrote her parents in 1894, complaining that the university had adopted the rhythms of industrial life. She observed, "The professors in this university treat us like a lot of machines."[44] Two years later, another Cornell student complained, "All Sage has developed into a mill, which is grinding day and night. . . . Every other door almost is adorned with a sign 'Busy' and 'Please do not disturb.' . . . In the evening, Faith, Agnes & I went to the Library where we could get quiet."[45] Harold Gulvin, a student at Cornell in 1927, wrote to his girlfriend, "They have a system here of cramming knowledge into you in a short time."[46]

It was not just the workload that students found off-putting; it was the need for close attention. Many students found that it was harder to concentrate during lectures than they had expected. While educators might idealize focused intelligence, students themselves found it difficult to embody that ideal. Evan Randolph, a Harvard freshman, wrote on March 1, 1900, that he had started his day by attending his Government class. Despite the expectation that he would pay attention, he could not. The class "is given by Professor McVane. He is probably a very learned man, but a dam [*sic*] stupid lecturer, so I wrote my daily theme there instead of listening." The rest of Randolph's day was filled with classes and meetings with professors, as well as a trip to the gymnasium. He confided, "I don't remember studying so hard since I have been at college."[47] A more cynical student defined the lecture as "a system of education whereby the professor talks while the students do the *Sun* crossword puzzle, read alien textbooks, sleep, write the next hour's assignment, discuss past, present, and future dates, comb their hair, catch flies, knit, and as a last resort, take notes."[48] Such laments filled the writings of college students for decades. Nile Kinnick, a student at the University of Iowa, and a Heisman trophy winner, wrote his family in 1939, telling them he had been giving serious thought to "just how a man should study, etc." He had concluded that rather than be educated narrowly in a single profession, students should instead be taught to read widely. He railed against "too much detail, mediocre conservatism and directed study," believing they impeded "initiative, imagination, and ideas."[49] While educators might celebrate the intensive mode of education their universities offered, not all students shared their enthusiasm.

Neurasthenia and Worries about the Mind

Others watching the transformation of higher education as well as the larger economy harbored deeper misgivings, for many doctors believed that the overly focused intellectual work that the world of business increasingly required might actually carry with it physical harm. Focused attention seemed to be necessary for success in industrial society and corporate capitalism, but it was not without risks. Beginning in the 1870s, and continuing into the early twentieth century, physicians, educators, and psychologists in England and America documented physical ravages caused by what they variously termed "mental strain," "cerebral hyperaemia," or "mental fatigue." They suggested that individuals who spent too much time paying attention, too many hours immersing themselves in just one topic, risked damaging their brains.

Time and again, doctors documented dire medical conditions that they claimed resulted from students concentrating on the rigorous studies and intellectual labors demanded of modern mental workers.

In an 1878 work on insanity in ancient and modern life, British physician Daniel Tuke suggested, "Monotonous work long continued would seem to exert an unfavourable influence on the mind. . . . If diversified, hard work is much less likely to prove injurious." He presented case histories of students, school masters, accountants, and others engaged in tasks that required a narrow focus, and told of their unhappy outcomes. He pointed an accusing finger at "the psychological mischief done by excessive cramming both in some schools and at home," which he declared to be "reckless."[50] He recounted cases of students who had suffered profound physical damage as a result of such studies:

> A youth aged nineteen, studying for a certain status at Cambridge under a tutor was obliged to do a great deal of work in a short time. The first symptoms were inability to give sustained attention to his studies, and great listlessness. Then, after being in bed for one night for about an hour, he woke up raving mad, and required three people to hold him. The pulse was feeble and there was no evidence of inflammation. The brain was utterly exhausted. The following night he woke up again maniacal, and so again on the following morning.[51]

Only slowly did he recover.

Some doctors maintained that these deadly mental afflictions also struck younger children who were compelled to study intensively at a tender age. According to a report published in the British medical journal the *Lancet,* youngsters who engaged in rigorous study were dying of water on the brain at a far higher rate than in previous years. One had expired while crying, "I can't do it! I can't do it." Another slipped away while saying, "Governess, it's too hard; I can't learn it."[52] Doctors confirmed that American children's delicate brains were similarly fatigued, leading to a spike in nervous disorders among the nation's young.[53]

In fact, a number of medical authorities believed Americans were more prone to the condition than people of other nations. Doctors such as Charles K. Mills documented the "mental over-work and premature disease among public and professional men" and noted that "extreme mental activity, overstrain, and excitement must be regarded as characteristics of American civilization." This sprang from the perception that there was endless opportunity and possibility for all—and the ironic result was that "our very liberties and opportunities become sources of peril to health and life."[54] George Beard, author of *American Nervousness,* suggested that the speed of American life and its embrace of new technologies could also be blamed for mental and nervous stain. Beard believed that his generation was living through a revolutionary time, where new knowledge was being generated and adopted at an astounding rate. "The experiments, inventions, and discoveries of Edison alone have made and are now making constant and exhausting draughts on the nervous forces of America and Europe, and have multiplied in very many ways, and made more complex and extensive, the tasks and agonies not only of practical men, but of professors and teachers and students everywhere."[55]

This overload of information made it necessary for individuals to think more and to be ready for all contingencies. In his 1875 treatise, *On Mental Strain and Overwork,* Doctor Frederick MacCabe of Ireland discussed the financial world that businessmen must navigate, the risks they must assume, and the faith in and knowledge of distant governments and economies they must cultivate. "All these conditions," he concluded, "constitute a web of inextricable complexity surrounding the enterprising commercial man of the day. . . . Telegrams arriving at all hours with fluctuating quotations, and producing rapid alterations of hope and fear, must preclude all possibility of mental repose."[56] Such an expenditure of energy ultimately might deplete the resources of the modern mental worker, leaving him exhausted.

Technology and urban industrial society placed unnatural demands on human physiology, requiring the body to do things for which it was not constructed. Doctors believed such unprecedented mental exertions drained blood, depleted molecules, stretched vessels, and caused other harm to the fragile, finite brain. Beard described "the heightened activity of the cerebral circulation which is made necessary for a business man since the introduction of steam-power, the telegraph, the telephone, and the morning newspaper."[57] William Hammond, a former Surgeon General of the United States, and a member of several East Coast medical school faculties, offered a new diagnosis for the many overtaxed brains he had encountered. He labeled the condition that plagued them "cerebral hyperaemia," and described it as a problem that arose when "the cerebral vessels" had "become permanently over-distended" from excessive mental work. In extreme cases, the condition caused "the decomposition of brain substance."[58]

Hammond provided case studies, writing of a certain F. H., "engaged in a manufacturing business which required all his attention to make it profitable." When F. H. learned of a problem with his supplies, he "experienced an intense sensation of vertigo, a sharp pain in the head, palpitations of the heart, . . . a roaring sound in the ears and flashes of light before the eyes." These symptoms persisted for several months. Hammond reported on a bookkeeper, who, after a day of taxing work, "was seized with vertigo and fell unconscious on the sidewalk." When he regained consciousness, he suffered from "noises in the ears and indistinctness of vision. . . . Mental application was impossible without leading to an aggravation of all of his symptoms." More dramatically, a prominent banker tried to commit suicide because he could not tally a column of figures correctly.[59] These were not isolated cases but, according to Hammond, alarmingly widespread. Those who were prone to such conditions found it "difficult to fix the attention," and when they tried, this "almost invariably increases the pain or uneasiness felt in the head."[60]

Dr. Horatio C. Wood, of the University of Pennsylvania medical school, also observed such problems but offered a slightly different explanation, arguing that "thought . . . marks the death of protoplasmic molecules." While usually the brain could replenish such molecules, when "thought follows thought with such instant rapidity," one might use up too many molecules before they could be naturally replenished, and the brain might never again fully recover."[61]

These explanations for why people were wearing out in the new industrial society were built on the idea that brains were finite in their capacity.[62] They could do so much, and no more. Telegrams and telephones required unprecedented and unhealthy expenditures of energy, cells, and blood. It was violating rules of nature to demand of mortal minds so much more than they had been built for.

While all humans might be taxed by the demands of the age, some were at a greater disadvantage than others because of differences of race or sex. For instance, medical authorities frequently argued that African Americans were only rarely afflicted with the nervous disorders that sprang from mental work because they were supposedly incapable of doing it. George Beard asserted that African Americans were instead more likely to suffer from hysteria—a condition that was a sign of "an excess of emotion over intellect." The disorder could be found among "stout Irish servant girls, . . . Southern negroes, and among the undisciplined and weak-minded of all races and classes and ages."[63] Charles Burr, a Philadelphia physician, maintained that the condition was rare, but not completely unknown, among African Americans. "The American negro never suffered from neurasthenia till recently. But now it is of not infrequent occurrence, especially in half-breeds." Burr alleged that racial inferiority was the cause. "[A] lower race is trying to compete with a higher, and is unfit to do so. The strain is too great. . . . [T]he mere struggle of the best specimens of an inferior race to attain the plane of a superior leads often to their downfall."[64] Mental work was draining for all, but turn-of-the-century doctors believed that brain power was not evenly distributed across the population and those with the least amount of brain power might be taxed the most.

For this reason, they believed there were stark contrasts between men and women as well. While men might fall victim to mental overstrain and experience symptoms such as vertigo, hallucinations, and palpitations, women faced a far grimmer fate. After all, they supposedly were doubly hampered in their efforts to assimilate into the new information economy of the nineteenth and early twentieth centuries: not only was there more information than ever before, but their brains were far smaller than men's, which made them particularly maladapted to the demands of modern education and training. Hammond claimed that women were especially prone to cerebral hyperaemia when doing arithmetic, for their brains were not large or strong enough for such an endeavor. He claimed he had seen many a female sufferer of the malady in whom the condition "was directly induced by the study of calculus, spherical trigonometry, and civil engineering." He continued in a resigned and glum tone, predicting, "I suppose civil engineering will be responsible for many hyperaemic brains in young girls."[65]

Edward Clarke, a member of the Harvard Medical School faculty, believed women also might ruin their reproductive systems by engaging in mental activity during sensitive times of their menstrual cycle—after all, women's natures were characterized by "rhythmic periodicity," and if they did not work in concert with their biological tempo, they faced grave dangers. Women who ignored their menstrual cycles and did focused intellectual work at college straight through a month, rather than resting for a week, might divert to their brains blood that was vital to their reproductive systems. For many such young women, the result was sterility and the withering of the ovaries.[66] Other experts, including Silas Weir Mitchell and pioneering psychologist G. Stanley Hall concurred, warning women that their reproductive systems unfitted them for intense mental endeavor, at least during some days of the month.[67]

While men and women might be harmed in different ways by intense mental labor, the cure for both was the same. The best way to remedy the condition was to do exactly what leaders of the new industrial economy maintained was ill advised. One should rest and gain hobbies to divert the mind away from the specialized knowledge that could injure brains. Women were sometimes prescribed a complete "rest cure" in which they lay abed and did little, while men were given slightly more active regimens, which nevertheless still required a mental vacation.[68] Hammond wrote, "If the patient is unable to take an interest in anything but his business, the case is indeed a bad one, and the possibility strong, that some more serious affection than cerebral hyperaemia will assuredly make its appearance." To use the brain for anything strenuous while recovering from an attack of hyperaemia was equivalent, he said, to walking "three or four miles while suffering from an inflamed knee-joint." Patients must remember that "the first requirement in the treatment of the diseased organ is almost always *rest*." The way to treat it was "to afford rest to the brain by cessation or diminution of mental work and by diversions of various kinds."[69] Horatio Wood recommended rest and recreation but warned patients that even these could be too taxing. Patients must choose their amusements carefully. For instance, he particularly warned against the game of chess, arguing that it was too intellectually intense an activity. "It is absolutely to be condemned as a recreation to those whose life-work requires long-continued hard thinking." He warned, "Recreation should not involve mental labor."[70] Overtaxed brains deserved a true reprieve from concentration and intense attention.

To turn-of-the-century Americans confronted with the vast web of complexity created by industrialization, by telegraph wires crisscrossing the nation, by sprawling railway systems, the world often seemed overwhelming. Some believed the only way to grapple with all that modern society demanded was to narrowly focus on one small part of it. Others believed that such a narrow focus brought with it dangers of mental strain. Perhaps minds were just not made to assimilate so much information. The suggestion that individuals should take a rest cure, avoid excessive work, at least for a time, illustrates one difference between nineteenth-century and modern approaches to attention, mental effort, and intelligence. Earlier generations believed the mind was a limited resource with finite supplies of blood, molecules, and energy. It might need a break from constant activity, and should not always try to focus on the new information coming in, whereas today, with Ritalin and Adderall, the goal is to train the mind to accommodate the many demands that the information economy puts on it.

Manual Work and Minds

While concentrated study and focused *intellectual* activity excited commentary, so too did the rise of focused *physical* labor on factory floors. Commentators suggested that factory work could be demoralizing to workers because of its boring repetitiveness and mental emptiness. An early critic, the Unitarian minister William Ellery Channing, objected to industrial labor as one "monotonous, stupefying round of unthinking toil."[71] An Episcopal minister and social reformer, R. Heber

Newton, testified to a congressional committee that the factory worker had become simply a "tender upon a steel automaton, which thinks and plans and combines with marvelous power, leaving him only the task of supplying it with the raw material, and of oiling and cleansing it. . . . What zest can there be in the toil of this bit of manhood?"[72]

Medical experts catalogued the effects of this labor, believing that it exerted a mental toll on manual laborers just as it did on the more elite mental workers. Beard described this problem in *American Nervousness,* claiming that because of specialization, industrial workers' activities were "restricted to a few simple exiguous movements, to which they give their whole lives—in the making of a rifle, or a watch, each part is constructed by experts on that part. The effect of this exclusive concentration of mind and muscle to one mode of action, through months and years, is both negatively and positively pernicious. . . . Herein is one unanticipated cause of the increase of insanity and other diseases of the nervous system among the laboring and poorer classes."[73]

Yet while Beard conceived of narrow focused work as dangerous to menial laborers, some commentators believed that manual workers were, in fact, suited by nature for their repetitive, monotonous jobs. As a proponent of factory labor intoned, "The majority of human minds are weak, and slow, and could do little in the world but for simple tasks adapted to small and barren brains. . . . Monotonous toils suit them exactly. . . . The exact and punctual habit, which the machine engenders, trains careless minds with a discipline most wholesome, . . . and the better minds quickly rise out of that class to something better."[74] Harvard's president, Charles Eliot, also expressed sanguine attitudes about the state of blue-collar work. He saw the benefits of intellectual specialization and celebrated its virtues in higher education. But he was relatively blind to the plight of lower classes who complained of the stultifying narrowness and mental emptiness of manual work.[75] During an anti-union rally in Boston, he attacked unions for regarding "labor as a curse." In a withering review of Eliot's speech, a labor activist wrote that Harvard's president was oblivious to the fact that factory work meant that "man is compelled to toil over one kind of work, with its wearying monotony—work that is frequently rendered soul-deadening in influence." In contrast, the union activist claimed that Eliot's work (specialized though it might be) was "of such a character that the imagination is constantly fed, the brain stimulated as well as fatigued, while the body is not kept on a strain that produces constant weariness."[76] Yet complaints about the tedium of manual labor generally fell on deaf ears, for like Eliot, many elite and middle-class Americans seemed unmoved by the possibility that specialization might be affecting industrial workers' mental acuity and well-being.[77]

The most extreme, famous, and unapologetic proponent of the idea of focused manual labor was the efficiency expert Frederick Winslow Taylor, who, as noted in Chapter 3, also played a role in the emergence of boredom. In his 1911 *Principles of Scientific Management,* he argued that workers should specialize, and that their tasks were to be determined by their intelligence and their abilities to focus their minds. With stopwatches and time-motion studies, Taylor believed he could streamline industrial processes and increase the productivity of people of all ranks, talents, and classes. He wrote that scientific management would foster "the development of each man to his state of maximum efficiency, so that he may be able to do, generally speaking, the highest grade of work for which his natural abilities fit him, and it further means giving him, when possible this class of work to do." Workers' intelligence, or lack thereof, was innate, and little could alter it. Of men loading pig iron, he wrote, "Now one of the very first requirements for a man who is fit to handle pig iron as a regular occupation is that he shall be so stupid and phlegmatic that he more nearly resembles in his mental make-up the ox than any other type. The man who is mentally alert and intelligent is for this very reason entirely unsuited to what would, for him, be the grinding monotony of work of this character." Taylor believed that the industrialized economy was so complex that such mentally addled workers could not possibly understand the larger structures and forces that shaped their work. "The workman who is best suited to handling pig iron is unable to understand the real science of doing this class of work." Other, smarter workers would teach him how to work "in accordance with the laws of this science."[78]

Some workers might protest such divisions of labor, Taylor explained: "As the workmen frequently say when they first come under this system, 'Why, I am not allowed to think or move without someone interfering or doing it for me!'" This, however, was not just the workingman's complaint, for "the same criticism and objection . . . can be raised against all other modern subdivision of labor." All workers needed to realize that "this system is not intended or designed to develop geniuses. . . . Geniuses . . . are not wanted in every-day work."[79]

Workers of all sorts therefore must subsume themselves and their minds to the larger system. As Taylor famously declared, "In the past the man has been first; in the future, the system must be first."[80] Employees with wide-ranging intellects were not needed; instead, a coordinated team of narrowly trained, tightly focused, highly synchronized minds, capable of deep concentration on a single thing, was what the new industrial economy required.

Taylor denied that manual workers had significant mental capacity; others argued African Americans or women of all races lacked brain power. And those who worried about mentally fatigued white men also believed there were limits to what their brains could do. While some believed elites were more suited to intellectual work than other groups, experts still conceived of their mental energy as limited. The doctors who worried about cerebral hyperaemia, neurasthenia, and mental fatigue contended that human brain power was ultimately a finite resource. Individuals had only so much energy and capacity in their heads. When they worked too hard, they eventually ran out of fuel. The industrializing society was requiring men and women to exceed their natural limits, to do unnatural and reckless things with their brains.

That view—of the brain as finite—began to change in the early twentieth century. Central to this transformation was philosopher William James. He believed that nervous exhaustion was an actual condition, but also that it could be overcome. In a 1906 speech that was later reprinted and widely circulated, James suggested that in the United States "few men live at their maximum of energy."[81] But he believed that with proper training, Americans could use all of their mental and physical energies far more efficiently and effectively than they currently were.

Out of this speech slowly emerged a new mythology about the brain—that humans used only 10 percent of their true mental powers, and that the brain might do ever more. James himself had offered no such numerical estimate; instead, that ratio was attributed to him in the preface to Dale Carnegie's 1936 best seller, *How to Win Friends and Influence People,* a guide for those eager to rise in the world.[82] The claim of untapped mental power suggested that individuals need no longer fear mental exhaustion from too much brain work; instead, they could accommodate all that came their way. There need be no natural, biological impediments to success. Everything was possible. This new folk conception of the brain gradually gained traction over the course of the twentieth century, and despite the dubious science behind it, still continues to inform the way many see the brain, attention, and intelligence.

Media and Minds

Perhaps this new vision of the brain was compelling because many of the technologies that were invented at the turn of the century seemed to require increased brain power. While machines and the corporations that owned them were placing new demands on Americans, requiring them to concentrate with greater intensity, many of the inventions of the nineteenth and twentieth centuries were also producing new distractions that undermined concentration. These technologies seemed to demand of their users more focused intellectual activity, yet they also seemed to be the very things that interfered with focus.

For instance, the telephone brought people more information, but its ringing bell could jangle the nerves and interrupt concentration. In 1881, a journalist for the *Omaha Daily Bee* detailed the advantages and disadvantages of the telephone, describing typical annoyances that it brought. Among them

> The accountant is half through with a column of figures when the bell rings; the dentist is pulling a tooth and the bell rings; the

> clergyman is sending up a fervent prayer when this terrible din destroys all thought of devotion. The poet may have an inspiration, the merchant a bargain, the sick man a taste of sleep; the remorseless bell spoils them all and throws each disappointed listener into an hour of agony and worry. It is very wonderful that you can hear a man talk who is a hundred miles away, but it is infinitely discouraging to have him talk at all sorts of hours and times when you are not prepared for him.[83]

The new interdependent, highly technological society required a delicate balance. It asked people to concentrate, but it also provided new distractions. One needed to be hooked into the system, but also able to ignore its excesses and interruptions.

As a result of such connections, the American at the turn of the century might be better informed but also more distracted than ever, and this conundrum only grew as new modes of communication were introduced. In the 1920s and 1930s, many grappled with the meaning and place of the radio in their mental lives. Could they take in all it had to offer without sacrificing other mental powers? Was it a force of enlightenment or a source of distraction and dissipation? At various times, listeners took both sides of the issue. Annie Maude Dee Porter, a native of Ogden, Utah, reported in 1936 that the radio interrupted serious thought. She wrote in her diary, "Spent a stupid & useless morning at home did not even get the papers read. The radio interferes with my intellectual life very much."[84] On the other hand, it sometimes was enriching, as it seemed to be in the fall of 1940. "We live on the radio War & politics (& football) give a lot to hear & think about."[85]

Some suggested that in addition to providing news, the radio honed mental powers, for its very noisy intrusiveness made listeners exercise their brains and taught them to focus their attention in the face of distraction. In their 1935 study of the psychology of radio, Harvey Cantril and Gordon Allport wondered "whether such persistent use of the radio is having an effect upon our powers of concentration, upon our habits of listening, and upon our nerves." They believed that the radio might be inculcating in Americans habits of careful listening:

"Radio is probably improving the capacity of the average man to listen intelligently to what he hears." And they suggested that learning to close out distractions might enhance concentration. For instance, students might study with the radio on, and learn to let its tunes not intrude on their homework. But Cantril and Allport believed that, nevertheless, the practice of listening to the radio while doing something else was costly in terms of psychic energy. "Even when distractions are inhibited they may be nerve-racking in the long run. A selection between competing stimuli can be made only at the cost of effort." They concluded that "working 'against the radio' may enhance the degree of attention given to a chosen task, but only at the cost of strain and fatigue."[86]

Certainly this was something that worried educators. In a survey of university and college deans, the *New York Times* inquired about campus regulations for radios. Should colleges promote radio, establish their own stations, or ban sets from dorms? The answers varied. Clarence Mendell, the dean of Yale University, explained that radio was discouraged: "We have, at Yale, no central radio for broadcasting to the student body, nor do we encourage private sets. I believe that life is already too complicated and noisy for the best results without introducing any further disruptions." Lieutenant Colonel S. Whipple of the United States Military Academy shared many of these hesitations. Radios were not permitted in individual rooms, for they would be "a hindrance to the concentrated study required." On the other hand, they were allowed elsewhere, for they gave students "a necessary connection to the outside world which cannot be maintained through the medium of newspapers."[87] Students needed to know what was going on in the world, but they also needed to be protected from too much distracting knowledge that might impede their concentration.

Most students did not agree, however. Cantril and Allport pointed out that "those who have walked past college dormitories and heard the rhythmic strains of jazz orchestras or the voices of crooners issuing from numerous loud-speakers have wondered what effect radio is having on the scholastic habits of the younger generation. A questionnaire was given to 200 students of both sexes to see whether they actually try to combine listening with studying. Seventy-one per cent of these students reported that they had radios in their rooms." Sixty-eight percent of students admitted that they thought they studied less effectively with the radio on.[88] Yet nevertheless they continued to tune in while studying, reflecting a growing faith in the possibility of dividing attention, of trying to juggle more mental tasks simultaneously. Whereas an earlier generation of doctors had conceived of the brain as finite in its powers, Americans, in their listening habits and study routines, were expressing greater faith in the expandable capacity of their own brains.

Educators also worried about younger children, who seemed to be powerfully drawn to the radio and its offerings. In a debate with George Gershwin over the virtues of radio, Blanche Colton Williams, a Hunter College professor, took the role of critic, noting, "Let's observe a few practical examples of Radio vs. The American Family: Children trying to study over the clamor, people attempting to sleep while competing loud-speakers blare in neighboring apartments. In many families, the instrument is left on from the time papa comes home, sometimes from the time mama gets up."[89]

Azriel Eisenberg backed up this claim as he described the children he had observed in New York City. "For many hours each day, the youngsters gather around the radio to listen with rapt attention to the thrilling adventures of their beloved comic strip heroes and heroines. So enthralled are they with the 'air-and-ear' show that they have developed the habit of dividing their attention between the humdrum preparation of their school assignments and the compelling excitement of the loud-speaker." Yet Eisenberg suggested that the radio was actually teaching new cognitive skills—the art of splitting one's attention. "With the introduction of the radio, both adults and children have learned to divide their attention. The danger of developing a tendency toward poor concentration is not so great if the listener's attention is divided between a musical program and whatever else he may be doing (although it is true that his appreciation of the music will be impaired). Should he, however, be listening to a narrative or dramatization and attempt to do something else at the same time, one or the other must suffer. Probably children give all their attention to the radio when listening to their favorite programs. To programs that are secondary in their preference they most likely give only part of their attention."[90] Interestingly, he reported that children who were rated more intelligent were more likely to divide their attention and listen to the radio while doing other things, including their homework.[91] Whereas undivided attention had been the ideal for at least half a century, the rise of radio, and its entry into American living rooms, bedrooms, and dormitories, presaged an increasing tendency for Americans to split their attention among several activities and to believe they could do multiple things simultaneously.

The rise of radio represented an important turning point in other ways as well. As was the case with loneliness, radio lowered people's tolerance for solitude, making it more difficult for them to be alone. The ability to experience solitude as a positive state was not only necessary for those dealing with aloneness; it was also central to ideas of focus and concentration. The radio jeopardized the idea of the single, sometimes isolated, autonomous, thinking individual, and began instead to link listeners to other places and groups with just the turn of a dial. To choose to be alone with one's thoughts came to be a more difficult challenge, for the radio made it easy to distract oneself from the task at hand.

Many of the concerns about the radio were also echoed in people's reactions to the television. One fear was that it captured the mind in an almost hypnotic way and diverted people away from what they ought to be doing. Viewers were paying attention, but to the wrong things. Orville Prescott, the *New York Times'* book critic, worried about what he saw around him. Writing in 1950, early in the television era, he noted, "I have frequently gone into the homes of my friends who have young children," he said, "and it is an alarming sight to see those kids, particularly on a Sunday afternoon from about three-thirty until about eight, sit with a glazed stare in their eyes and look at that television screen. There is no intelligent reaction at all, and if you go into the room this little child of five who has been taught manners for five days a week, says, 'Shut the door or get out!' It is not only corrupting their minds, I think, but ruining their manners."[92]

A multitude of studies produced in the 1950s indicated that as TVs arrived in living rooms, Americans put their books aside.[93] In 1959, a critic complained that "with few exceptions, all American homes are wired to TV's increasingly imbecile radiation. Statisticians estimate that 'the average American' between six and 90 spends an average of 20 hours a week staring at the TV screen—one half as much time as he works for a living, two-thirds as much time as he passes in schools, 20 times more than he meditates in church."[94]

Many feared that the "imbecile radiation" was particularly dangerous to the intellects of young children. TV viewing seemed to be positively correlated to an average 15 percent decline in grades.[95] And some worried that rather than replacing books, the TV was introducing a new kind of distracted reading, for a growing number of students were doing their homework while watching their favorite programs. By 1963, one study indicated that 16.4 percent of students in second and third grades and 18.3 percent in fourth, fifth, and sixth grades watched TV and did homework frequently.[96] While this was a cause of alarm for many, it also reflected the growing folk wisdom that the brain could do anything, that attention could expand.

Attention in the Postwar World

If radio and television divided attention and provoked debates about distraction, so too did other mid-century developments. While psychologists had neglected the subject of attention during the 1920s and 1930s, the topic was revived during World War II. The military wanted to help its personnel concentrate when performing jobs that required extreme forms of multitasking as well as work like radar watching and aircraft piloting, which demanded critical concentration for extended periods of time.

In prosecuting the war, the Allies manufactured increasingly more sophisticated aviation technology as they attempted to outpace the efforts of the Axis powers. But as aircraft became more advanced, aviation personnel not only had to perform challenging cognitive tasks, but also had to perform them in situations where mistakes had costly, sometimes fatal, consequences. These were epitomized by air traffic controllers who could not keep up with all of the information on their radar screens or by pilots who would mistake the levers that controlled their flaps with the levers that controlled their landing gear.[97]

Following the war's end, British psychologist Donald Broadbent (who had been a Royal Air Force recruit during the conflict) began experiments that examined how humans selected what to pay attention to and what to ignore in the process of selectively focusing the mind. In 1958, Broadbent published *Perception and Communication,* which quickly became a seminal work in cognitive psychology and attention studies. He suggested that attention depended on being able to inhibit some responses to stimulation, to filter out irrelevant information, and to ignore distraction. His work was subsequently refined and sometimes challenged by a host of other scholars who began to assiduously study the nature of attention, distraction, and multitasking. While the scholars did not always agree with each other, on one matter there was a consensus: in the post–World War II era, it had become imperative to fashion a formal study of attention, and that study was spurred on by the challenges associated with the increasingly powerful machinery of modernity.[98] Recalling these times, Paul Atchley, a University of Kansas psychologist, reflected that "we had these highly motivated individuals—radar operators and pilots—who would miss attacks or drop bombs on the wrong cities. Why did they fail?. . . . We could quantify the machine, but not the human. That's where cognitive neuroscience really started."[99]

Like their nineteenth-century predecessors, postwar psychologists took an interest in attention and the mental challenges that modernity imposed. Some things had changed though. Most notably, words like "cerebral hyperaemia" and "neurasthenia" fell into disuse, as did the suggestion that the mentally fatigued and overstimulated should seek relief for their tired brains by cloistering themselves away from too much sensory input or by reducing their workload. Rest cures were a thing of the past, as were the fears that one could damage the brain by overwhelming it with too much information. Instead, psychologists placed renewed emphasis on finding ways to expand Americans' cognitive capacities. As Matthew Smith argues in *Hyperactive: The Controversial History of ADHD,* this was especially manifest after the launch of Sputnik, when Americans began to doubt whether they were producing a workforce with enough brain power to compete in the space race against their Soviet rivals.[100]

One of America's first attempts to redress this worry was to pass the one-billion-dollar National Defense Education Act in 1958. As envisioned by its proponents, the NDEA was designed to improve teaching in English and foreign languages, science, and math, and to reduce attrition rates. While it had notable effects on these disciplines, one of its ancillary consequences was to reform how less attentive students were treated in the education system. Prior to Sputnik, these students might have dropped out of school and gone directly into the workforce. With the new emphasis on graduating a more educated populace, however, teachers found themselves pressured to keep less-attentive students in school. These pressures, in turn, were compounded, as educators, under the sway of the NDEA, assigned more homework and increased the number of hours of study that students were expected to do each day. All of this led to a subtle but pervasive shift in how psychologists and educators regarded the relationship between students and the information stream. They believed that one way or another, students must learn to absorb the torrent of information headed their way, for new geopolitical demands imposed new learning imperatives. In the post-Sputnik school system, students were encouraged to increase their capacity for focused study.[101]

In 1969, America celebrated a successful moon landing and declared victory in the space race. And with the race won, there was at least one less nationalist pretext for pressuring American youth to stay focused in the classroom. Yet the reaction to Sputnik was emblematic of how modern Americans would conceive of attention, information, and intelligence in the years to come. While information overload continued to be a concern, and while critics sought ways to stem its flow, in the latter part of the twentieth century it became increasingly incumbent on people to adjust to the information stream rather than to adjust the stream to their own cognitive dispositions and capacities. The key was to learn how to channel one's mental power and potential, so that it would not go to waste.

This modern perspective was evident in the addition of attention-deficit disorder (ADD) to the *Diagnostic and Statistical Manual of Mental Disorders* (*DSM-III*) in 1980 and its eventual redefinition as attention-deficit/hyperactivity disorder (ADHD) in *DSM-III-R*, published in 1987.[102] In Chapter 3, ADHD was described as "a search for stimulation disorder," for sufferers of ADHD also report higher incidences of boredom. Boredom, however, is only an incidental feeling in the ADHD diagnoses—a symptom, but not the root affliction. As the name denotes, the diagnosis is more directly related to attention and implies that those who cannot concentrate have a codified medical disability.

Whereas in the past, illnesses like cerebral hyperaemia were the result of paying too much attention, by the late twentieth century, doctors regarded as ill those who paid too little attention. Too much focus was the medical problem of the nineteenth century, and too little was the concern in the twentieth—a sign of how Americans have internalized both the imperative to focus and the belief that this is something their brains should be able to easily do, given its ostensibly limitless nature.

Since their creation, the ADHD and ADD diagnostic labels have seen heavy use, with a steep escalation in the number of Americans formally and informally diagnosed in the last three decades.[103] When the Centers for Disease Control and Prevention (CDC) first started measuring the prevalence of the condition in 2003, an estimated 7.8 percent of American youth had been diagnosed. By 2012 (the most recent year for which the CDC has data), the figure had climbed to 11 percent. The CDC has not formally measured rates among adults, but some researchers estimate that about 5 percent of adults have also been diagnosed.[104]

Concomitant with this growth have been large increases in the number of Ritalin, Adderall, and other psychostimulant prescriptions dispensed for ADHD conditions. In 2010, $7.42 billion worth of ADHD medications were sold. In 2015, sales approached $10 billion and by 2020 are likely to reach $17.5 billion.[105] Moreover, while the rise in the diagnosis and treatment of ADHD indicates that psychologists and educators have been paying more attention to attention, so too have pundits and social critics. For Richard DeGrandpre, who wrote *Ritalin Nation* in 1999, ADHD was caused by a "rapid-fire culture," in which the information flow had sped up. Once people had become accustomed (and sometimes addicted) to this speed, they had difficulties adjusting to classrooms and other environments where information flowed less rapidly. In DeGrandpre's view, Ritalin provided "a potent backdrop of stimulation," which freed sufferers from having to act out or seek stimulation when they were put into slower-paced environments.[106] While the drug seemed to have a "calming" effect, he believed that Ritalin users would be better served by weaning themselves from a hectic lifestyle and revisiting what Milan Kundera once called "the pleasures of slowness."[107]

Other social critics, however, question whether the speed of modern life is to blame for Ritalin use. Malcolm Gladwell is one. He cites the work of Rosemary Tannock, a psychiatrist at the Hospital for Sick Children in Toronto. Tannock had studied how ADHD and non-ADHD boys played video games. While the ADHD boys enjoyed playing the games, their reaction times were much slower than their counterparts, suggesting that sufferers of ADHD were unable to function well even when presented with abundant external stimuli. In Gladwell's view, these results hardly seemed to describe "a child attuned to the quicksilver rhythms of the modern age," depicting instead youngsters who were unable to screen out distractions or navigate "the rising complexities of modern life." In a conclusion that seems to counter DeGrandpre's, Gladwell posits that Ritalin should be welcomed as a valuable and increasingly indispensable tool as other proven cognitive aids like nicotine fall into public disfavor.[108]

Many Americans have recognized it as just this, for a surprisingly large number take the drug without a prescription. As *Time Magazine* reported, a significant portion of Ritalin and Adderall users have not actually been diagnosed with ADHD and take the drugs to enhance mental performance.[109] Estimates vary, with researchers suggesting that between 25 percent and 34 percent of college students use the drug without a prescription, primarily to improve their ability to concentrate;

the drugs hold such appeal because, in the words of Alan Schwarz, a *New York Times* reporter, they offer "tunnel-like focus."[110] Christian Wutz, the technology entrepreneur we spoke with, said that many people he knew—young and eager to succeed in business—believed it to be a necessity. "They definitely use a lot of Adderall. . . . It's a fact that it's going to make you perform better." Did he know lots of people who used it? "Oh yeah, I think it's almost a given."[111] Similarly, in the classroom, educators have contended that the drugs can create a more harmonious and productive student body. Matthew Crawford, who worked as a teacher for a year before going on to write a number of acclaimed books on learning, work, and distraction, recalls that he "would have loved to have had a Ritalin fogger" in the "classroom for the sake of maintaining order."[112] Even some bio-scientists, ethicists, and psychiatrists endorse the drugs for those with no diagnosed condition, regarding them as a useful way for individuals to enhance their cognitive performance and powers.[113]

While fans of Ritalin and Adderall might consider the stimulants to be godsends, the drugs also attract criticism. As psychologists Thomas Hills and Ralph Hertwig argue, their use is based on an unsubstantiated belief—"that cognitive traits conform to a linear model in which more means better: More memory is better, more focus is better, more self-control and willpower are better. . . . Just as we cherish faster processing speed and larger memory in our digital electronics, we may assume that boosting a particular cognitive trait will bring better mental performance and affective well-being." They suggest though, that this is a flawed assumption. Their research indicates that rather than more being "inescapably better, . . . optimal control of attention represents a delicate balance between too much and too little focus."[114]

Other critics of these drugs consider them an abdication. Rather than addressing the underlying environmental causes for why distraction occurs—the hectic nature of contemporary culture and the constant distractions—Americans seek medical palliatives endorsed by a billion-dollar pharmaceutical industry with vested interests in the subject. These differences delineate the broad outlines of the "Ritalin

Wars," which have simmered with the growth of ADHD diagnoses and have polarized the public. On one side are doctors, pharmaceutical companies, and national support and advocacy groups such as CHADD (Children and Adults with ADHD), who maintain that ADHD is eminently real and that the medications genuinely help people diagnosed with the condition. On the other side are critics who claim ADHD is overdiagnosed and ADHD drugs are overprescribed and have adverse effects. These differences gained national attention when the Church of Scientology weighed in on the side of the skeptics and filed a spate of class action lawsuits against psychiatrists and pharmaceutical companies. While the suits were dismissed, they inspired critics like Peter Breggin who, in his books *Talking Back to Ritalin* and *The Ritalin Fact Book: What Your Doctor Won't Tell You,* says that stimulant drugs are much more hazardous than advertised. He was joined by others who maintained that the diagnosis of ADHD was often just masking poor parenting or poor schooling, or that ADHD was simply a way of pathologizing children who were maturing in a delayed fashion.[115] These debates reflect a divide over what should and could be asked of the brain. How much focus is enough? What should people expect of their mental powers?

If drugs have represented one (much debated) way to expand mental capacity, some Americans have tried other means, relying on brain games or brain exercises that are available as apps on their phones. Over the last two decades, programs like Lumosity, Cogmed, Brain Games, Brainiversity, NeuroNation, Mind Quiz, and many others promise to enhance working memory, sharpen focus, increase fluid intelligence, and improve neural plasticity.[116] Lumosity describes its mission as "helping people keep their brains challenged," and it promises that its "simple online tool" will "allow anyone to train core cognitive abilities."[117] Meanwhile, Brainiversity reminds you that "to stay fit and healthy we should regularly exercise our bodies. And the brain is no different. According to research . . . the use of brain training programs and games can have a positive impact on your mental fitness. Brainiversity is a brain training game designed to stimulate your brain with 24 different activities covering Language, Memory, Math and Analysis."[118] Such programs reflect the popular wisdom that has sprung up in the twentieth and twenty-first centuries that there need be no limits to mental power, and that with hard work, cognitive aids, constant exercise, and new technologies, our minds can be infinitely powerful, and our ability to assimilate new information unlimited.

Endorsing these optimistic views on human cognition, media theorist Clay Shirky, in his book *Cognitive Surplus,* argues that the web offers a way to expand brain power. He maintains that this is a departure from the past, for during much of the twentieth century, humans wasted large stores of cognitive capacity viewing television shows like *Gilligan's Island* and consuming other forms of mind-numbing mass culture. However, the development of the Web and social media allowed the public to create their own content, rather than just passively consume pablum, and this Shirky considers a better and more productive use of mental energy. Moreover, he believes the Web has offered a way to access untapped cognitive capacity.[119]

He is joined by those who celebrate the "internet of things" as an extension of consciousness. Consumers can buy cameras that enable them to watch their cats while at work, or an app that gives minute-to-minute updates on the geo whereabouts of family members, or a refrigerator that sends a text message when the milk is running low. Millions of sensors are being embedded in clothes and carpets and a multitude of other things that can expand humans' ambient awareness.[120] They seem to offer a pain-free way to know more and attend to more.

The logical extension of such optimism about technology and mental powers can be found among advocates of artificial intelligence (or AI), who believe that human brains can create artificial brains that in turn can enhance human intellect. Ray Kurzweil, the chief futurist at Google, regards AI as a set of technologies that soon will extend "the most important attribute we have, which is our intelligence."[121] After AI has augmented humans' cognitive capacities, people will be not only more intelligent, but also "more profound," "funnier," and "sexier." He spells this out in a *Playboy* interview:

> By the 2030s we will have nanobots that can go into a brain non-invasively through the capillaries, connect to our neocortex and basically connect it to a synthetic neocortex that works the same way in the cloud. So we'll have an additional neocortex, just like we developed an additional neocortex 2 million years ago, and we'll use it just as we used the frontal cortex: to add additional levels of abstraction. We'll create more profound forms of communication than we're familiar with today, more profound music and funnier jokes.

Kurzweil imagines what an encounter with his boss might be like after AI becomes available:

> Let's say I'm walking along and I see my boss at Google, Larry Page, approaching. I have three seconds to come up with something clever to say, and the 300 million modules in my neocortex won't cut it. I need a billion modules for two seconds. I'll be able to access that in the cloud just as we can access additional computation in the cloud for our mobile phones, and I'll be able to say exactly the right thing.[122]

For Kurzweil, as for many of his acolytes, future generations can look forward to almost limitless increases in cognition.

This is not to say that everyone is willing to jump on Kurzweil's AI bandwagon. In a blog post reviewing the *Playboy* interview, Nicholas Carr takes a skeptical view:

> I think there's a flaw in Kurzweil's logic here. He fails to anticipate the inevitable arm's [*sic*] race in cleverness. Larry Page is going to be plugged into that enormous cloud neocortex, too, so surely Page's standards for what qualifies as a clever remark will have gone up exponentially. Even with his new brain, Kurzweil will still be in exactly the same boat, floundering to muster the wit necessary to impress the boss. Funny and sexy are relative terms.[123]

More generally, many psychologists critique the optimism that undergirds not just Kurzweil's efforts, but the apps, brain games, and drugs, all of which promote the belief that one can expand one's mental powers. For example, neuroscientist Adam Gazzaley and psychologist Larry Rosen, in their book *The Distracted Mind,* start from the premise that "the human mind has fundamental limitations when it comes to our ability to use our cognitive control to accomplish our goals." In a similar vein, the new mysterians, a group of writers, psychologists, and philosophers that coalesced in the early 1990s, argue that there are certain mysteries that will forever remain unfathomable to human consciousness.[124]

Yet despite such criticisms, many today hold great faith in the infinitely expandable nature of human cognitive capacities. Indeed, a recent survey by Harris Interactive found that 65 percent of Americans continue to believe that they are using only 10 percent of their brains—that there is great, excess capacity yet to be tapped.[125] They believe that through drugs, games, iPhones, and perhaps Kurzweil's nanobots, their mental powers can grow, that there are no limits.

Paying Attention: Productivity and Presence

Perhaps it is because of this popular faith in the power of our own brains that many worry when they do find evidence of cognitive limits, when they discover that they are not always able to pay attention to all things, or the right things, with sufficient intensity. As Charles Duhigg, author of *Smarter, Faster, Better,* notes, Americans worry about their productivity and focus "because our experience matches so poorly with our expectation. We're living through this age where they keep on telling us, 'Look, we have all these devices for you now.' We have e-mail, and we have a communications revolution, and we have computers in everything that you can possibly touch, and the idea should be that life gets easier. And instead, it's just getting harder and harder. And that doesn't seem like how things are supposed to go."[126] Indeed, despite all of these new aids designed to heighten brain power, many today are anxious about the distractedness of the American people. Pills, smartphones, and brain games promise to improve our efficiency and focus and raise our hopes, and yet their very presence in our lives creates more layers of complexity and often undermines our concentration.

This is an increasingly common problem. In the mid-twentieth century, psychologists believed that distraction was something that pilots and radar technicians needed to worry about; today that worry has spread more widely, for the complexity that challenged aviation personnel in World War II, and that spurred Broadbent to launch new research on attention, is now starting to affect the common office worker. As technology journalist Clive Thompson reports, "Office denizens now stare at computer screens of mind-boggling complexity, as they juggle messages, text documents, PowerPoint presentations, spreadsheets and Web browsers all at once. In the modern office we are all fighter pilots."[127]

Some commentators and researchers point to particular pieces of technology as the source of distraction and sensory overload. At Microsoft's labs, cognitive scientist Mary Czerwinski and her colleagues examined the way that attention was being reshaped in the modern digital office. When Microsoft popularized the Windows operating system in the 1990s, it facilitated multitasking in unprecedented ways. Where earlier systems had confined workers to single applications, Windows enabled users to write a report, while also browsing the Web, while also reading their email. Microsoft had eagerly marketed the multitasking capabilities of its new product, but the success of Windows also multiplied the number of interruptions that workers were subject to in the workplace. As Czerwinski began to suspect, workers in modern office environments who were using Windows did not stay on one task for very long before allowing themselves to be interrupted—often by a different task in a different window on their desktop.[128]

With the advent of smartphones, this problem was only compounded, for texting and social media were even easier to access. In a recent survey, more than 50 percent of employers reported that cell phones were distracting their workers, and 44 percent listed the internet as a workplace distraction.[129] One estimate is that workers waste two to three hours per day because of such disruptions.[130] And workers themselves admit their concentration is often broken. According to a survey by Virgin Pulse, "[Fifty-two] percent of . . . survey respondents say they're distracted between 1 and 20 percent of the time during the workday. Forty-three percent of respondents say they get distracted between 21 and 75 percent or more of the time during their workday." They pointed the finger at social media, texting, and the internet more generally, as well as "chatty colleagues."[131] Many employees worried about distraction and lamented the decline of their attention span. In a 2015 Harris poll, 73 percent of the more than two thousand people surveyed said that technology "has become too distracting." However, only 8 percent of those surveyed suggested that such distractions were decreasing their productivity.[132]

Given the widespread incidence of distraction, some observers have characterized the search for deep attention as an elusive quest, born of nostalgia for a different time and different economic system. A number of scholars have recently argued that our idealized cognitive modes reflect the modes of production of the era. For instance, Cathy Davidson, in *Now You See It,* maintains that the esteem for deep, undivided attention reflected the needs of twentieth-century Taylorized factories that thrived on laborers concentrating on single tasks.[133] Literary critic N. Katherine Hayles argues in "Hyper and Deep Attention: The Generational Divide in Cognitive Modes" that the celebration of deep attention and the denigration of multitasking have been further reinforced by the American education system, which has created "environments conducive to deep attention . . . combining such resources as quiet with an assigned task that demands deep attention to complete successfully. So standard has deep attention become in educational settings that it is the de facto norm, with hyper attention regarded as defective behavior that scarcely qualifies as a cognitive mode at all."[134]

Yet, while academics have sustained and propagated these norms for decades, Hayles and Davidson think that these assumptions are being challenged by a younger generation with different habits of attention who cannot sit still and who "squirm in the procrustean beds outfitted for them by their elders."[135] Young people have grown up in an age of

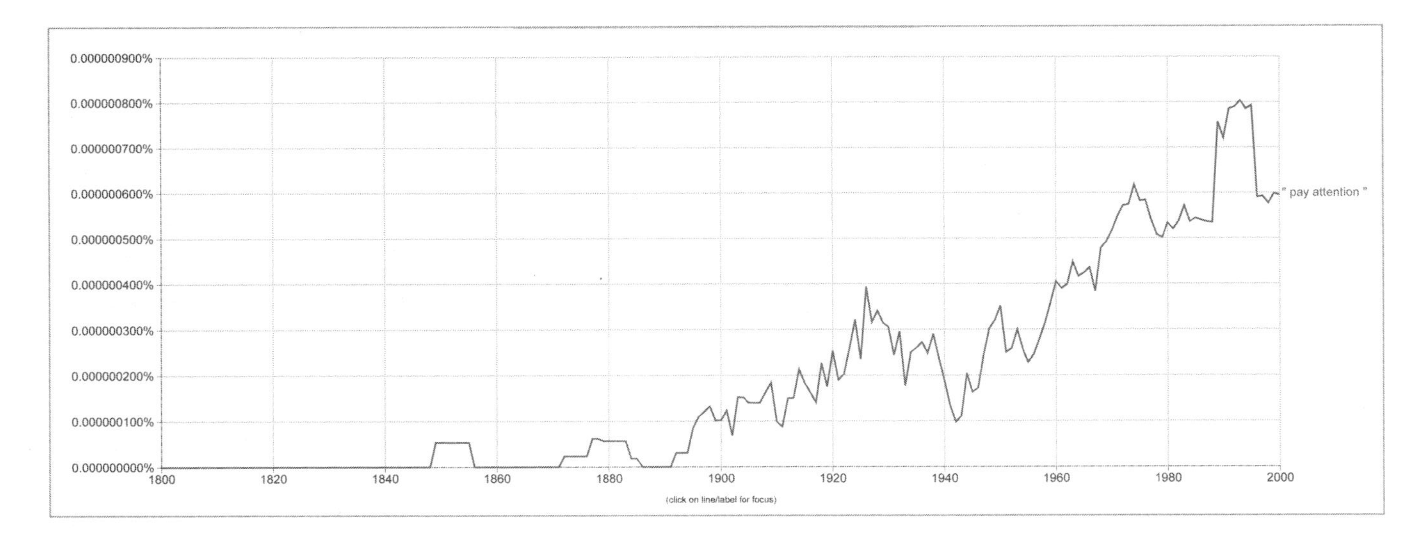

FIGURE 4.1 This Google Ngram traces the increasing use of the phrase "pay attention." *Google Ngrams: http://books.google.com/ngrams*

smartphones, Windows, and social media, so they have internalized the syncopated rhythms of attention. Moreover, changes in the mode of production demand laborers who can divide attention among rapidly changing tasks. Hayles argues that "hyper attention is more adaptive than deep attention for many situations in contemporary developed societies. Think, for example, of the air traffic controller who is watching many screens at once and must be able to change tasks quickly without losing track of any of them. Surely in such a job hyper attention would be an asset." She maintains that the number of such jobs is "increasing more rapidly than those that call for deep attention, from the harassed cashier at McDonald's to currency traders in the elite world of international finance."[136] Hayles and Davidson therefore worry less about the decline of deep attention, suggesting that that cognitive mode carried in it the oppressive norms of industrial capitalism. In contrast, they paint multitasking as potentially liberating.

But darker visionaries regard multitasking and frenetic online activity as just the latest extension of capitalism, a new mode in which users are being turned into both consumers and products. As Jonathan Crary, the art historian, writes, "one crucial aspect of modernity" is "an ongoing crisis of attentiveness, in which the changing configurations of capitalism continually push attention and distraction to new limits and thresholds, with an endless sequence of new producers, sources of stimulation, and streams of information, and then respond with new methods of managing and regulating perception."[137]

In Crary's view, the individual who fits best into market economies of today is the one "who is constantly engaged, interfacing, interacting, communicating, responding, or processing with some telematic milieu."[138] Distraction is indeed a reflection of the new mode of production, Crary suggests, but rather than seeing it as liberating, he regards it as a sign of just the latest intrusion of capitalism into human consciousness.

Modern commentators, then, attach a host of meanings to attention and distraction. Perhaps concentration is liberating, a sign of autonomy, as Carr suggests in "Is Google Making Us Stupid?" Or perhaps concentration is a relic of an industrial order that controlled individuals and their intellects, and multitasking is therefore liberating. Or, alternatively, perhaps distraction is merely a practice that plays into the hands of online merchants eager to sell products and commodify minds.

Our interviewees are not as doctrinaire as these social critics and philosophers, and they seemed to have mixed feelings about attention. They continued to hold up deep concentration as an ideal—after all, for generations, Americans have been taught its virtues in schools, churches, and libraries. Yet they also realized that they could not always achieve this ideal. They multitasked, even though they sometimes doubted their efficiency when they did so. They were aware that companies wanted to distract them and grab their attention, even while they themselves hoped to attract the attention of others through their status updates, posts, and feeds.

At one end of the spectrum were "techno-optimists," like Warren Trezevant of Oakland, California, who celebrated "the fact that we each carry humanity's knowledge in our pocket." Rather than technology making people stupid, he believed that computers and phones offered the opportunity to "offloa[d]" our brains so that "we can store that knowledge" on devices, which "frees us up to do other things and think about other things."[139] Greg Bayles, the app developer, described phones as a type of "neural augmentation," and he believed such cognitive enhancements would only proliferate. When asked whether there was a limit to what the brain aided by technology could do, he appealed to Moore's Law: "there's some law in tech where every year the storage capacity of a device goes up by two times and the cost goes down by half or something like that." While there was "some limit way far off," at the present, he concluded, "it's not a super-foreseeable one."[140]

Many shared this faith that properly managed, digital devices expanded mental capacity. Barry Gomberg, the attorney from Ogden, Utah, was similarly optimistic: "I don't think that [technology] makes us more stupid. I think that [it] has the potential to connect us to a phenomenal level of information and awareness of the world around us." He admitted that at times he had a hard time focusing, but for him that had been a perennial problem. He had always had distractions;

before the advent of the Web, television had been "an enormous invasion" in his life. In contrast, modern technology allowed him to work more effectively. For instance, Microsoft Windows allowed him to keep multiple projects open at once and this actually engaged his attention even as it indulged his penchant for distraction. As Barry put it, "I think the ability of the technology to hold my attention makes up for whatever the interruptions cost me moving . . . back and forth."[141]

Often our interviewees celebrated their ability to multitask. Will Xu, a Delaware native attending Grinnell College, explained that dividing his attention made him more efficient. He used two monitors and often had "fifty tabs open," whether at work or at play on the computer. He explained, "I feel like all the time I'm multitasking. . . . And it's just a thing that I think I've developed naturally. . . . I have the ability to search a billion things at the same time and be more efficient with my time." Will believed that multitasking offered him a new way to make more time in the day. "Even . . . during my leisure time, I feel like I'm trying to maximize my leisure time; in a way, . . . doing two things, I feel like I'm doing that. Twenty-four hours in a day doesn't feel like enough, so I'm trying to get more hours out of the same hours." The ability to divide his attention multiplied his time. It also reflected his faith in his own unlimited mental power. If he didn't do two things at once, if "I just do one or the other, . . . there will be moments when my brain is like 'we're not doing anything.'"[142]

Others celebrated the fact that such divided attention provided breaks during the work day. Cooper Ferrario readily admitted that he used three screens at his job, and that while two were dedicated to work, the third streamed entertainment. He believed that the third screen actually enhanced his efficiency by providing him with a quick way to take a break: "I think that that does help my productivity."[143]

In contrast to Will, who was writing college papers, and Cooper, who was working as a summer intern, Anna Renkosiak, a museum vice president, split her time between the boardroom and her screens. Yet she too could see merit in the occasional distraction, considering it something that could enhance rather than diminish productivity. Her job required her to work at a "high level" and to be focused "all the time."

And yet she still saw the occasional text message interruption as a welcome interlude. As she put it, "If you get that little like 'Hey babe, love you' text message, it's like 'aww' . . . you can . . . daydream a half minute [and then go back to budgets]. . . . It lets you take that momentary break and then come back to the task at hand." She contended that it could be a good thing to be temporarily distracted.[144]

Yet the ideal of deep attention had not disappeared, and many continued to aspire to it. They had learned in school that they should strive to sustain their focus, and while they might not be able to achieve this goal, they still measured themselves and their performance against it. While Hayles and Davidson might celebrate multitasking, many people trying to do their work were far more worried about their scattered thoughts and interrupted attention. For example, Stephanie Christiansen, who worked for a Utah nonprofit, thought digital distractions were reducing her productivity. She observed, "I think it's creating an absolute ADD world. I see it in myself. . . . I could sit and look at numbers forever and ever and ever. It would never bother me. And now I can't." She described her office:

> I probably have four or five or six [tabs] open all the time. Some of them are work related and some of them are [not]. . . . If I get a little "ding," . . . I'm immediately over there going "who sent me something?" . . . I don't know if it makes me more or less productive. I don't. I honestly don't. There are tools that make us work easier . . . but then there are all the distractions that come with it.[145]

Others articulated reservations in terms of dissolution, waste, and the loss of control over time. Sometimes these dissatisfactions were expressed without any references to cognitive decline. For example, Jack Pleger, a contractor in Phoenix, observed that the computer had been brought into his office in the hopes that it would save time, but that paradoxically he felt increasingly "nailed" to it.[146] Similarly, K. described an experience familiar to anyone who has used Facebook, Twitter, or other "bottomless" social media applications with an "endless scroll":

> For Facebook, I say I'm going to go in for thirty minutes. That's what the plan is. You see different posts and you see different cat videos and different things going on. . . . You kind of laugh. You keep scrolling, scrolling, you see . . . a message pops up and someone's talking to you. And so you kind of get carried away because you lose track of time very easily in Facebook.[147]

In K.'s case, social media, irresolution, and an absurd attraction to cat videos lead to moment after moment of inconsequence. Alta Martin confessed to this tendency as well. But for her it was not merely comical. It made her anxious about her productivity and her ability to get homework done:

> It's just a horrible thing. Especially with Facebook. . . . If you post something, you want to make sure people are responding to it. So then it's a constant distraction. . . . I don't want to worry all day long whether people are liking what I've put up or not. . . . During homework it's a horrible, horrible distraction. . . . It postpones your homework, which is stupid, 'cause you still have to end up doing it. So it's not helpful. . . . In that context, it's more hurtful than helpful.[148]

Retired librarian Allen Riedy had seen technology's effects on library patrons at the University of Hawaii firsthand. He observed that the internet "makes people stupid, maybe lazy too." Unable to focus for a long period, students were "skimming" rather than reading. He thought diligence and focus brought their own rewards. People learn things, he said, through "digging." "Slogging through something to find something . . . is instructive in itself," but to his mind, individuals willing to dig and slog were becoming a rarity.[149]

Katie Schall, the social media manager for an artisanal cheese company, expressed some of these anxieties about herself, and she looked back to her past mental abilities with a sense of nostalgia. As an undergraduate in the early 2000s, she had double-majored in philosophy and political science at Fresno State. Back then, she recollected, "I could sit and read fifty pages of . . . a philosophical text." Five years later, she returned to pursue a master's degree in public administration, and she simultaneously served as the university's director of social media relations. In that role, she "built their Facebook page from zero to twenty thousand people" and managed their Twitter feed, but it came at a cost: "I found that I could not sit and read a text for longer than . . . five minutes without checking Facebook. . . . I just couldn't sit and read things like I [once] could anymore, and so it felt like in a way it was kind of making me more stupid. And even now . . . this BuzzFeed-type stuff like 'The top 10 celebrities most likely to look horrible in a bikini this year'—I find myself clicking on that stuff more . . . than . . . [on] what's actually important. . . . Social media and technology is bringing me more things that maybe aren't making me a better person, living a more examined life."[150]

Many found fault not only in the technology, but also in their own irresolution and imperfect ability to use these tools consistently for nobler and more consequential purposes. They blamed their devices, but perhaps even more they blamed themselves for the way they used their devices. In doing so, they clung to the ideal of focused attention, even as they lamented their failure to achieve it.

People frequently made conscious efforts to avoid interruptions and distractions. Ana Chavez, the immigrant from Mexico who now was a college student, erased her apps at the start of each school year.[151] Christian likewise worked to fight distraction. For instance, he blocked his Facebook feed "because it's a distraction. . . . Browsing on Facebook is a luxury that I don't want . . . and that I think detracts from my ability to perform for both my team and my company and myself."[152]

Some avoided signing up for social media in the first place but this was seldom sufficient to shield them from distraction, for they still found themselves flooded with new and unwanted electronic communications. N., a lawyer in Honolulu, who only used email and texting, observed how these messages interrupted his concentration. He was not in control of his work time or his attention, because it was often compromised by others who sent him messages and expected instant replies.

> Sometimes I'll be doing something and somebody will actually call and say "Did you read my email?". . . . People are expecting me to respond right away all the time, because they think . . . I am watching my email and even moreso . . . my messaging. . . . I get a lot of interruptions like that. Suddenly I have to drop what I'm doing and it breaks my concentration.[153]

Even if one took measures to close out distractions, to turn off apps or never to get them in the first place, there was no true retreat from the constant hubbub. In the modern workplace, it was hard to find a cloister.

There was, then, a wide range of responses to digital technology. Some observed that new digital technologies were helping them to work more efficiently, increase their brain power, or maximize their time. Others judged it "more hurtful than helpful," describing these new technologies as a "time suck," a "distraction," a form of "mind pollution," or a "waste" of time. These different evaluations should not obscure a fundamental commonality. The majority of our interviewees were people with middle-class means, students who aspired to find jobs with middle-class salaries, or middle-class professionals who worried about whether they could hold on to their current status. Victor, a software developer living in the Rocky Mountains, described this anxiety in a particularly compelling fashion, for he feared falling behind at work and falling down the class ladder:

> I'm an information gatherer. . . . I'm trying to learn. . . . It's kind of worrisome because there's definitely the thing of, "Do I know enough? No, I don't; I need to learn more. . . . The internet makes you realize how many extremely qualified . . . experts there are . . . and to see that level of expertise definitely brings out a lot of anxiety in me because I actually wonder if I'm an impostor at times. . . . Do I actually know enough? . . . I think there's definitely a fear of me regressing toward the kind of working class I grew up in and that makes me always concerned about what's going on in the field.[154]

To Victor and many like him, paying attention was a crucial economic necessity, but it seemed as if there was ever more to know and absorb. Caught up in a high-speed digital economy, these Americans used "attention" and "focus" and "distraction" to describe the demands they were feeling. When a device or social media platform was distracting, it was distracting because it compromised productivity. Conversely, if a technology served to focus attention, they regarded this as a way to enhance productivity. The words our subjects chose, and the meanings they attached to them, reinforced the economic imperatives of the society in which they were socialized or (in the case of many students) into which they aspired to become socialized.

Some found other reasons to lament their declining attention spans. For example, Xavier Stilson, a student, at first offered a pretty conventional account of the relationship between technology and distraction: "There are more distractions and they're coming up more often. . . . I know that I need to study . . . so, yeah, I feel anxious that I need to get that done. . . . It does take a lot of willpower to turn off the cell phone and put it away and just focus on what you're supposed to be doing." After talking with us further, however, he started speaking of digital technology as something that was not only thwarting productivity, but also interfering with people's ability to tell right from wrong. Asked who and what were threatening the ability to focus, Xavier replied, "Being religious, I . . . honestly think it could be people who don't want us to do what is right. . . . They try to take your focus off of what's most important."[155]

Others worried that their electronic devices were distracting them from human interaction. Stephanie observed, "Do I think that . . . [technology is] making us stupid? I think it is making us extremely disconnected." For that reason, she limited the news feeds on her phone, because "I don't want to be walking around with my face down in my phone."[156] Camree Larkin, an avid user of social media, nevertheless worried that it distracted her from important relationships. "I think I spend way too much time on it. It takes away from calling people to talk to them."[157] While concerns about attention were often expressed as fears about productivity, they could also be expressed as concerns about moral corruption and the degradation of friendship and social life.

This was most evident in Eli Gay's comments. Eli, the tax preparer in the San Francisco Bay Area, had been taking an extended sabbatical from Facebook and other mainstream social media sites. At first, her reasons for doing so seemed to be related to productivity, when she observed that Facebook was "such a time suck." But even more, she found that social media distracted her from more spiritually meaningful activities. She talked of the way it created distance at social events, as friends became distracted from conversation and looked down to their phones mid-sentence. Eli, who had a master's degree in sociology and women's studies from Brandeis, could have explained these incidents as moments of rudeness or as evidence of a new set of norms for socializing. Instead, she resorted to a framework inspired by her interest in Buddhism:

> The idea . . . of being present, or . . . in Buddhism, being awake, and being conscious . . . that is one of my stronger values. . . . I think there has to be a degree of sleepwalking or unconsciousness when you are sitting at a table with someone and you stop and look at your phone. . . . For the subject who's doing it, it doesn't feel like sleepwalking but in the bigger picture objectively, that's what it looks like. . . . It would be like a narcoleptic . . . who just fell asleep for a moment.[158]

As Eli saw it, much of the problem of distraction stemmed from the fact that people were forgetting how to be "present," to pay attention to the moment. She talked about FOMO, the "fear of missing out," which she had been trying to conquer so that she would not be distracted by regrets about what might have been, or questions about what she should be doing.[159] In Eli's view, FOMO was at the heart of distractedness; it explained why people could not pay attention and felt compelled to check their phones while engaged in some other activity. As a way of transcending FOMO, Eli had learned of new and more useful attitudes to cultivate, all with their own acronyms. She described

JOBI and JOMO—mental states that kept one focused on the present. "I was in a whole Burning [Man] camp called JOBI which is the Joy of Being In [the moment] as opposed to the fear of missing out. . . . But now I'm more about JOMO, the Joy of Missing Out and just really knowing that there's . . . only ever where we are, and there isn't this other reality."[160]

People needed to take pleasure in their present circumstances rather than worry about alternate realities. Moreover, the incapacity to experience and enjoy the moment—to pay attention to the here and now—was actually impeding an important form of self-awareness. Eli said,

> I think . . . it's . . . important . . . for people to know themselves and to be in touch with what makes them feel good. . . . And all the technology is . . . interfering with that and taking the people away from that sort of self-knowledge . . . and that's where I sort of make fun of people when they are on their phones. It's like "what's happening that's more important than this moment of you sitting here with me at dinner?" . . . And even if it's just a second . . . it's seeing a little bit of that FOMO: "Who's on the other line? Who's on the other end of this call?" . . . I think working a lot to again just transcend that . . . feeling of "there's an alternate reality." It's been a good few years of that [being] . . . my mantra. . . . There's just no other reality. There's just this. . . . Everything has led up to this moment. And this is where I am.[161]

Of course, Eli also qualified this advice as behavior that not everyone could practice in equal measure. She was self-employed and living comfortably enough in one of the nation's most costly metropolitan regions. Unlike the students we interviewed, she had already managed to fulfill many of her income aspirations. Issues of productivity were no longer an overriding concern for her, and as a result, she was more apt to describe questions about attention as issues of face-to-face connectedness, self-knowledge, and spirituality.

But whether people felt distracted from spiritual or material goals—and regardless of whether they felt distractions were good or bad—no one could deny that they were omnipresent.

The Profits of the Attention Economy

Admittedly, Americans have worried about distractions for over a century, believing that as the world has become more complicated and as information has proliferated, the need for concentration has in turn become more pressing.[162] Yet there is something new about the technological distractions that contemporary Americans face. In the past when a locomotive whistled or a dynamo whirred or a telephone rang, the distractions were largely incidental: the sounds were bothersome intrusions but they were not fundamental to the mode of production at the time. Today, in contrast, distractions have become increasingly purposeful—companies like Google and Facebook and Twitter, which offer their services for free, recoup their expenses and make profits by distracting their customers with ads and links designed to sell them goods.[163]

In the attention economy, the chief means for making profits is through distraction. Because these distractions have become integral to the business process, they are at once vastly more ubiquitous and much more entrenched. They are embedded in smartphones, computers, and the webpages that one surfs through. Technology companies, web designers, and advertisers are intentionally manufacturing what Adam Alter, a New York University marketing and psychology professor, calls "irresistible" and "addictive" digital experiences.[164] These, in turn, are creating what Tim Wu, in his book *The Attention Merchants,* terms an "attentional crisis": "Ours is a time afflicted by a widespread sense of attentional crisis, at least in the West—one captured by the phrase 'homo distractus,' a species of ever shorter attention span known for compulsively checking his devices. Who has not sat down to read an email, only to end up on a long flight of ad-laden clickbaited fancy, and emerge, shaking his or her head, wondering where the hours went?"[165]

Similar observations were made by our interviewees. Christian, for instance, realized that companies tailored the pop-ups and ads based on the data they had accumulated on different consumers. Ads were customized to individual interests based on browsing history and they were "designed to distract you—just you."[166] They worked too. Q. reported, "Yeah, I'm definitely a victim of that. . . . I mean you go on to CNN you got things blinking at you. It's trying to grab your attention, it's competing for your attention."[167] In the same vein, albeit with a more critical tone, Emily Meredith, the nurse in Oakland, California, condemned "the content creator people . . . [who] just create content which isn't really content . . . clickbait. . . . There are so many people who are just employed to create this stuff out of nothing, right that will then force you to click on it . . . because it's . . . titillating or something. . . . That blows my mind. That just seems like the most unbelievable waste of human potential that I can ever imagine." She lamented "all the human energy and potential that's gone into just creating nothing."[168]

When Emily spoke to us in Oakland, she was only a short drive from the Google headquarters in Silicon Valley, the heart of the new attention economy. Her critique of clickbait resonated with a more well-known polemic that Carr has leveled against Google in *The Shallows:*

> Google's profits are tied directly to the velocity of people's information intake. The faster we surf across the surface of the Web—the more links we click and pages we view—the more opportunities Google gains to collect information about us and to feed us advertisements. . . . Every click we make on the Web marks a break in our concentration, a bottom-up disruption of our attention—and it's in Google's economic interest to make sure we click as often as possible. The last thing the company wants is to encourage leisurely reading or slow, concentrated thought. Google is, quite literally, in the business of distraction.[169]

This was true not just of Google, but of many other businesses as well. Tech entrepreneurs we interviewed recognized the financial value that distraction could offer. Christian's company sold Wi-Fi to businesses, which in turn might offer it free to their customers in exchange for watching a video or signing up for promotions. The success of his company—and the companies he worked with—depended on engaging internet users' attention. He noted, "We're an advertising company in many ways. What's the goal of an advertisement? To sell, right. And how do you measure that? . . . Ultimately how well did it distract you from what you were doing and motivate you to take action . . . so I would say 100 percent that they're in the business of distracting. . . . I think every company does that." He was about to make a presentation to a restaurateur later in the day and he was pretty sure the question that was foremost on his client's mind was "how effectively are you distracting? Which then means how well are you generating value for us?"[170] While concentration had value for oneself, so could the distraction of others. One person's distraction might be another's financial gain.[171]

This view of distraction is one sign of how conceptions of attention have changed over the course of two centuries. People often forget this and take the particular ways contemporary Americans channel their mental energy as natural and universal, as an innate part of the human condition, when in fact they are products of history. Norms about how to pay attention, and how much attention to actually pay, are different in today's world than they were in America's past. So are the hopes and anxieties that are invested in attention.

In the eighteenth century, the ability to "fix" one's attention on God was a sign of virtuous self-control; it reflected one's moral state. In the nineteenth century, as America became an urban, industrialized society, many began to place an increased premium on deep concentration because the economy demanded they be focused producers. The development of a large, complicated and interconnected economy spurred Americans—and aspiring professionals especially—to develop the capacity for sustained attention on single things. But as they confronted these new demands on their intellects, many also believed there were limits to what they could absorb and know. The human mind was finite, not infinite; humans could not take in everything.

That belief began to change over the course of the twentieth century. Americans, shaped by the demands of the military-industrial complex, believed they needed to concentrate for the sake of national defense. With expert psychological advice, they tried to find a way to maximize focus and brain power, to exceed natural limits with the help of pharmaceuticals and training.

In the twenty-first century, these imperatives still exist but have been joined by others. Now Americans feel they need to focus as well as to attend to multiple stimuli, multitask, and attract the attention of others. Today we are paying attention in order to take it all in—and we often believe we can and should be able to do so, because we have faith that our brains are infinite in their power and potential. And because there are now all sorts of cognitive enhancements—from brain games to iPhones, to the internet of things—both our sense of what intelligence is and can be and our hopes for what we can know and can do, have grown impossibly large.

CHAPTER 5

Awe

In a recent TED talk, comedian Jill Shargaa implores, “Please, please, people. Let’s put the ‘awe’ back in ‘awesome.’” She complains that people use the term “awesome” to describe completely mundane occasions and objects. Instead she says it should be reserved for more extraordinary events and experiences—the creation of the great pyramids of Giza, the moon landing, or a glimpse of the Grand Canyon. Contemporary Americans are cheapening the feeling, applying the term to just about everything.[1] Her interest in awe and her concern that the emotion is being diminished are shared by members of the general public as well as by technology evangelists and mental health experts.[2]

These commentators worry that a growing number of Americans are becoming incapable of feeling awe, and that social life is becoming shallower as a result. As a play on the familiar abbreviation ADD (attention-deficit disorder), one evangelist of the feeling uses the acronym to refer instead to “awe-deficit disorder.” Along similar, albeit more serious lines, psychologists Paul Piff and Dacher Keltner recently observed that many in modern society have become “awe-deprived,” because they are not afforded the same opportunities to witness starry skies, nature’s beauties, or transcendent works of art that earlier generations enjoyed. The result is that twenty-first-century Americans have become more focused on themselves and less connected to the grandness of the universe.[3]

But while evangelists of awe and psychologists who study it worry that contemporary Americans are awe-deprived, so far only cursory attempts have been made to show how the emotional experience has changed over time. Although the word is still commonplace in the American vocabulary, its connotations have undergone a transformation. At one time, awe was framed religiously—as a feeling that illuminated the relationship between humans and Gods. As such, it often served to outline the absolute power of the deities in relation to the limited power of humans, as well as to clarify the particular technological limits that mortals should not trespass.

These religious frameworks have become more attenuated over the course of the last two centuries. While nineteenth-century Americans experienced the feeling and expressed it freely, to many twentieth-century psychologists, religious awe seemed like a primitive, backward emotion, a vestige of earlier, less enlightened times, and they discouraged it. Today, the word "awe" is still used, but it has been divested of its older meanings, and in the twenty-first century, it often carries a smaller and less transcendent meaning. Contemporary Americans are awed less by God or by nature and more by their own technology. To the extent that they experience the emotion at all, they are awed by themselves and their own creations.

A look back at the original meaning of the word highlights the contours of this change. "Awe," which can be traced to the ninth century, once suggested a mix of feelings. The *Oxford English Dictionary* reports that it denoted "dread mingled with veneration, reverential or respectful fear; the attitude of a mind subdued to profound reverence in the presence of supreme authority, moral greatness or sublimity, or mysterious sacredness."[4] The feeling reflected individuals' sense that there was something larger and grander than themselves, something so vast and powerful that it might even provoke fear.

Earlier generations of Americans also used a number of related terms to describe their feelings about the grandeur of the universe and God. They often employed "wonder," to describe both an astonishing act or occurrence, as well as the feelings that it evoked. First used circa 700 AD, the word meant "something that causes astonishment. . . . A marvelous object." It could also denote "a deed performed or an event brought about by miraculous or supernatural power." Wondrous objects and marvels in turn produced the feeling of wonder, which was defined as "the emotion excited by the perception of something novel and unexpected, or inexplicable; astonishment mingled with perplexity or bewildered curiosity."[5] Americans also sometimes used the word "sublime" to describe their feelings at encountering "a feature of nature or art . . . that fills the mind with a sense of overwhelming grandeur or irresistible power; that inspires awe, great reverence, or other high emotion, by reason of its beauty, vastness, or grandeur."[6] Related feelings, like "astonishment," also evoked the grandness of forces outside oneself. Defined as a "mental disturbance or excitement due to the sudden presentation of anything unlooked for or unaccountable; [a] wonder temporarily overpowering the mind," astonishment also carried the meaning "to shock one out of his self-possession, or confidence; to dismay, terrify." The term's origins were the French words for being lightning- or thunder-struck.[7] All of these feelings denoted the experience of humans coming into contact with forces larger and more powerful than themselves.

Talk of awe and wonder abounded in earlier centuries, particularly when Americans encountered new technologies.[8] In 1858, in a sermon to mark the laying of the transatlantic telegraph, Presbyterian minister William Sprague termed the cable "the wonder of wonders," and he celebrated the fact that "human thought flashes in lightning all through the ocean's depths." It was now possible to "converse with those who live on the other side of the globe almost as if they and we occupied adjoining habitations." Sprague thought that contemplating the telegraph might fill his congregants' minds "with nobler thoughts of God," and their hearts "with more reverent and devout affections towards Him." It would teach mankind "to bow his spirit in humility in the reflection that his faculties can grasp so little, that the limit of his knowledge is so quickly reached."[9]

A century and a half later, in 2015, Allen Riedy, the retired librarian in Honolulu, reflected on his feelings about the technology of his day. "I've more than once felt awe when something first comes out. Some new technology. . . . Wow, . . . but it fades pretty quickly." He speculated that the "awe factor" might be decreasing because new technology "is so quickly integrated. . . . [It] becomes so commonplace so quickly."[10]

The gulf between these Americans' reactions to technology is profound. The awe that Sprague articulated reflected his belief that there was a spirit in the machine, that it was miraculous. Such attitudes were common, for in the nineteenth century, new modes of communication sparked astonishment in many Americans who regarded the technologies as harnessing the divine and mysterious forces of the universe. Gradually, however, their mental world became "disenchanted," as the sociologist Max Weber put it—or at least enchanted in a different way.[11] Today, many Americans are astonished when presented with new devices and contraptions, yet that awe has taken on a new shape and meaning. It is less religiously inflected and communal, and far more individualistic—and, perhaps indirectly, less capable of muting hubris and narcissism, since it no longer serves as a feeling that connects people to something larger than themselves.

The Myth of Icarus

Feelings of awe, wonder, fearful reverence, and terror—which highlighted both the marvelous power of forces grander than oneself, and one's own smallness in comparison—were commonplace when earlier generations contemplated new technologies. For centuries, inventors often faced resistance to their new devices, precisely because they seemed so extraordinary and supernatural. Some wondered if humans were becoming godlike by making these new contraptions. Should they be creating them? Were mortals even supposed to possess powers and forces that were traditionally the province of the Gods? Western societies have wrestled with such questions at least since the story of Genesis: Adam and Eve were punished for overstepping the limits God had placed on their knowledge, while those who tried to build the Tower of Babel in order to ascend to the skies were prevented from doing so when God divided them by creating a multitude of languages. Such episodes served to inculcate an awe of God and to remind humans of their place in the world.

Ancient Greeks told each other the myth of Icarus to teach similar lessons. Icarus tried to fly with wings made from feathers and wax. With them, he flew around like a bird. But eventually he soared too close to the sun, which melted his wax wings and caused him to fall to his death in the sea. For the Greeks, the tale served as a depiction of a tragic human condition, in which mortals aspired to transcend their current powers through invention, only to see those aspirations dashed by the limits of the invention and the limitations of being human. The myth of the Titan Prometheus, likewise, recorded his efforts to bring fire to humans, and his subsequent and painful punishment for doing so. These warning tales, well known in Europe and America, suggested that there were boundaries that human beings should not transgress.[12] To aspire to augment one's capacities was a sign of overreach, impertinence, and arrogant disrespect for divinity. These tales reflected the fearful awe with which humans regarded the powers of God (or the gods), as well as the limitations they believed to be inherent in humans' capabilities.

Such misgivings about human innovations did not quickly abate. They were present in medieval Europe and migrated to colonial America. Pious New Englanders, for instance, voiced similar worries after the invention of the lightning rod and, in doing so, revealed their feelings of awe. In the wake of an earthquake in 1755, Reverend Thomas Prince, minister of the Old South Church in Boston, suggested the rods were to blame. Benjamin Franklin, whom some dubbed "the American Prometheus," had created and popularized them. In Prince's mind, these new and popular devices interfered with God's power. Prince claimed that the lightning rods drew electricity from the air and brought it to the earth, leading the ground to quake. To divert electricity represented dangerous interference with divine plans, for many believed God used lightning as a weapon against sinful man. As Prince intoned, "O! there is no getting out of the mighty Hand of God! If we think to avoid it in the Air, we cannot in the Earth." In a subsequent letter to the *Boston Gazette,* he explained that while he did not oppose lightning rods per se, they must be used "with a due Submission to the sovereign Will and Power and Government of GOD in Nature"; he warned that human interference might lead "the offended Deity" to "make that in which we trust for Safety to be the very Means of our Destruction in a

Moment."[13] To this generation, technology was awesome, but frighteningly so, for it might intrude upon realms of power not intended for humans.[14] Prince's fearful awe at the lightning rod reflected a mode of feeling that was several centuries old. It was based on the belief that God jealously guarded his vast and mysterious powers and did not want humans to possess them.

A new, less fearful sense of awe developed during the nineteenth century, however, for many Americans came to argue that it was acceptable to harness the powers of the universe with new machines. They believed they were capturing supernatural forces but maintained that their technologies were created in accord with divine plan rather than in rebellion against it. The mere fact that they could invent something meant that God approved of their actions. Given this perspective, by the mid-nineteenth century, there was far less concern that technology interfered with God's will, and the feeling of awe became less inflected with dread.

During the twentieth century, awe fell out of fashion, for psychologists considered it to be a backward, premodern feeling. Today few worry that by "tooling up," their new inventions will intrude upon divine domains. Admittedly, many Americans may be a little reserved about their capacity to always use technology wisely in the wake of man-made, technology-enhanced disasters, including the Holocaust, Hiroshima, Thalidomide, and Bhopal.[15] And there are plenty of popular representations of technology that reflect dystopian outcomes—for example, films like *Frankenstein, The Terminator,* and *The Matrix*.[16] Nevertheless, Americans continue to demonstrate a faith in their own inventiveness and its power to bestow benefits to humanity. Much of the time, they are even willing to believe such inventiveness will help them transcend current circumstances and take them to a higher plane of existence.[17] Today they display a faith in the power of human—not divine—agency to find ways around the problems of being mortal, for over the last two centuries, through their inventions, they have acquired some of the awe-inspiring attributes that earlier generations considered divine.

Modern individuals have incorporated these new abilities and powers into their definition of what it means to be human. As a result, the feeling of awe—which once people felt only when they regarded the Divine—has been transformed. Whereas originally it made people aware of their limits and their smallness, now it does not. Today awe is a weaker emotion and when contemporary Americans express it, it is often in reaction to themselves and their own creations. Its former power, as a check on hubris and vanity, has largely disappeared.

Dreams for the Telegraph

Contrast that to 1844. When Samuel Morse gave a public demonstration of his telegraph in that year, he sent the now-famous message "What Hath God Wrought." His statement, with its suggestion that God's hand was at work in the construction of the telegraph, encapsulated the way that many would think of it for the next several decades. Observers quickly cast it as a device imbued with divine powers, and this view spread rapidly. Within just a decade of its invention, many in the United States came to regard the telegraph, as well as other contemporaneous technological improvements, as gifts from God.[18]

This vision of the telegraph, as something divinely ordained seems to have originated with one of the telegraph's early promoters, who was eager for profits. Even before it was completed, a principal investor in the telegraph, Congressman Francis O. J. Smith, postulated that it would offer humans the "high attribute of ubiquity," an attribute that historian Richard John notes was "an awesome power that had previously been the exclusive prerogative of God," a power of "awful grandeur," worthy of "religious reverence." Once the telegraph was functional, Smith waxed poetic about the "almost super-human agency" of "Professor Morse's lightning."[19] Smith had financial interests in the telegraph and every reason to hype its potential in overblown rhetoric. He wrote about it in such terms with the hope that the federal government would buy the patent in which he held a stake.[20] Despite its capitalistic and very worldly origins, Smith's formulation struck a chord with many Americans who had no financial interests in the telegraph's success and who latched on to the idea that the telegraph was divinely inspired.[21]

Admittedly, there was still some resistance to the wires spreading across the land, precisely because they seemed to be so awesome, wondrous, and mysterious. Those who feared the telegraph expressed the older, fearful type of awe, more prevalent during the colonial era. Some ministers, sounding much like the Puritans discoursing about lightning rods, suggested that to use the telegraph (which was popularly referred to as "the lightning line") would be to disrespect the Lord, for it was tapping into divine powers, but doing so illicitly, without God's consent.[22] Amos Kendall, an investor in Samuel Morse's telegraph company, and also its business manager, reported in a letter that "the North Alabamians continue to cut down the telegraph almost every day. At first it was under protest that it produced dry weather. . . . It is said they were encouraged by a preacher who proclaimed from his pulpit that God had sent the dry weather to punish the world because man had got to be 'too smart' in reducing his lightning to human uses."[23] The belief that the telegraph might anger God and cause bad weather was not isolated, for, according to an early historian of telecommunications,

> A Hard-Shell Baptist preacher in southern Kentucky proved to be another menace. "See that, my brethren!" he cried out to his hearers one day during a disastrous drought. "Out along the road thar, a set of ongodly men have dared to interfere with the Almighty's lightnings, and what, my brethren, is the upshot of it? They have robbed the air of its electricity, the rains are hendered, and ther ain't been a good crop since the wire was put up, and what's more, I don't believe ther ever will be." The result was that the congregation went forth from the church, stirred up their neighbors, and collectively cut down several miles of poles and carried off the wires.[24]

The ministers' speeches reflected an older vision of human limits and the fearful, awesome grandeur of divine power, but such visions were increasingly challenged by those who celebrated rather than feared the telegraph, and who remade awe into a less humble and frightened feeling.

Indeed, the general reaction to the telegraph was far more positive, with onlookers perceiving it as an inspiring part of God's plan. For instance, when the transatlantic cable to England was laid in 1858, numerous ministers preached sermons about it. Reverend Joseph Copp declared that, through the telegraph, God's "agency unveils itself to a[n] . . . awe-struck world."[25] The Presbyterian minister William Sprague preached that the telegraph represented part of the "unfolding of the Divine perfections." He reminded his listeners that "all the power that is displayed or exercised in the Universe, is but an emanation from Omnipotence."[26] In honor of the same event, the Reverend Rufus W. Clark, in a sermon to the Brooklyn South Congregational Church, described the "amazement and joy" that Americans felt about the telegraph. He proclaimed that God was "stretching, from continent to continent the chords of a universal harp, to the harmonious notes of which the anthems of all nations will be sung in honor of the high and Holy one." He rhapsodized:

> We stand upon the shores of the proud Atlantic, whose storm-tossed waves have so long defied the power of man, whose wild waste of waters have so long kept asunder England and her fairest daughter . . . and lo! The monarch is subdued. Far down in quiet depths, amid the wrecks and cavern sepulchers of past generations, lies the bond of union between two great Christian nations, the highway of lightning thought. . . . In considering this event—the laying of the Atlantic cable—we would call your attention to it as a development of Divine Providence, as a new force for civilization and as another mighty agency for extending the Gospel over the earth.

Clark's enthusiasm was typical of the clergy. He himself had received news of the transatlantic cable while at a meeting with some four hundred other ministers. "The effects of the news was [*sic*] electric. Never was a body of men more deeply thrilled. . . . They then united in singing 'Praise God from whom All Blessings Flow.'"[27] Reverend Henry Field (whose brother, Cyrus Field, had funded the Atlantic telegraph) declared, "The more we think upon it, the more we are thrilled with secret awe and wonder."[28]

Sixteen-year-old Caroline Cowles Richards, living in upstate New York, wrote in her diary about the completion of the transatlantic cable, sharing her response and that of her grandfather, a minister: "There was a celebration in town to-day because the Queen's message was received on the Atlantic cable. Guns were fired and church bells rung and flags were waving everywhere. In the evening there was a torchlight procession. . . . Grandfather says he thinks the 19th Psalm is a prophecy of the electric telegraph. 'Their line is gone through all of the earth and their words to the end of the world.' It certainly sounds like it."[29]

Such optimists must have been disappointed—and such prophecies seemingly undermined—when the transatlantic cable stopped working after only a month in service. It would take another eight years until a functional cable was established. Yet, during that hiatus, the overland system of telegraph wires expanded, and many continued to describe the invention with wonder and amazement.

Often, nineteenth-century Americans regarded the telegraph as Reverend Clark had: the telegraph and associated inventions were "developments of Divine Providence," a sign that God wanted humans to expand their powers and to communicate with others in both earthly and spiritual realms. In that spirit, many focused on the idea that the telegraph was akin to prayer. Sometimes they used the telegraph as an analogy, as when, in an 1858 letter to her son, Elizabeth Stuart quoted with approval a fellow parishioner who had proclaimed "Prayer is God's Telegraph," or when an anonymous poet scribbled,

> What wondrous methods God has given
> Salvation wires from Earth to heaven
> The spirit's current runs up there
> I'll send a telegram of prayer. . . .[30]

This tendency to link the telegraph to prayer crossed denominational lines. Sister Saint Xavier, a French nun living in the wilds of

Indiana, far from her family and native land, wrote to her mother in 1855:

> We often receive news from Europe by the telegraph. . . . [I]n less than half an hour we can have information regarding what has happened fifteen hundred miles away. . . . There is talk of having a telegraph which shall pass through the ocean. The other day I thought what a happy invention it would be if a bridge could be made to extend from earth to heaven. But on reflection I recalled that Our Lord has made such a bridge—the Cross.[31]

In some ways, these comparisons diminished the importance of the telegraph as an invention. After all, for Sister Saint Xavier, prayer and the cross seemed to offer the true means for communing with a greater power. Yet on a tacit level, something else was expressed: that the telegraph, like the cross and prayer, was conferring astonishing powers that would enable everyday people to transcend the quotidian, to be more than mortals, and that the invention was graced by God.

Some observers went beyond analogy and suggested that the telegraph wires were not just analogous to prayer but in fact partook of the spiritual and even the supernatural essence of prayer. The *Telegrapher,* a newspaper published by the National Telegraphers' Union, supported this interpretation. The paper had as its motto "Is it not a feat sublime? Intellect has conquered time." In 1866, it suggested that the telegraph could prove the existence of heaven and the efficacy of prayer, claiming that nature as well as man-made objects offered "spiritual lessons." The telegraph had prompted "great thoughts of the invisible communication between heaven and earth," and from it, Americans might "gather fresh lessons from its teachings. Prayer has been called the telegraph line from earth to heaven. Is not the invisible fitly typified by the visible?" The author claimed that individuals who received telegrams from afar over the telegraph should not be dubious about messages sent through prayer. "Should not every skeptic stand rebuked when he receives a message over these wires? He believes in the message just received from a distant city, but he does not believe in messages sent from the heavenly world."[32]

With the new, awe-inspiring, and mystical powers afforded by the telegraph, perhaps this generation could do what humans had longed to do for millennia—communicate with the great beyond. Some embraced this hope as a result of the new fad of Spiritualism, which began to spread across the country. While the yearning for communication with heaven and the dead was long-standing, Spiritualism—which had first emerged in 1848, just four years after the telegraph's debut—seemed to promise a more effective way to reach through the veil. Jeffrey Sconce and John Durham Peters have shown that there were many links between the technological media and the spiritual mediums who contacted the dead in séances. Perhaps most noteworthy was that Spiritualists relied on knockings to spell out a code alphabet, much akin to Morse code.[33] Some nineteenth-century Americans made explicit connections between the two modes of communication, perceiving them to be essentially one and the same. Julia Ellen LeGrand Waitz, a resident of New Orleans displaced by the Civil War, wrote in her diary in 1863 that, despite her despair at her present circumstances, she took great hope from the technological changes she saw about her. She reported that she had just had "a long talk about spiritualism. . . . I feel sometimes almost persuaded that we are on the eve of some great change which will affect men both physically and spiritually. I have long held a notion of my own about electricity—it is the spirit, the soul of the world. . . . When my undefined hopes in . . . future revelations flag, I think of the telegraph. One by one the mysteries of creation are unfolded."[34] There were many like Waitz, some Spiritualists, others not, who believed the telegraph could span not just the miles that separated those living on earth, but also span the chasm between the living and the dead.

If talk about the telegraph was conjoined with hopes for the unification of the sacred with the quotidian, the heights of heaven and the depths of the ocean, the human with the superhuman, the living and the dead, that talk was further complemented by people who hoped the telegraph would finally bring world peace by uniting people across the globe. In 1846, a Philadelphia newspaper called the telegraph an "extraordinary discovery," and predicted that it would "make the whole land one being—a touch upon any part will—like the wires—vibrate over all."[35] A few years later, in an 1852 issue of the *American Telegraph Magazine,* a poet intoned,

> Lo, the golden age is come!
> Light has broken o'er the world,
> Let the cannon-mouth be dumb,
> . . . .
> God hath sent me to the nations
> To unite them, that each man
> Of all future generations
> May be cosmopolitan.[36]

Provincial and national identities would be no more, and all people would become citizens of a perfectly peaceful world. Abolitionist Frederick Douglass was likewise optimistic that the telegraph and other emerging technologies would lead to a new age of harmony. He celebrated the fact that "growing inter-communication of distant nations, the rapid transmission of intelligence over the globe . . . cannot but dispel prejudice, . . . bring the world into peace and unity, and at last crown the world with just[ice], liberty, and brotherly kindness."[37]

For many nineteenth-century Americans, then, the telegraph was more than a wire and a key. It was instead a powerful instrument that seemed to permit communication with new realms of the universe. God had seemingly shared some of his mysteries with them. They were awed by the telegraph's powers to make them part of something larger, to connect them to divine powers and to each other. Collectively, the dreams that Americans vested in the telegraph revealed deep longings to escape the inherent limitations of being human, of being separate, apart, sometimes lonely, and tragically finite. They optimistically believed that the telegraph, through its invisible but powerful connections, could join people across great distance, create a new community that was unbounded by the constraints of time and space, and break barriers between man and god, living and dead. The awe they felt expressed their belief that there were forces larger than themselves in the universe, forces that might bring true reunion and communion. Whereas colonial Americans had stood in fear of such forces, many nineteenth-century Americans believed it was their right to share in them.

Disappointments

In 1874, poet Anne Lynch Botta, on one side of the Atlantic, wrote to James Froude, residing on the other, and summed up the excitement of the era. She declared, "The telegraph is the miracle of the age. That you can sit in your study in London, and I here in mine, and carry on a conversation limited only by our pounds, shillings, and pence to pay for the privilege; that the message which goes does to the depths of the Atlantic's unexplored water, reaches you almost as soon as if spoken across the table,—this puts necromancy to the shade."[38]

Botta believed the telegraph was miraculous. But she also admitted that there were limitations placed on its use by price—by how many pounds, shillings, and pence one possessed. Indeed, she transmitted her sense of wonder in a letter rather than a telegram, for it would have cost a small fortune to wire her long missive across the ocean. While Americans' hopes for transcendent uses of the telegraph were fantastical, they were made even more so by the limits of their pocketbooks.

High prices therefore soon tempered many people's hopes for the telegraph and sobered their expectations. The diary of Delia Locke, the California rancher's wife, offers a window on the evolving meaning of the telegraph, initially as a symbol of hope and then eventually of disappointment. On September 18, 1858, she wrote of a letter she had just received: "We have again received the Eastern mail. And the joyful tidings it brings sends a thrill of pleasure through every heart. The Atlantic Cable is laid—the mighty Telegraph enterprise has been successful! . . . [T]hirty one nations could now converse with each other with the rapidity of thought! . . . God be praised for all his rich gifts to rebellious man!"[39]

There are many things interesting about Locke's diary entry. First, since the transcontinental telegraph—which would eventually connect the East Coast with the West—was not yet built, she received news of the transatlantic telegraph's completion by letter rather than by cable, and a month after it had been laid. Second, on the very day she recorded the letter's arrival, the transatlantic cable ceased to work (although, of course, she had no way of knowing that yet). Finally, 1858 was the last time she spoke hopefully of the telegraph, for her actual responses to telegrams she herself received in later years were far different from these initial expectations.

Like Delia Locke, when Americans considered the telegraph in the abstract, as a system, a network that spanned miles and continents, they predicted it would bring a joyous and providential reunion of the human race, of the living and the dead, and they spoke of it as a gift from God and regarded it with expectant awe. Yet when they considered actual telegrams they themselves received, they had different, and generally less optimistic, feelings about them. While the telegraph promised to bring a communion of souls, telegrams often did just the opposite. Telegrams merely communicated in new form the quotidian, and often reinforced a sense of the reality of distance and, more profoundly, the limits of life. Their arrival made clear that the earlier awe-filled hopes—that one could transcend the human condition and solve the problem of mortality—were empty ones. While Locke may have thought the telegraph was a gift from God "to rebellious man," the actual telegrams she received brought with them the recognition that in fact she still did not possess the powers of the divine.

Consequently, over the next several decades, when she wrote in her diary about telegrams she herself received, Locke struck a much different tone. On a few occasions she mentioned the telegraph as bringing neutral news, but most of her references to it were far more freighted. In 1864, she wrote it was a "sad day—Aunt Hannah died about three o'clock this afternoon. . . . A telegram was sent to Mr. & Mrs. Read this morning."[40] In 1871, Delia Locke herself fell so ill that she wrote she was "brought upon the borders of the 'valley and shadow of death.'" A relative telegraphed her brothers of her illness, from which she soon recovered.[41] A few months later she reported that news of the devastating Chicago fire had come over the wire.[42] Later that year, there were cables predicting bad weather.[43] When her mother was ill, another relative sent a telegram for the doctor and kin to gather.[44] The following year, in 1875, she reported that the telegraph had brought news of an infant's death. "This is sudden news, for we had not heard of its sickness. . . . The funeral is to be on Tuesday."[45] And during the next years, when telegrams arrived, they carried news of other deaths in the family and little else.[46]

The reason that Locke and most other Americans encountered mostly tragic news when they opened a telegram was because the price was so high relative to income. For many of modest means, the expense was justified only in extreme—and often dire—circumstances. Consequently, telegrams generally brought more fear and sadness than joy to the average American, a fact widely acknowledged in the nineteenth century. Writing during the Civil War, Mary Boykin Chesnut reported grimly on what telegrams had come to mean to her. She confided to her diary, "A telegram reaches you, and you leave it on your lap. You are pale with fright. You handle it, or you dread to touch it, as you would a rattle-snake; worse, worse, a snake could only strike you. How many, many will this scrap of paper tell you have gone to their death?"[47] The telegram's fearful connotations did not quickly abate. A 1906 newspaper reported, "An Atchison woman claims to have well-controlled nerves because she can open a telegram with no more emotion than she would show in opening a can of corn."[48]

While Americans had thought the telegraph would give them the awesome powers of divinity, help them overcome distance and death, in reality, telegrams did no such thing. They only reminded men and women of their own mortal limitations. Those who received telegrams no doubt wished to know about their family and friends' conditions, illnesses, and deaths, but those messages brought proof that the latest technologies could not transcend the human condition. They could neither stop death nor open up communication with the dead. That telegrams arrived unexpectedly and brought with them a sense of shock made clear that the connection the telegraph wire established between individuals was not a constant one, but a sporadic one. Locke had hoped the telegraph would allow people to communicate joyous news to each other "with the rapidity of thought." The fact that telegrams were few and far between, and carried news of the unexpected, the unsuspected, and the tragic, revealed the limits of this technology, limits exaggerated by high prices.

Then too the telegraph fell short of expectations because it was also part of a problem many hoped it would solve. The telegraph was one development in the emergence of modern, capitalist markets. It simultaneously promised to connect people, but also justified and enabled their separation.[49] The longing for communion and connection—with kith and kin far away, with God in heaven, with the dead—seemed to grow as Americans scattered and hustled and bustled. New communication technologies like the postal service and the telegraph promised that there would be some possibility of reunion should Americans wander far from home and the familiar in search of profits and opportunity. In doing so, it helped set many in motion. The telegraph, like the train whose tracks it often paralleled, was simultaneously an instrument of communion and dispersal, a source of awe as well as disillusionment.

"There's No Telephone in Heaven"

If the invention of the telegraph and its extension around the globe brought with it both awe-filled hopes for transcendence and salvation as well as eventual disappointments, other new media inventions of the nineteenth and early twentieth centuries excited many of the same dreams and then dashed them as well. The telephone, phonograph, and radio all seemed to hold within them unknown potential that might offer humans new powers. Americans, however, soon discovered their limits. Yet even when their fascination with one device waned, they quickly vested their hopes in the new ones that emerged. Historian David Nye writes of this tendency, noting, "Despite its power, the technological sublime always implies its own rapid obsolescence, making room for the wonders of the next generation. The railway of 1835 hardly amazed in 1870, and most Americans eventually lost interest in trains (though that particular 'romance' lasted longer than most). . . . During each generation the radically new disappeared into ordinary experience."[50]

However it was not merely that Americans transferred their excitement from one technology to another, it was that the emotions they felt about these inventions changed dramatically over time. Twentieth-century awe differed markedly from nineteenth- or eighteenth-century awe. Technologies had changed; so too had emotional culture. Gradually, Americans became convinced that awe—with its connotations of superstition, fear, and submissiveness before a powerful God—was an outdated feeling, one that earlier, more primitive people had felt. Even as early as 1869, signs of such attitudes were emerging. A poet celebrating the telegraph described it as the work of Morse, "our modern Prometheus," and declared it had ushered in a new age: "No more the hours of awe and gloom, Which filled our childish hearts with dread. . . . Now science grabs the lightning's fire."[51] Unlike the mythological Prometheus, the American Prometheus of the nineteenth century no longer need stand in fearful awe of the gods.

Poets were joined by moral philosophers who questioned the value of awe and wonder. The influential English philosopher, educator, and psychologist Alexander Bain, whose work was read on both sides of the Atlantic, wrote of wonder and the sublime, noting, "In matters of truth and falsehood, wonder is one of the corrupting emotions. The narrations of matter of fact are constantly perverted by it." Wonder, he believed, was on the decline: "The discovery of uniform laws makes wonder to cease in one way by showing that nothing in nature is singular or exceptional. . . . [S]cience and extended study naturally bring a man more or less to the position of 'nil admirari,' depriving him of the stimulating emotion bred of inexperience. Even the unexplained phenomena can be looked at with composure by the philosophic mind."[52]

Reflecting this new appraisal of awe, Dr. David Inglis, in 1898, reported on what he termed a "Remarkable Exaggeration of the Sense of Awe" in the *New York Medical Journal.* There he told of a patient who experienced too much awe. He described a well-educated woman who felt "intense awe" whenever she saw a rainbow, red sunsets, or the aurora borealis. Inglis believed her case illustrated "the type of mind of the early ancestors of our race." He continued, "My patient seems to me to represent the probable mental state of early man before the intellect had reasoned out the causes and relations of natural phenomena. The average man reasons these things out, and, coming of generation after generation which has reasoned them out, he is born with a mind in which the fear of the great and strange forces of Nature is relatively small." As Inglis concluded, because of her sense of awe and mystery, "this patient is a reversion to an early type. She is an aboriginal."[53]

Many psychologists concurred that some religious feelings were useless vestiges of the past. James Leuba, a psychologist of religion, described the fear with which many regarded God as representing "the lowest form of religion" and a "survival of a by-gone age." Awe, he believed, was more evolved and noble than fear, though it too contained an element of "arrested fear." He suggested that "the stage of culture at which awe can be the dominant religious emotion is also passed"; it was becoming "obsolete." Those who still felt fear and awe when they contemplated nature and God, he wrote, were not to be celebrated, for the emotions were "in no way praiseworthy." He concluded, "The powerful support which traditional Christianity—and of course, other forms of religion also—receives from the emotional reaction in question is due to the fact that both are survivals of an earlier age. . . . [T]he lapse of intelligence induced by emotion brings man down to the level of antiquated religious beliefs."[54] In 1917, social psychologist William McDougall likewise described "the long persistence of fear and awe in religion," suggesting that it was fading away among the "more highly civilized peoples at present time."[55] Across the Atlantic, a year later, Max Weber famously observed "that there are no mysterious incalculable forces that come into play, but rather that one can, in principle, master all things by calculation. This means that the world is disenchanted. One need no longer have recourse to magical means in order to master or implore the spirits, as did the savage, for whom such mysterious powers existed. Technical means and calculations perform the service."[56]

Awe and wonder were becoming outdated emotions, ill-suited to the modern, rational world.

If awe was no longer welcomed in modern society, neither was the tendency to regard machines as providentially designed. Many Americans continued to hope these inventions would offer new possibilities for communion with the dead and connection to the greater forces of the universe, but they eventually came to regard the machines themselves as the product of human effort rather than divine inspiration. This change did not happen overnight. Some still continued to invoke God or mysterious powers when talking about new technologies. Ministers still sometimes claimed them as God's creation.[57] Yet the frequency of such pronouncements gradually declined, just as excited celebrations of human ingenuity began to emerge.

Many observers of turn-of-the-century inventions focused more on their human creators than on their relationship to divinity. Americans still regarded these new devices as holding the potential to tap into strange and mystical forces of the universe, but it was humans, not gods, who were doing this. For instance, to many the phone seemed magical but not quite divine. As a newspaper reported in 1877,

> Neither in the picturesque phrases of half-civilized man nor in the boldest flights of fancy or tradition is there anything quite so weird as the speaking telegraph. In all the Eastern legends of magic, people who are placed wide apart never communicate directly with each other by speech. After the magician has drawn his circles in the sand, and lighted the mystic fire, and spoken the cabalistic words, he may perhaps summon the distant one by occult influence or through the agency of a genie. It is a thousand times more astounding as a mere conception that the voice, the tones, the very utterance of a friend who is miles on miles away, may be distinctly heard by the listener who holds to his ear the trumpet of the telephone.[58]

Yet by 1901, some had become so accustomed to the telephone that the *New York Times* declared, "This is an annihilation of time and space which would belong in the realm of magic if it was not a commonplace of daily experience."[59]

Even if phones were becoming commonplace, marketers nevertheless continued to focus on their magical qualities. "The Magic Flight of Thought" was what the telephone offered, according to a 1914 AT&T advertisement. It declared that while humans had long hoped to move across time and space instantaneously, now it was possible, thanks to the technical expertise of the phone company: "The flight of thought is no longer a magic power of mythical beings, for the Bell Telephone has made it a common daily experience. . . . [T]houghts of people are carried with lightning speed in all directions."[60] The phone was magical and awe-inspiring, but the Bell Company, not God, had created it. And in doing so, the company had given what once were considered divine powers to mere mortals.

The long-standing hopes that these new inventions would help men and women transcend their mortal limits did not completely disappear either; however, many of them came to rest on the technical skills of the human creators rather than on divine intervention. Shortly after inventing the phonograph, Thomas Edison explained that it offered the possibility of immortality. He declared that one of its key uses would be to create a "Family Record" which would preserve "the sayings, the voices, and *the last words* of the dying member of the family."[61] Elsewhere, he elaborated on this use, noting, "Centuries after you have crumbled to dust, [the phonograph] will repeat again and again to a generation that will never know you, every idle thought, every fond fancy, every vain word that you choose to whisper against this thin iron diaphragm."[62] These were not merely the wild promotions of the phonograph's inventor—instead, they revealed a widely shared desire for communion with the dead, for a guarantor of immortality. *The Farmer's Cabinet,* a New Hampshire newspaper, spoke with a similar reverence about the phonograph and its powers to reshape the boundaries between mortals and immortals:

> Nothing could be more incredible than the likelihood of once more hearing the voice of the dead, yet the invention of the new

The Magic Flight of Thought

AGES ago, Thor, the champion of the Scandinavian gods, invaded Jotunheim, the land of the giants, and was challenged to feats of skill by Loki, the king.

Thor matched Thialfi, the swiftest of mortals, against Hugi in a footrace. Thrice they swept over the course, but each time Thialfi was hopelessly defeated by Loki's runner.

Loki confessed to Thor afterward that he had deceived the god by enchantments, saying, "Hugi was my thought, and what speed can ever equal his?"

But the flight of thought is no longer a magic power of mythical beings, for the Bell Telephone has made it a common daily experience.

Over the telephone, the spoken thought is transmitted instantly, directly where we send it, outdistancing every other means for the carrying of messages.

In the Bell System, the telephone lines reach throughout the country, and the thoughts of the people are carried with lightning speed in all directions, one mile, a hundred, or two thousand miles away.

And because the Bell System so adequately serves the practical needs of the people, the magic of thought's swift flight occurs 25,000,000 times every twenty-four hours.

AMERICAN TELEPHONE AND TELEGRAPH COMPANY
AND ASSOCIATED COMPANIES

One Policy *One System* *Universal Service*

FIGURE 5.1 Into the twentieth century, advertisements for the telephone continued to invoke the idea that the phone had magical qualities. This 1914 AT&T ad suggested that the magic of the gods had become the possession of humanity. *Courtesy of N. W. Ayer Advertising Agency Records, Archives Center, National Museum of American History, Smithsonian Institution*

> instrument is said to render this possible hereafter. It is true that the voices are stilled, but whoever has spoken or whoever may speak into the mouth-piece of the Phonograph, and whose words are recorded by it, has the assurance that his speech may be reproduced audibly in his own tones long after he himself has turned to dust. A strip of indented papers travels through a little machine, the sounds of the latter are magnified, and possibly centuries hence hear us as plainly as if we are present. Speech has become, as it were, immortal.[63]

In sum, as Evan Eisenberg points out, the phonograph offered a form of séance.[64] Yet it was a séance created by Edison, not God, and one at which the dead said nothing new or unexpected. Their voices were frozen in the past rather than communicating new thoughts in the present.

When radio, or wireless telegraphy as it was often called, was first invented, it too excited hopes of connections to the otherworldly. And some observers still discussed their feelings in terms of the religious awe they felt. Historian Susan Douglas maintains that the radio's rise and spread during the 1920s was a sign "that people were hungering for otherworldly contact, for communion with disembodied spirits, for imaginative escapades that affirmed there was still wonder in the world."[65]

A case in point was the popular belief that the radio would allow listeners to tune in to the ether and happen upon voices talking long after the speakers had died. The *New York Times* carried word that "the voices of famous men who have spoken over the radio are still wandering in the ether and if wireless development continues at the present rate they may be picked up a hundred years hence, according to engineering experts of the Marconi Company." Engineers told the reporter that radio messages "were never lost" and might "go on forever." They claimed that they had "actually trapped messages after they have passed a third time around the earth. It is not impossible, they say, that fifty years hence the voices of men long dead may be still wandering about and be picked up by sensitive instruments."[66] Anne O'Hare McCormick wrote that it was both "interesting . . . and appalling, to realize that only the turn of a knob stood between you and all the voices, human and instrumental, vibrating at that moment throughout the world." She claimed that "every ripple of the surrounding ether was soaked with transmissible sound—always had been, from that matter, so that in the air that beat upon the ears of the aborigines camped on the banks of the Potomac (if they had but known how to tune in) were hidden the words of Socrates discoursing in the Agora of Athens, of Cataline declaiming before the Roman senate, of the Sermon on the Mount."[67]

Yet such speculations revealed the limits of the machine and the limits of faith. Listeners were not going to pick up any new sermons from Jesus; nor were they going to actively communicate with him, or Socrates, or Cataline. They might hear echoes of their words, spoken long ago, but the telephone, phonograph, or radio could not rouse the dead. Hearing these voices from the past might be thrilling, but it was not the true or complete connection for which earlier generations had longed.

And there was a rising tide of skepticism about the devices as well. For instance, many Americans scoffed at the idea that one could telephone to heaven, though their grandparents had believed they might send a telegram there. In August 1890, the *New York Times* reprinted news of what a London paper termed called "Telephone Insanity." "In the infirmary connected with the Central Police Station the doctors have received to-day a curious case of what they call 'telephonic madness.'" A young lady had stopped in the middle of the streets and "shouted at the top of her voice, 'Hallo! Hallo!' the preliminary words used here when a person wishes to converse with another through a telephone. A crowd at once gathered around the young lady, who put her hands to her mouth and ears in telephonic fashion. 'Is that you, Saint Peter?' continued she, as if speaking into a tube. 'Right, give me my keys? What? You cannot be bothered! Then send your commissionaire. I must get home!'" After watching the woman do this for some time, the crowd gradually concluded "that she was wrong in her mind. A constable took her to the police station, where she went on in the same way, declaring that she heard distinctly through the telephone the celestial music of Paradise; that she could hear Saint Cecilia playing the piano, and that the chorus was composed of cherubim." She was sent to an infirmary by the police.[68]

Such delusions apparently were quite common at the turn of the century. For instance, a Lexington, Kentucky, newspaper, the *Blue-Grass Blade,* reported in 1905 that "a doctor of this city named Lillokrone developed religious mania and rigged up his boarding house a telephone with which he imagined he could communicate with heaven. He landed in Bellevue Hospital, psychopathic ward."[69] "Was Telephoning to Heaven, but Policeman Got Him before Central Makes Connection" was the headline in a Spanish Fork, Utah, paper in the same year.[70] Whereas few had mocked those who imagined they might telegraph to heaven and communicate with the dearly departed, a new skepticism was visible by the late nineteenth century. Such beliefs were coming to be regarded as signs of madness or "religious mania," not religious awe and reverence.

That skepticism was visible even in popular songs of the era. "Hello Central, Give Me Heaven" told of a "tearful little child" who tried to call her mother on the phone, proclaiming "call her; won't you please; For I want to surely tell her, We're so lonely here." The child's request was poignant precisely because listeners knew such a feat was impossible.[71] Those who heard the song might have held out hopes for communicating with the dead, but they no longer seemed to believe telegrams or phone calls would be the means of conversing with them. A number of newspapers carried similar accounts of small children longing to call heaven to talk with dead loved ones. The *Valentine Democrat,* for instance, carried a "Child's Query" in 1911:

> "He is five years old, and his brother, two years older, had just died at the family home on G. Street. He was talking it over with his grandfather.
> 'Say, grandpa,' asked the little fellow, 'where has Roger gone?'
> 'To heaven,' answered his grandfather.
> . . . .
> 'Have they got a telephone in heaven?'
> But there was no reply."[72]

FIGURE 5.2 The 1901 song "Hello Central, Give Me Heaven" described the long-standing hope of contacting heaven with new technologies, while also underscoring the futility of such hopes. *Courtesy of David M. Rubenstein Rare Book & Manuscript Library, Duke University*

By the early twentieth century, the desire to phone heaven was portrayed as a sign of youthful innocence and naïveté, described as a longing that would eventually be outgrown. Whereas in the nineteenth century, adults had hoped to telegraph to heaven, in the twentieth century such dreams had become a mark of an immature intellect.

FIGURE 5.3 Reflecting a growing cynicism toward those awestruck by new technologies, this 1923 comic parodied the belief that one might tune in to heaven with a radio. *Courtesy of George H. Clark Radioana Collection, Archives Center, National Museum of American History, Smithsonian Institution*

In fact, some pundits openly mocked the hopes for connection with God, heaven, and the dead. A 1923 comic in the syndicated series *Somebody's Stenog*, depicted a woman tuning in to her newly purchased radio. A man, watching her tune her set, asked what she was able to hear. She replied, "I'm liable to get Mars or Juniper [*sic*] the way this little box is working–Hush I hears Harps playing and sounds like wings flapping and people singing hymns. . . . Oh Ted–I bet I got Heaven–" Her friend replied, "Babe, you're a nut–it's that revival meeting in the church next door!"[73] The dream of using technology to connect with heaven, once at the mainstream of American thought in the nineteenth century, was increasingly marginalized in the twentieth.[74] There were newly imposed limits to the religious awe one should feel and express in the face of new technologies.

Finally, while nineteenth-century observers believed the inventions of the age reflected God's glory and were part of his plan, many twentieth-century Americans saw the new devices as reflective of the genius of individuals. A century before, God had been in the machine; now men and women had created machines that were Godlike. A poem from the 1930s, entitled "Radio," illustrated this new viewpoint. It began,

> There is no land so barren, bleak
> That I will not abide therein;
> There is no storm that wrecks and maims
> But I survive its direful din.
> . . . .

The poet then suggested the radio was enormously powerful:

> For, under God, no power lives
> But hides in fear before my presence.

Yet lest anyone be confused, the poet quickly cleared up the nature of radio's power, as well as its origins:

> "Radio," they name me now—
> Product of research and science.[75]

This trend of celebrating technology's power but desacralizing its origins became even more pronounced over the course of the twentieth century. Often Americans regarded their televisions, computers, and phones with astonishment, thought of them even as magical, but did not express the same religious awe of a century before. These were the products of "research and science," artifacts of human ingenuity rather than divine inspiration. To think otherwise—and to express religious or spiritual awe—was to be seen as backward, primitive, and to risk mockery.

Digital Dreams and Their Half-Life

Today the discourse about technology is far more secular than it once was. On the one hand, this is not surprising. Since the early twentieth century, a number of social theorists, most notably Max Weber, have suggested that in an age of rationalizing bureaucracies, enchantment has "retreated from public life."[76] Yet while modern Americans might be skeptical that they can communicate with heaven or God, they nevertheless still sometimes sound awed when they describe their feelings about their technologies. Indeed, today some speak of "the cloud" in tones their grandparents reserved for the ether. However, their awe is not the awe of their nineteenth-century ancestors. The awe that twenty-first-century Americans express has been redefined; it is a weaker and far more secular feeling, and therefore less likely to take them outside themselves, for the things that awe them today are their own creations.

What seems to be the case for many is that they attribute the transcendent, enchanting possibilities that technology offers today to the ingenuity of man, not to God or divine will. They still hope for immortality, a connection with the dead, and a true global community, and they harbor the desire to be part of something grander, but they believe that it is men and women, and the wireless devices, apps, and software they create, not God's hand, that will ultimately produce it. It will be achieved during life rather than after death. As Joel Dinerstein writes, in the secular and modern era, "the future replaced heaven as the zone of perfectability."[77]

What has also changed is what objects actually provoke awe. The technologies that gave rise to the feelings and gave hopes for transcendence in earlier times often tended to be large civil engineering projects like the telegraph network, the transcontinental railroad, the telephone system, dams, bridges, and dynamos that served to draw Americans of disparate faiths together in a way that no sectarian religion could have. As David Nye has argued, these technologies and the civic feelings they evoked were wrapped up in the so-called technological sublime, which he described at length in *American Technological Sublime.* Published in 1994, at the very nascence of the World Wide Web, it begs the question, what has happened to awe and the sublime with the growth of the cyber sphere? Sociologist Vincent Mosco contends that the twenty-first-century version of the technological sublime is now "the digital sublime," and he argues that, in contemporary society, people have vested the internet with hopes for perfection, for an end to history or to politics and geographical distance, much as their ancestors did with the new inventions of the nineteenth century.[78]

However, like the hopes for sublime transcendence that were inspired by previous technology, most of these dreams have been only imperfectly realized. No better illustration of these initial hopes can be found than in the ebullient celebration of the World Wide Web that Thomas Friedman offered in his 2005 runaway best seller, *The World Is Flat.* As Friedman saw it, new digital technologies would break down barriers between global powers, would lubricate markets, and also diminish discord, as nations found their futures better forwarded through trade and communication on the internet than through war. Later pundits picked up the theme, suggesting that perhaps the Web was a harbinger of better things to come, as phrases like "the Twitter revolution" and "sharing economies" were bandied about. For a while, it seemed like such technological hopes were being realized. But contemporary world events continue to remind the public that they have not yet transcended history or mortality. Despite technological innovations, our boots remain literally and metaphorically on the ground.[79]

Yet, nevertheless, some Americans, when they are online, are experiencing awe, albeit in a newer, weaker, and more mediated form.[80] One might not be expecting people holed up in their basements playing video games to be experiencing any form of the emotion, but Jane McGonigal, in *Reality Is Broken,* reports that this is exactly what some gamers feel when they play *Halo,* the military, first-person shooter made by Bungie for the Microsoft Xbox. For the uninitiated, it might be hard to believe. But for *Halo* gamers who are placed on a battleground where they are both hunter and hunted, the game can provoke "spine-tingling chills" that produce the feeling. McGonigal explains that *Halo* also offers players an opportunity to contribute their services to a human army that is waging "The Great War" against an invasive confederation of aliens known as "The Covenant." The game, of course, is a fiction, but in casting players in the role of soldiers who are risking their virtual lives for humanity, Bungie has created an "epic" where gamers can imagine that their gameplay is a part of something much larger than themselves.[81] As McGonigal puts it,

> Not every game feels like a larger cause. For a game to feel like a cause, two things need to happen. First, the game's story needs to become a collective context for actions—shared by other players not just an individual experience. . . . And second, the actions that players take inside the collective context need to feel like service: every effort by one player must ultimately benefit all the other players. . . . *Halo* is probably the best game in the world at turning a story into a collective context and making personal achievement feel like service. . . . Thanks to Bungie's exhaustive

> data collection and sharing, everything you do in *Halo* adds up to something bigger: a multiyear history of your own personal service to the Great War.[82]

Of course, *Halo* is a fantasy. But as McGonigal observes, the emotions it triggers—awe, "reverence," and a sense of solidarity—are real. "The Great War isn't real, but you really do feel awe when you think about the scale of the effort so many different people have made to fight it. In the end, as one player sums it up 'Halo proves that you can have a shooter game with a story that really means something. It draws you in and makes you feel like you're part of something bigger.'"[83]

This sentiment is common among gamers. Greg Bayles, a native of Las Vegas, testified to the awe that video games might induce. He was a project manager at a research lab that produced therapeutic apps for medical conditions, but he had trained as a video-game designer and had written and thought a great deal about awe.[84] He saw many opportunities for astonishment in those games. For instance, meeting others from around the world who had played these games produced a feeling that was "really unifying," for, as Greg put it, the games had "invited us into a new headspace and they made us think about ourselves and about the world in new ways . . . and I think that's some of what awe is." In the nineteenth century, Americans had hoped for reunion and communion in the clouds of heaven; for Greg and many who played the same games, that reunion was happening on earth, in a shared "headspace."

In fact, he saw religious parallels between the awe available online and the past transcendent experiences people had had in churches. Religion had offered entry into another world; he believed games and apps did the same.

> I was thinking about the . . . architecture of a cathedral—how the doors are often so small so that as you step through them you are required to hunch down, and as you come into the grand vestibule . . . it seems that much bigger. . . . I think there's something really interesting about having entire worlds contained in

> this little smartphone that you carry around in your pocket, having entire galaxies, . . . just presented to your senses in ways that is impossible otherwise. . . . That's really something that invites me into an experience of awe. . . . It's just mind-blowing. . . . I can't possibly experience these things outside of this context and yet . . . I can pay $100 for this phone and suddenly I am invited into this unimaginable experience.[85]

Greg's enthusiasm about the awe-inspiring nature of digital technology was palpable.

He was not the only one to be so excited. Others have found awe online as well. An example is visible—and audible—in a symphonic internet choir that Eric Whitacre, a Grammy-winning composer, describes in his TED talk "A Virtual Choir 2,000 Voices Strong." In front of a rapt audience, Whitacre tells how he had solicited thousands of a capella recordings from separate individuals scattered across the globe, who had sung the various parts of his composition "Lux Aurumque." After merging all the contributions into one choral piece, Whitacre felt that his compositional talents, in concert with the internet, and the contributions of his devotees, led toward a transcendent, religiouslike experience. As he describes it in a separate interview:

> There's just this incredible energy when you have that many singers, all of them singing and breathing at the same time. . . . [A]s a composer I want so desperately to make this a communal experience, so that the performers and the audience members are all sharing the same ecstatic experience at the same time. . . . [W]hat I'm describing, basically, is a religious experience, and I don't know that I'm setting out to do that. But it sure sounds like Mass, doesn't it?[86]

For Whitacre, there was a spirit in the network, and it offered his acolytes, as well as his audience, a place of connection and communion with others that might not be available elsewhere. The network, or at least Whitacre's network, seemed to redress at least some forms of spiritual emptiness.[87] Listeners to his concert on YouTube left comments testifying to the music's effects on them: "Whitacre's 'Lux Aurumque' brings me to tears every time. Totally awe-inspiring," posted Valerie Ferguson, while another viewer, Anna Ray, wrote, "Wow!!! This is amAZING!! This technology creates an awe within me! These emotions send hope to my soul!! Someday, in heaven every nation will join to one choir and we will all sing together. :)"[88] Yet unlike the communion nineteenth-century Americans hoped to achieve—enabled by divine powers—this kind of spiritual meeting was wholly engineered by humans.

There were some we interviewed who shared Whitacre's and his fans' amazement at the seemingly magical powers summoned by modern technology. Those who produced digital artifacts professionally described the relationship between magic, awe, and technology with excitement. This was the case with Warren Trezevant, the San Francisco Bay Area resident who worked as a product manager for a company that created animation software. Warren landed his present job after seventeen years at Pixar Studios where he had worked as an animator for *Toy Story 2, Cars, Finding Nemo,* and many other films. He held the experience of wonder in high regard and it motivated his current work: "I do it because I felt it when I watched *Star Wars* [as a child]. . . . I experienced movie magic. So part of my dream is how do you create that awe in other people? . . . How do you become the magician?"

Warren believed that most magical moments happened as a result of individuals' deep investment in presentation. When people confronted technology and regarded it as magical, it was because others had spent a tremendous amount of time and energy in making it appear that way: "It's something that I've learned both at Burning Man and doing tricks. . . . Animation is magic. . . . It's an illusion you are presenting to somebody, much as technology is presenting an illusion. And there will be some people who want to understand its component parts like 'what is the screw tolerance that put it together?' but most of the time people want to know what's the magical experience."

Warren worked in a field more self-consciously devoted to the creation of magic and illusion than most other professions. But he felt that comparable techniques were being employed in the creation of phones and other personal technologies: "With the iPhone . . . or even the iPod . . . what a magic trick that is. . . . That they could figure this all out, put it together. . . . It's not new. It's never new technology. It's like magic tricks are never new. It's about the deeper presentation. . . . All of a sudden, you touch things and move things and swipe."

When Warren talked about awe, his frame of reference came from how he himself consumed and produced magic and animation. If he spoke fondly of the way he had been amazed by *Star Wars* as a child, his wife, Harriet (who we interviewed at the same time), spoke with similar enthusiasm about the time one of Warren's zoetropes had entranced Steve Jobs and impelled him to stare fixedly at the magic that Warren had spent a good deal of time creating. But Warren's definition of awe was not confined to his work at Pixar. Now that he was creating software that helped animators do their job, he was thinking a lot about how tools could help or hinder work. During our interview, he and his wife brought out two knives that they had painstakingly forged during a weekend workshop. For Warren, the point was that all tools matter, and that whether people were making knives or software they should be seeking to find out the "deeper truth about how people want to work." Awe emerged when that truth was found and when it was embedded in good design that replaced an older, more cumbersome, approach. The goal in his current job was to design software so that animators could "work at the speed of thought." The experience of work became "awe inspiring . . . when things become easier." And it was indisputably the result of human ingenuity and industry.[89]

Greg likewise found awe in the technology he used and produced. He had trained as a video-game designer and he saw many opportunities for astonishment in those games. He confided, "In terms of the video games that I make, I really am trying to kind of leverage awe. . . . I think video games especially . . . have an interesting potential, because they pull in so many different media types"—from art to programming to music to philosophy and storytelling. Together, "they have a lot of potential as far as transcendence goes."[90]

Others involved in creating technology also found that when they had built something powerful it could feel like magic to themselves or to others. For instance, when programmers create features that seem to inexplicably save users time or reduce a complex sequence of tasks to a simpler one, those features are sometimes described as behaving "automagically." On Stackoverflow (a social media site for programmers), a subscriber collected the best comments that programmers had left in code. (Comments are a bit like annotations in a book. They are inserted for other programmers to read but they are not formally part of the program flow.) Among the most notable are the following:

> //Magic. Do not touch.

and

> <!–Here be dragons–>[91]

While programmers often reveled in the beauty of their code for its own sake, there were also business advantages to be found in magic. As one CIO noted, there was a "mystery/mastery principle" in technology. It boiled down to this: "who creates mystery, gets mastery."[92] Adam Kaslikowski, the high-tech entrepreneur in Venice, California, was acutely aware of the market value of awe. He had served as COO, CEO, or founder of a number of lifestyle brands that were primarily Web-based. In addition, he had recently headed a company named Authorly, which had transformed books into apps. In Adam's business, magic featured prominently "at the design stage." Designers would often say, "I want it to be amazing; I want it to be magic; I want it to be a delightful experience." Moreover, magic was a "minimum bar for entry . . . in app or game development. . . . It has to feel magic. If it doesn't, you are relegated to the also rans."[93] Likewise Greg believed that technology producers should strive to create "awe-inspiring experiences." While some might lament that the emotion was being created for a commercial purpose, he observed that this had long been the case. After all, he noted, Michelangelo had painted the Sistine Chapel on commission.[94]

Today, awe and magic are commonly harnessed in the service of selling. For example, Christian Wutz, the twenty-three-year-old chief marketing officer of a wireless company, declared, "You want to motivate people to be astonished." If they were, it "results in more value for us or for the investor we're talking with, the venture capitalist we're talking with." One of Christian's mentors had told him, "The goal of the founder is to create believers," and he said, "That really stuck with me. . . . That's what your goal is . . . to create that awe," for one could create customers by eliciting the feeling.[95] This is a process that sociologist George Ritzer describes as the "rationalization" of "enchantment," by which companies make awe repeatable and purchasable.[96] Twenty-first-century Americans are very aware of the human origins of awe—as well as its commercial potential. It was not created by a force larger than themselves. Instead, they themselves might produce it—and profit from it, as well.

Awe and Transhumanism

Awe is also being evoked by a group of people who believe that technological innovations are on the verge of enabling human immortality. This is the hope of Ray Kurzweil and other transhumanists. Kurzweil calls his path to immortality "the singularity," and he believes it will be possible to transcend mortality and the human condition more generally by 2045. The singularity is not exclusively a transition from a biological to a nonbiological self. Instead, according to its proponents, technological advances will allow people to move seamlessly between biologically and non-biologically based consciousness. In his 2005 book *The Singularity Is Near,* Kurzweil explains, "humans will develop the means to instantly create new portions of ourselves, either biological or nonbiological. . . . [A] biological body at one time and not at another, then have it again, then change it." In this nonbiological embodiment, they will become "software-based humans . . . [who] live out on the Web, projecting bodies whenever they need or want them, including . . . holographically projected bodies, foglet-projected bodies and physical bodies comprising nanobot swarms."[97]

If Kurzweil and his singularity seem to be promising awe, it's ramped up even more by Jason Silva, a self-proclaimed "media artist, futurist, philosopher, keynote speaker and TV personality," who, like Kurzweil, holds out hopes for immortality. Silva developed a YouTube channel titled Shots of Awe, where he gives more dramatic expression to the way transhumanism and other embodiments of high technology spark the emotion. He proclaims that Shots of Awe "offers a series of reflections on my current thinking on the human condition, the way we use technology to transcend all previous limits. . . . [T]hink of them as inspired nuggets of techno rapture."[98] In keeping with this promise, Silva offers a series of videos with titles like "To Be Human Is to Be Transhuman," "Technologies of Ecstasy," and "We Are the Gods Now."[99] In "We Are the Gods Now," he explains why the future inspires awe:

> I do feel ecstatic when I contemplate these possibilities. Just reveling in those possibilities gets me off. I have a mindgasm, literally. . . . With the revolutions in biotechnology, nanotechnology, the free exchange of information is allowing us to conceive of radical new things. . . . It's a universe of possibility. It's gray infused by color. It's the invisible revealed. It's the mundane blown away by awe. . . . What I want to communicate to you guys is a sense of awe and a sense of wonder. . . . And this idea of awe as a kind of reassuring quality and ability to contemplate our own existence and marvel at ourselves.[100]

To "marvel at ourselves" is the new awe of the twenty-first century, a different feeling from the humbler variety felt in earlier years.

The awe that transhumanism and technology promote is also less enveloping, less grand in scale. While Kurzweil and Silva might be ecstatic at the thought of what technology can do to life and death, their vision is not quite as mystical as that of nineteenth-century Americans. There is no glorious communion with divine powers, or complete reunion with friends and family after death—no spiritual transcendence.

What they mostly offer is the continuation of the self through a melding of biological and digital material.

While Kurzweil and Silva are awed by future hopes, and today's movie and app producers are likewise astonished by the digital world they helped create, we often discovered a more tempered experience of the feeling, and a reticence to discuss it among those with whom we spoke. Awe seemed like a naïve feeling, out of step with the worldly wise emotional mood among many technology users.[101] The prejudice against deep awe, which psychologists first enunciated a century ago, still lingers. In the twenty-first century, people can Skype or Facetime friends and family across the globe; see videos of long-dead kin and celebrities from when they were alive; and hear any sound, symphony, or Whitacre choir they want. While nineteenth-century Americans were effusively enthusiastic about these possibilities, their twenty-first-century descendants are much less impressed. Rather than seeing such powers as divine, people now regard them as part of the human condition. As Greg noted, "I see phones as a . . . part of the modern human being." He explained that if you "don't have it on you, suddenly you can't navigate, you can't communicate with the social circles you typically communicate with, your information bank is limited. . . . In a real sense we have assimilated smartphones into our bodies."[102] As people are becoming accustomed to having such power at their fingertips, they are coming to consider it as just part of being human.

They occasionally still see these powers as amazing, but rarely consider them divine. When we asked our subjects whether they were astonished or saw anything magical in their devices, they often assented. But when we asked them whether there was anything divine, they were reluctant to adopt that language and steered clear of framing their experience with technology in overtly religious terms. For example, Skyler Coombs, the student who had recently returned from a Church of Jesus Christ of Latter Day Saints mission in Washington State, said of his new smartphone, "It was pretty amazing, everything that it could do. When you get to have everything you need right at your fingertips. . . . Sometimes it does feel like magic. Like the internet—where is it actually at? Like, how is it getting to all these devices that everyone is using?" But, he admitted, at other times, he took "it for granted," remarking that most of the time the phone "just kind of seems normal."[103]

Likewise, Xavier Stilson, who had recently returned from a mission to Brazil, was awed "at some points," but when asked if the phone was divine, he replied, "I wouldn't say that." He then observed, "I think I definitely do take it for granted. . . . It's something that's always there. . . . I went to Brazil [from 2013 to 2015] and there they don't have all of the technological advances that we do and we definitely take it for granted." He explained how he was able to have both of these somewhat contradictory reactions, noting that when a product was new, "It's there and you want it . . . but as time goes on . . . it's always there. . . . You just take it for granted. . . . It no longer awes you until something new comes out."[104]

This mix of feelings was widespread, evident in interviews with people from a range of backgrounds. For example, Anna Renkosiak was reserved about framing her encounter with technology in explicitly religious terms. She said there were digital effects that amazed her, such as holograms performing at concerts, but such inventions were "not necessarily divine." Instead, she credited humans, thinking it was "neat that somebody had the brain power to actually come up with this stuff . . . but I don't know that I'd ever be 'oh, this is from a higher power.'"[105]

Yet if an overt reference to divinity was avoided, people still used terms like "magic" and "amazement" and, on occasion, "awe" to describe their feelings. Moreover, the common tendency was to vacillate between feelings that technology was embedded in the ordinary texture of life, and a sense, often fleeting, that it had transcendent qualities. Interviewees would say that they were not astonished by innovations because they had come to expect updates every day; or they were jaded because new technologies were constantly being superseded by even newer ones; or they took these technologies for granted because they had not bothered to think about them. However, in subsequent observations, they would say that when they did think about the innovations in their midst, they were astonished.

For example, Cooper Ferrario initially declared that digital technology was "science . . . not magic." But a moment later he observed, "If I take a step back and think about . . . this [phone,] . . . it is awesome. It is totally incredible and amazing that I have this in my hand. And that everyone has this. And it's affordable."[106] In a similar vein, Barry Gomberg, the attorney, observed, "I don't generally feel that in social communications the platforms that we use . . . have ennobled our interactions," and he added that "the routine communications" they enabled are "mundane pablum." Yet in the very same interview he observed that social media was allowing his mother-in-law to "watch her granddaughter growing up," and that he was "way awed" by his iPhone.[107] In general, the people we talked with had mixed feelings. They occasionally hinted at magical, awe-inspiring qualities in their technology. In comparison to their nineteenth-century forebears, however, they talked about it in far more secular terms.

There also seemed to be a rhythm to people's emotions. They felt awe at a first encounter with a new technology, and then it would quickly diminish. They came to expect constant innovations and refinements to the technologies of daily life as a matter of course. Ana Chavez had initially been amazed when she first came to the United States from Mexico, because she saw a range of new devices and appliances: "When I got here . . . I thought everything was awe." She explained, "A toilet . . . [a] shower head was awe." But once she became accustomed to these things, her astonishment diminished. "So now that I have phones and people are coming up with new things, . . . it's no longer . . . awe to me, it's like holy crap, humans are so smart. . . . It's a big deal . . . but it's not . . . total shock."[108]

Gregory Noel had felt a similar ebb and flow of awe. When he had first begun to use social media, he thought "MySpace was mind blowing . . . and Facebook just . . . elevated that. . . . I got a chance to reconnect with people I haven't seen in a long time in Miami. . . . That opportunity alone was just amazing, . . . mind blowing." Yet ultimately, his awe diminished dramatically, and he came to conclude that he was better off without social media.[109] Emily Meredith, the nurse in Oakland, California, suggested that her wonder at new devices was lessened because changes were "so incremental. . . . It's not like all of the sudden a computer appeared. . . . It was sort of slow the way it was introduced." As a result, she noted, "I don't really have a spiritual, divine" feeling about it.[110]

Those over thirty suggested that the younger generation, completely accustomed to new innovations of the digital era, might feel such inventions were just natural rather than astonishing and awe-inspiring. Born in 1980, Katie Schall believed that younger people in particular had become so used to technology that they took it for granted. "I feel like the generations coming up are less impressed by things. . . . I think people just expect electronics and things [to] just work perfectly all the time and have very little patience when things don't work out right and so I think they're more hard on technology now than they were before. . . . I just think people have . . . higher expectations and they're more quick to be annoyed with things. . . . So their expectations are higher. . . . I think it's entitlement; I think it's technology entitlement. . . . So things must work perfectly and must be seamless and intuitive and if it doesn't, then yeah, you get really irritated."[111] Camree Larkin agreed: "I think that people my age and including myself often times take it all for granted." Older generations, however, reminded her of what she had. "My dad never lets me forget how amazing it is [when he says,] 'Do you realize how incredible this is?'"[112] But without such reminders, many regard their devices—and the powers they conferr—as natural and unremarkable.

Another way that contemporary Americans' feelings differed from their forebears was that an overwhelmingly optimistic view of technology was no longer pervasive. Emily suggested that Americans' sense of awe might be tempered by a greater knowledge of technology's dark side. "Maybe there were naysayers" in the nineteenth century, she observed, but perhaps the general mood of the era was "it's all good. . . . Everything about the telegraph is positive; there's no drawbacks to that." In contrast, contemporary Americans recognized "drawbacks," and as a result, "there are so many reservations about technology now."[113]

Not only did people evince less faith in technology, but some suggested that technology itself was interfering with awe. Eli Gay, the tax preparer in the Bay Area, thought that smartphones and computers were in fact the "antithesis" of a "spiritual place":

> Most of my . . . spiritual ideology would be . . . Buddhist . . . you know, just about "presence". . . . A lot of what technology does is actually the opposite, where . . . it's just making people unconscious or taking them out of the moment. . . . I'm just always aware of that and . . . especially with the phones. . . . I was out with a friend the other night. It's just acceptable now . . . it's in his pocket and . . . every five to ten minutes it's like this. [She pretended to check a phone.] There wasn't a lot of stepping out and texting. But just every time that happened . . . it's like "What's happening here? You're stepping out of a moment with me to go be in this other moment." And that feels sort of like the antithesis of . . . a present, . . . grounded, spiritual place.

She recalled a moment when the power of technology to distract from the spiritual had been particularly apparent and concrete. "This one time at Brandeis when the Dalai Lama came . . . the Dalai Lama was speaking and we were sitting there, and someone's phone rang. . . . And the person answered the phone and starts walking up the aisle! That for me was just this moment of '*What* or *Who* is on the other end of the line?'" She imitated the man answering his phone and saying, "'Oh, hold on. I'm just listening to the Dalai Lama.'" Eli continued: "I thought that was really crazy, but I saw it with my own two eyes." She intimated that moments of awe-filled transcendence were to be found in the here and now, and that technology threatened to take people away from those moments. Unlike nineteenth-century Americans, who wanted to use technology to find a distant God, Eli suggested that modern devices distracted from the holiness of the present, the sacred that was embedded in daily life.[114]

FIGURE 5.4 Caspar David Friedrich was a German romantic artist. His 1818 painting, *Wanderer about the Sea of Fog*, is an iconic representation of the human encounter with nature and the sublime, and the powerful feelings evoked by this encounter. In Kim Dong-kyu's modern rendition, a smart phone is added, thereby sparking new questions: Does the introduction of the new technology mute or amplify the sublime? And is the wanderer still awed by nature? Or is he awed by his device?
Courtesy of Kim Dong-kyu

Eddie Baxter, the social worker, also recognized that technology often took "people out of the moment." He had recently returned from a trip to the Galapagos where he felt deep awe and had also taken many pictures that he had posted on social media. Yet, nevertheless, he noted, "I feel like there's a sort of a sickness in . . . I feel like it kind of takes me out of that present moment. So I'm no longer really being mindful of . . . what's actually going on."[115] Eddie had sometimes worried about being "braggy" online, and his concerns about being taken "out of the

FIGURE 5.5 Instagram post by Eddie Baxter, of Ogden, Utah. Eddie expressed awe during his trip to the Galapagos, but he worried that taking the pictures might interrupt the emotional experience. He also worried that he might come off as "braggy." *Courtesy of Eddie Baxter*

present moment" compounded those feelings. He did not want picture-taking to interfere or mediate the powerful feelings he was having as he took in nature in the Galapagos. At the same time, he wanted to share his awe with less privileged friends back home and by so doing, he hoped to inspire them to find awe in their own lives. As Eddie put it,

> A lot of my posts . . . I'm hoping that it inspires and urges folks to get involved themselves. Like "Hey, Eddie did that and . . . Eddie was a drop out of high school." . . . Here I am doing this thing that I truly thought I would never have access to. . . . Hopefully that would spark some awe in other folks and definitely not in a braggy way. In a way "Oh my gosh, . . . I know Eddie, . . . I know him, we've hung out together. Maybe I can do that too." So I try to share that a lot with the kids that I'm working with because working with at-risk youth—that was me, that was 100 percent me.[116]

Eddie's experience is emblematic of the ambivalence that many of our subjects felt as they tried to sort out their feelings about technology, narcissism, and awe. They recognized that social media could be construed as a vehicle for sharing awe with others and as a means for common uplift. But as often as not it could be interpreted as a tool that catalyzed narcissism and inhibited people's ability to seize the wonder of the present moment.

The Decline of Awe?

While Americans might continue to use the word "awe" in casual conversation, lately there have been worries about the decline of a richer, deeper sense of awe, as well as fears that the overengineered, digitalized world interferes with the experience of the emotion. These concerns have gained visibility only recently. Of the 694 articles about awe published in psychological journals and books since 1795, 530 of them have been published since the year 2000.[117] The new interest in awe represents a change for the field of psychology, for traditionally much research in the discipline has focused not on the emotions individuals feel in relation to a larger universe, but instead on more interior feelings. Additionally, early twentieth-century psychologists were critical of the emotion, believing it to be an atavistic feeling, left over from earlier phases of human history. They had little interest in reviving it.

Since the start of the twenty-first century, however, a number of psychologists have begun to reexamine the feeling, arguing that it is declining and that its disappearance represents a loss. They have tried to measure it, find its causes, as well as understand the impediments to it. For example, in 2006, the, psychologist Dacher Keltner of the University of California, Berkeley, and his colleagues developed a survey instrument called the "Dispositional Positive Emotions Scale (Dpes)–Awe Subscale" that, much like the survey instruments used to measure narcissism, loneliness, and boredom, presumes to measure awe.[118] In 2015, Keltner along with his colleague Paul Piff celebrated the feeling in a *New York Times* op-ed titled "Why Do We Experience Awe?"[119] And in June 2016 Keltner organized a conference titled "The Art and Science of Awe," which assembled scholars, artists, and educators to further explore the emotion.[120] Awe has also been taken up in a series of articles by Andy Tix in *Psychology Today* and in his own blog, titled "Reflections on Mystery and Awe."[121]

For Keltner, Piff, Tix, and other psychologists who take an interest in it, awe is understood as an overwhelming emotional experience typically triggered by encounters with religion, nature, music, and art.[122] The feeling takes people outside themselves. Because awe disrupts conventional notions of the individual self and encourages people to see themselves as part of a much larger social and natural landscape, Keltner sees it as an experience that has "prosocial" benefits. And Keltner and Piff demonstrate this in another study they conducted: when awe was induced in research subjects, they displayed more ethical behavior and were more generous to others than those in control groups.[123] At the same time that awe helps develop prosocial behaviors, Keltner speculates that it might also curb narcissistic tendencies, because narcissists typically disregard the welfare of others. While this is a more secular version of awe, lacking the dread and fear of a deity that earlier generations experienced, it nevertheless has a similar effect of minimizing individuals' sense of personal grandiosity.

In the psychologists' view, awe promises to help enhance people's civic and communitarian dispositions; however, Keltner and Piff worry that this promise might be difficult to realize because the older sense of awe may be less accessible than it once was, and they wax nostalgic for the feeling in its earlier forms. Perhaps its decline is due to a drop in formal religious participation, to the possibility that twenty-first-century Americans are less exposed to nature than their forebears, or to the fact that by some measures they are more isolated from each other than past Americans. At the same time that awe might be in decline, the way it is being induced could also be changing. Instead of being triggered by visits to the church or nature, the emotion is increasingly induced by encounters with technology or with religious and natural experiences mediated through technology.[124]

While video-game players value the awe they feel while playing, one can wonder if it is not a much-cheapened digital version of the feeling that their ancestors experienced. For while *Halo* players might feel they are saving a made-up world in a video game, nineteenth-century Americans hoped the telegraph could truly accomplish that goal for the real world in which they lived. Although twenty-first-century Americans could feel a moment of transcendence when they participate in a virtual world with others, earlier generations hoped for a true, global reunion and communion with others. Their awe at technology and their hopes for transcendence were, perhaps, greater. Their sense of their own limitations was more acute; their longing and reverence for something larger than themselves more apparent.

Significantly, as this deeper, older form of awe may have declined and been replaced with a less transcendent feeling, narcissism may have been on the rise, signaling a shift in the way people regard themselves and the world around them. Perhaps they no longer carry a small sense of themselves in the world; perhaps instead, with the help of new technologies, they feel extremely powerful and significant, that they need not accept the limits of being mortal. As they contemplate living longer, thanks to a variety of new devices, as they see their powers to move, to talk, to create, enhanced, they may feel like they have unprecedented capabilities. Many feel the limits on their powers receding, as well as the social mores that discourage the advertisement of those powers disappearing.

The Myth of Icarus Revisited

Yet if the decline of awe is a mere speculation to psychologists whose views are bounded by the present, and rendered less plausible by our interviews with people like Greg Bayles, who likened staring into a smartphone to entering a cathedral, it is clear that the meaning and experience of awe have indeed shifted across time. Americans today might still feel it, but not in the same way that their ancestors did. In the colonial era, some believed that they should not try to capture the grand and awe-inspiring forces of the universe, for they might bring divine wrath down on their heads. They learned such cautious attitudes from myths like those of Icarus and Prometheus and from sermons delivered by worried ministers. These tales and lectures carried a warning: humans overreach when they try to gain heaven through the use of technology. To do so is folly.

However, by the mid-nineteenth century, many believed that it was perfectly acceptable to try to harness the forces of the universe and to flout such warnings. Ministers of the era often described the telegraph as a tool that would lead to world peace and that would put individuals in closer touch with the divine. Their sermons offered glowing portraits of technology far different from that described in the myth of Icarus. In fact, many seemed confident that their actions and inventions were in keeping with divine plans. Unlike ancient fears about humans offending the gods, Victorian Americans felt they should innovate and invent.[125]

In the early twentieth century, people continued to embrace new technologies as their right, but psychologists and educators encouraged them to overcome their awe, which they believed was a premodern feeling, one that interfered with clear reasoning. Gradually over the course of the twentieth century, such transcendent awe was less frequently expressed aloud, and was eventually replaced by a less enveloping, more worldly feeling, one that celebrated human grandness and potential.

By the late twentieth century, Americans manifested a faith in the power and promise of human invention, and they displayed few fears about transgressing divinely set limits. No longer scared of overstepping sacred boundaries, they demonstrated a belief that it was their right to innovate and create. The civil engineer Henry Petroski expresses that faith in his 1992 book, *To Engineer Is Human,* in which he explains to the lay public why engineering projects sometimes fail. He uses the myth of Icarus to illustrate his argument. But he no longer casts it as a comment on human limitations. In Petroski's view, Icarus fell to his death because his wings were not designed to cater to his uses. The lesson for the engineer (and for Americans) is not that humans are fated to never transcend their existing condition, or that they should be awed by the powers of the universe, which they themselves could neither access nor equal. Rather the lesson is that transcendence is an iterative process that takes place through trial and failure and through the refinement of design. Astonishment and awe wax with each new successful design and then wane as the limitations of the new design become apparent. But transcendence remains a hope even if it happens in smaller leaps than are conventionally imagined. Humans might yet have it in their power to fly to the heavens.[126]

By the late twentieth and early twenty-first centuries, then, few worried about treading on sacred ground when they made new discoveries or created new machines. Americans were sometimes awed by novel inventions, but they were much more likely to see in them the human touch rather than divine fingerprints. The feeling such contraptions provoked in them was dramatically different—and somewhat weaker—than it had been in earlier centuries. And in contrast to the past, when Americans today feel this weaker sense of awe, it often gives them a larger sense of self rather than a small one, for they experience it when they witness their own contrivances.

Americans' sense of their own potential has grown, and the definition of what it means to be human has expanded. Today they are awed by their own creations. In the twenty-first century, powers that once were seen as belonging only to God—omniscience, control of the elements, the triumph over time and space—are now seen as the proper possession of mankind. And while none of these new powers have yet conquered death or offered true transcendence, people continue to strive to overcome the limits of being human.

In Whitacre's online symphonies, listeners still aspire, and fleetingly obtain, feelings of awe that transcend the quotidian. And in a growing transhumanist movement, Kurzweil and his acolytes speak with awe of a future where through the melding of mind and machine one can leave behind the limitations of the corporeal self. These scenarios, along with a nascent and growing interest in once again studying awe in the field of psychology, suggest that Americans still value the feeling and hope for its resurgence even if it has been in decline. The hope for transcendent, awe-filled moments is still there, beckoning Americans into the future even as that hope connects them to earlier generations who expressed a more religious, more communal, and less self-aggrandizing version of the feeling.

Undergirding the evolution of awe is a parallel story about secularization. As America has become a more secular society, the definition of what it means to be human is no longer illuminated by a language that distinguishes between mortals and gods. Although a small British charity called the Daedalus Trust takes its inspiration from the Icarus myth and professes that its central mission is to "raise awareness about hubris," it represents an isolated effort. Contemporary Americans often no longer register the difference between themselves and gods because they lack a set of tools—like the myth of Icarus or the Tower of Babel—for apprehending limits.[127] Awe, however, insofar as it reminds us of both our limits and our dreams of limitlessness, might help to fill this cognitive gap. As such, the effort to create more opportunities for awe, whether through modern psychology, or through technology, nature, or a resurgence of religion, constitutes an important project.[128]

CHAPTER 6

Anger Rising

In Chapter 5, Greg Bayles enthusiastically celebrated the capacity of the internet to provoke awe, likening his smartphone to a cathedral: "There's . . . entire worlds contained in this . . . smartphone . . . that invites me into an experience of awe. . . . I can pay $100 for this phone and suddenly I am invited into this unimaginable experience."[1] These comments were striking when he confided them to us in the summer of 2016. If awe had all the "prosocial" values that psychologist Dacher Keltner and so many others claimed it had, and if these experiences were, as Bayles thought, available by dint of a mere $100 purchase, then perhaps people were at the dawn of a happier and more harmonious age where awe would mitigate the more negative experiences that often accompany narcissism, loneliness, boredom, and distraction.

Unfortunately, this was wishful thinking. When we visited with Bayles fifteen months later, in the fall of 2017, his views were no longer quite as celebratory. As he put it,

> Yes, the screen is a tiny cathedral but it's still not the cathedral. . . . OK, yeah, the reproduction still gives you those feelings of awe and the sublime but there's still something that's lost in stepping away from the original. So, yes, you have maybe 360-degree footage of the inside of the cathedral, but you don't have the walk up to the cathedral. You don't have that moment of stooping through the low door and then standing up into the majesty of the cathedral. Again. There's meaning lost and it's only those people who have the economic capital to be able to

> travel to the cathedral that actually get that full, sensory-rich, experience.[2]

This shift in Bayles's thinking was not entirely accidental. In the interim months, a presidential election had taken place, and fears about the power of technology companies, particularly the Big Five—Amazon, Apple, Google, Facebook, and Microsoft—had grown. Such events had revealed to him and other evangelists of high tech that not everything was quite as rosy as they had originally assumed:

> So pre-November 2016, you know, everyone thought the world was going in a really good place. Everyone was so hopeful. You ask any liberal and they say, "Well, of course we're going to win the election." You know; that kind of thing. And then we have this sort of . . . come to Jesus moment, where we said, "Wait, we were wrong about a lot of things. We've been way off base and listening to ourselves. We've been sort of patting our own head and 'good job liberals; we're carrying everyone forward,'" and it turns out no. We were carrying us forward. But there were a lot of people who were left behind.

These sentiments were of a piece with his more tempered views on technology and its capacity to bring people toward a more civil, more sociable, more awe-infused future. His new comments suggested that if smartphones offered access to some muted form of the sublime, it certainly was not powerful enough to act as a panacea for the problems of economics and politics. And if the election had led Greg to qualify his technophilia, it also provided an occasion to reflect on the relationship between the awe that technology had promised, and the anger and division that it had delivered in recent years. As Greg saw it, "For a time it felt like America as a whole was experiencing this limitlessness, this expansion of the self," but the election of 2016 "was a pretty clear indication that only one part of America was experiencing that, and the other part was experiencing tech-mediated obsolescence. You know, 'My job is disappearing, my livelihood is disappearing, I don't have the funds to be able to move somewhere where I might be able to find work, I've never had access to education,' and so, I don't know, I guess a lot of times we are predisposed to think that everyone is just like us."

For Greg, the election had revealed that half the nation was experiencing awe and "the expansion of the self, where you realize that you're actually part of something bigger," while the other half was experiencing anger. As Greg put it, "With anger you realize essentially that all those forces are pressing in on you. Trying to turn you into something other than what you are. So rather than being a moment of discovery, as it is with the sublime, it's this moment of sort of oppression or . . . defensiveness."[3]

The division that Greg was describing was rendered more obvious by the election of 2016, but it had already been foreshadowed by earlier events. Americans had been expressing anger through and about their digital technologies for several years. For example, in 2013, at a time when Google was already burnishing its reputation as the harbinger of limitless life and limitless search, Google's less wealthy neighbors in the San Francisco Bay Area were presented with the hard fact of real honest-to-god limits.[4] They were being priced out of the rental market by new Google recruits, with high-flying salaries, who were buying up what was left of the Bay Area's housing stock. These class tensions, which had simmered for years and risen with each new wave of high-tech recruits, finally exploded in 2013 when Google buses, which transport Google employees from the city to the company's headquarters, were attacked by demonstrators, who replaced the Google insignia with a new one that read "Gentrification and Eviction Technologies." And in another incident, protesters threw rocks at a bus, breaking windows and frightening the Googlers who were riding inside of it.[5] Such protests have continued in the Bay Area.[6]

The idea that technology might be dividing people and fueling anger is not the perception of only Bayles; it is widely shared. Nor is the Google bus incident, however emblematic, the only example of the way that technology has often led to disenchantment and disappointment rather than to limitlessness. For instance, the revelations that Facebook

FIGURE 6.1 Real estate prices in the San Francisco Bay Area have become some of the highest in the nation, due to a steady influx of professionals recruited by Silicon Valley tech companies. Disgruntled by rising rental costs and gentrification, some Bay Area residents have organized demonstrations that target the private shuttle buses that ferry tech-company employees to their jobs at Google, Facebook, Apple, and Genentech. *Photo courtesy of Steve Rhodes. Used by Permission*

shared data with Cambridge Analytica and that Russian operatives used social media to manipulate the 2016 election likewise have sparked popular ire toward technology companies.[7]

This wave of widespread anger was anticipated by some scholars and media critics who have long worried about the effects of digital technology on civic life and civility. Cass Sunstein's *Republic.com* and Eli Pariser's *The Filter Bubble* famously predicted the way that Facebook and other social media platforms could move Americans out of a common public square into segregated ideological silos where sensationalist fake news, and the anger that is often fueled by that sensationalism, could spread with relative ease.[8] Moreover, since many social media platforms are not tied to a particular physical place where people expect to meet

FIGURE 6.2 In May 2018, activists organized another round of protests directed at Google and high tech more generally. Here, they criticized the tech industry's effects on housing costs, the homeless, and the environment. *Photo courtesy of Ashwin Rodrigues. Used by permission*

each other face-to-face, there are often fewer incentives to maintain comity.

Americans today are therefore confronted with a torrent of online anger, and they wrestle with the question of whether they themselves should express their own outrage. News outlets have run a multitude of articles with titles like "How Digital Media Fuels Moral Outrage. . . ." "Angry Social Media Posts Are Never a Good Idea . . . ," and "The Internet Isn't Making Us Dumb. It's Making Us Angry."[9] Among those we talked with, the subject of online anger was also a fraught topic. Some of our interviewees thought that anger has not increased in the last few years, and that the Vietnam War era was a period when anger boiled at least as strongly as it does now. Others disagreed. What we discovered is that they worried a great deal about the social repute of anger and whether it was a sign of virtue or of turpitude. One interviewee kept an anger diary whose pages she regularly burned.[10] Another allowed that he was rightfully angry at the lies a litany of politicians had told. At the same time, he did not want to be labeled or caricatured as angry. He insisted that his Republican politics were grounded on long-standing political beliefs—not anger per se.[11]

The 2016 presidential election provoked similar debates about how (or how not to) manage anger. For a significant portion of the electorate, "no drama Obama," and his anointed successor Hillary Clinton, offered a model of how best to govern one's passions. These voters regarded Donald Trump as a demagogue, a firebrand, and a provocateur who won by engaging people's prejudices and sanctioning their ire. In contrast, a substantial number of voters celebrated Trump's emotional style, seeing him as more honest and authentic about his feelings—anger included.

This debate about the appropriateness of anger is not new, because modern views about the feeling did not emerge out of thin air. Americans have been debating whether anger should be repressed, managed, or indulged for centuries. Consequently, the emotion has a long and well-examined history.[12] That history is tied up with the idea of limits, for the question in European and American societies has long been just how much anger one should express. The word's history reveals this, for its origins lie in proto-Germanic terms for something "narrow" and "painfully constricted"—perhaps the feeling that people experience when they are so mad they feel they might explode.[13] Such an explosion might lead to the related feeling of outrage, which originally meant "the passing beyond reasonable bounds"—that is, an unconstrained eruption of anger that burst the limits and transgressed social boundaries.[14]

Anger was a feeling that seventeenth-, eighteenth-, and nineteenth-century Americans sought to limit and control but not completely eliminate, for many also regarded it as a socially useful passion and as an appropriate response to sin or injustice.[15] Anger was visible in parades, protest movements, "indignation meetings," workplace disputes, and politics. Over the course of the twentieth century, however, American psychologists, business leaders, and educators attempted to reduce anger and some, in fact, tried to banish it all together.[16]

Although it never disappeared from men's and women's inner lives, there were consequences if they expressed it in public, and, as a result, many learned to repress the feeling; over the course of the twentieth century, anger seemed in retreat, at least in public. But recently, there has been a change. Anger has become newly visible again, at least in some quarters. Today, in the age of social media, many Americans seem to feel there are few limits on how they should and can express their rage, rancor, anger, and indignation. Yet because it is expressed via phone or computer, it has at once become more visible, but often less effective as a prompt for change.

Colonial Anger

During the colonial era, Americans differentiated between types of anger. Some like John Robinson, a Pilgrim minister, argued that anger was always bad—he termed it "a hideous monster." A cure for the passion was "humility and meekness." Robinson suggested that it was the individual "that thinks not himself great," who would be able to control the feeling. Having a small sense of self—not a large one—was key to subduing anger.[17] Only God was entitled to be wrathful, not humans.[18]

Others in early America argued that anger had at least two faces. Sometimes it was necessary for humans to be angry, for the passion could be righteous, one that God wanted individuals to feel. Puritan minister Cotton Mather explained in 1717, "It may sometimes be . . . Needful, to be Angry. . . . [O]ur SAVIOUR Himself, when he had some Sinful People about Him, . . . He looked round about on them with Anger. Goodness it self may have Anger in it." But it could take other forms as well. Indeed, Mather offered these thoughts on anger just before the execution of Jeremiah Fenwick, who had killed a man in fury. Mather explained that while there was good anger, "'Tis UNGOVERNED ANGER that has brought all this upon him. UNGOVERNED ANGER; How Fatal a Folly! And yet, how Frequent a Folly!"[19]

Some colonists suggested that certain groups of people were particularly prone to ungoverned anger.[20] Historian Nicole Eustace has shown that in the eighteenth century, often the harshest condemnations of anger were leveled at poor whites and enslaved African Americans who expressed the feeling. Their anger was seen as without cause and their outbursts taken as a sign that they could not govern themselves and thus must be governed by others. Those at the bottom of the social hierarchy were supposedly prone to uncontrollable bouts of fury, untempered by reason. In 1747, for instance, Benjamin Franklin described "the wanton and unbridled rage" of "negroes, mullatoes, and . . . the vilest and most abandoned of mankind." In contrast, the upper classes believed themselves to be endowed with a more discerning set of sensibilities, which gave their anger—or resentment, as they often called it—a greater legitimacy.[21]

Colonial Americans accorded one particular form of anger—indignation—great respectability, seeing it as a sign of an individual's sense of justice and desire for virtue. For instance, colonists often proudly and publicly displayed their righteous anger as they protested British oppression during the American Revolution. In a 1774 oration that John Hancock delivered to commemorate the Boston Massacre and to rally opposition to the British, he invoked the righteous feeling, telling his audience to always remember the carnage. "Let not the heaving bosom cease to burn with a manly indignation at the barbarous story, through the long tracts of future time; let every parent tell the shameful story to his listening children until tears of pity glisten in their eyes, and boiling passions shake their tender frames."[22] Hancock invoked indignation to build national identities and bind communities.

Anger in the Nineteenth Century: Limits and Liberties

Righteous anger could indeed bind people together, but only some were entitled to express it.[23] During the nineteenth century, Americans developed new rules governing the expression of anger, rules that suggested that outrage and indignation were not the right of all. It was legitimate for white men to display, but it was a passion forbidden to African Americans and women, both of whom were expected to accept the limits on their lives with contented silence.

Those who broke such rules faced serious consequences. In 1849, James W. C. Pennington, born a slave in Maryland, recalled what happened to his father when he expressed anger toward their master. The master had suggested in "an angry, threatening, and exceedingly insulting tone" that he would sell some of his slaves because he believed he had too many and therefore none worked very hard. "My father was a high-spirited man, and feeling deeply the insult, replied to the last expression,—'If I am one too many, sir, give me a chance to get a purchaser, and I am willing to be sold when it may suit you.'" The master was outraged at his slave's show of temper and replied, "'Bazil, I told you to hush!' and suiting the action to the word, he drew forth the 'cowhide' from under his arm, fell upon him with most savage cruelty, and inflicted fifteen or twenty severe swipes with all his strength. . . . As he raised himself upon his toes, and gave the last stripe, he said 'By the ***, I will make you know that I am master of your tongue as well as of your time!'" Pennington reported that his whole family was indignant at this exchange and at the mistreatment of their patriarch.

> I was near enough to hear the insolent words that were spoken to my father, and to hear, see, and even count the savage stripes inflicted upon him. Let me ask any one of Anglo-Saxon blood and spirit, how would you expect a *son* to feel at such a sight? This act created an open rupture with our family—each member felt the deep insult that had been inflicted upon our head; the spirit of the whole family was roused; we talked of it in our nightly gatherings.

Unlike whites, however, they dared not display more anger or fight back against the master; instead they must swallow their rage, or flee, as Pennington ultimately did.[24] African Americans in slavery found it dangerous—and even life-threatening—to express the feeling.

Anger was also discouraged in women of all races, for it was supposed to be absent from the female character and, by extension, absent from the home life they created. According to ministers, magazine writers, family advisors, novelists, and artists, the home should be a paradise, suffused with love.[25] Historians Peter and Carol Stearns have shown that as these sentimental ideals of family life took hold, many nineteenth-century moralists preached that anger had no place in the well-ordered household. It was particularly taboo for white women, who were supposed to submit obediently to the conditions of their life and not manifest an autonomous will. A short story that appeared in the popular ladies magazine *Godey's* in 1853 laid out this ideal. It told the tale of "The First Quarrel," in which Flora Hastings, newly married, lost her temper at her husband when he arrived home late. Her punishment came when he soon fell gravely ill and almost died. After he recovered, whenever she "was tempted to give utterance to impatient moods, fretful and angry words died away on her lips, rebuked by the remembrance of that terrible agony, lest her husband should die with the words of forgiveness unspoken."[26] Such restraint was not always the lived reality in American homes, for many men and women did become angry, displaying the passion toward each other or their offspring.[27] Yet the admonitions against family discord grew even more common over the course of the century.

These rules limiting anger in familial and friendly relations also applied to the communications one had at a distance. As Americans became a nation of letter writers in the nineteenth century, they gradually adopted conventions designed to reign in anger. In an editorial reprinted across the country, readers learned the perils of long-distance anger. The piece, adapted from a passage in an Anthony Trollope novel, warned,

> An angry letter, especially if the writer be well-loved, is much fiercer than any angry speech, so much more unendurable. There the words remain scorching; not to be explained away, not to be atoned for by a kiss, not to be softened down by the word of love that may follow so quickly upon spoken anger. Heaven defend me from angry letters! They should never be written. . . . This, at least, should be a rule through the letter-writing world, that an angry letter shall [not] be posted till four and twenty hours shall have elapsed since it was written.

It was acceptable to vent one's feelings, but only in private and only for a moment: "Sit down and write your letter. Write it with all the venom in your power, spit out your spleen at the fullest. . . . Then put it in your desk, and as a matter of course, read it before breakfast the next morning. You will then have a double gratification."[28] Another advisor warned, "There is a certain indestructibility about letters. . . . It is best, therefore, to be careful what one puts in them. . . . Never send an angry letter the day it is written."[29] In correspondence as well as in face-to-face affectionate relationships, anger was to be largely absent.

There were places where it might be legitimately expressed, however. White boys and men generally were allowed to cultivate and display anger at school, at work, on the street, at the ballot box, in taverns, so long as they did not bring the feeling into home life. In fact, parents often actively encouraged it in their sons, seeing the passion as a necessary sign of strength and courage. John Allen Wyeth, born in 1845 in Alabama, recalled that despite prevailing gender roles, his mother (rather than his father) taught him to indulge in anger and fight other boys. When he was seven,

> a boy playmate lost his temper at something that happened between us, and in anger gave me a slap which I did not resent. At this juncture I heard a voice from a near-by window, and, turning, I saw my mother leaning out, her eyes flashing so that I could almost see the sparks flying and her cheeks as red as fire. In a tone about which there could be no misinterpretation, . . . she asked me if the boy struck me in anger; and when I told her he had, she blazed up and said, "And you didn't hit him back?" My response was that father had told me it was wrong to fight, and that when another boy gave way to anger just to tell him it was wrong and not fight back.

His mother was impatient with her husband's advice "and she fairly screamed: 'I don't care what your father told you; if you don't whip that boy this minute I'll whip you!' And she looked on, and was satisfied when it was all over." After this incident, Wyeth became convinced that it was acceptable to lose his temper.[30]

Indeed, for some boys, learning to express anger and to stand up for themselves was an important step on the road to manhood. Hamlin Garland, the celebrated chronicler of Midwestern life, found this to be the case. Born in 1860, he learned quickly that his father wanted him to act on his anger. He recalled that after he received a new sled, "I took it to school one day, but Ed Roche abused it, took it up and threw it into the deep snow among the weeds.—Had I been large enough, I would have killed that boy with pleasure, but being small and fat and numb with cold I merely rescued my treasure as quickly as I could and hurried home to pour my indignant story into my mother's sympathetic ears." Garland, however, did not tell both his parents, for, he admitted, "I seldom spoke of my defeats to my father for he had once said, 'Fight your own battles, my son. If I hear of your being licked by a boy of anything like your own size, I'll give you another when you get home.' He didn't believe in molly-coddling." Garland gradually internalized his father's endorsement of anger; as an adolescent, he came to see the feeling and his ability to act on it in fights as signs of his maturity and masculinity. Describing the "ending" of his childhood, he wrote, "I was rapidly taking on the manners of men. . . . I was able to shoulder a two bushel sack of wheat. . . . Although short and heavy, I was deft with my hands, as one or two of the neighborhood bullies had reason to know and in many ways I was counted a man."[31]

Given that anger and the fighting it often inspired were part of the way white men asserted their masculinity, many saw anger as a legitimate force in politics, in protest movements, and in labor strife. For instance, nineteenth-century men lived in a highly charged, partisan political climate. As historian Michael McGerr observes, party affiliations were key to social and civic life. Men wore their political affiliations proudly and publicly, seeing them as a fundamental part of their male identity, and they were often willing to aggressively display and defend this identity.

> Party encouraged an intense, dogmatic cast of mind. Men made little distinction between fact and opinion: one's beliefs . . . rightly conditioned one's perception. This was a subjective way of apprehending the world, with party at the center, the basic principle of public life. Mid-nineteenth-century partisanship was aggressive, demonstrative, contentious, and often vicious. Party membership was a part of men's identity; as such, their partisanship had to be paraded and asserted in public.[32]

Political life was imbued with spectacle, pageantry, and feeling. In this context, passions—like anger—were expected. McGerr explains,

> Spectacular campaigns and . . . election ritual helped to reduce the complex, daunting process of thinking through many issues . . . to a matter of black and white opposites. Newspapers and torchlight parades legitimized and even necessitated partisan commitment. . . . [But] Spectacular display was not simply emotional oversimplification. Mid-nineteenth-century campaigns tied parades and fireworks to long expositions of issues on the stump and in the press. . . . Popular politics fused thought and emotion in a single style accessible to all—a rich unity of reason and passion that would be alien to Americans in the twentieth century.[33]

As a result of this masculine, passionate political culture, angry Election Day brawls were not uncommon in nineteenth-century cities.[34]

Men also displayed their politics—and frequently their ire—through the newspapers they read. Newspapers were the mass media of the new republic; in Senator John Calhoun's words, "The mail and the press are the nerves of the body politic." But these nerves coursed with anger and rancor, because most of the papers were partisan and routinely vilified the opposing party in histrionic and indignant tones—and they did so to gain profit and maintain their readership. Historian David Nord, for instance, notes that, in the first fifty years of the new republic, "newspapers were unusually outrageously partisan, and factional," and "exacerbated *all* the lines of cleavage in the early republic." He explains, "To hear the newspapers tell it, traitors and seditionists lurked everywhere. . . . [N]ewspapers cultivated faction and dissension." He concludes that "when Americans in the early republic saw treason, sedition, fragmentation, dissension, disintegration, degeneration, disunion, anarchy, and chaos, they usually saw it first in the newspaper."[35]

A dip into papers of the day reveals plentiful examples of such partisan anger. For instance, in 1846 in Ohio, an editorialist in the *Kalida Venture* took aim at the Whig Party and its candidates. Indignant over recent bank failures, the editorialist wrote, "We have declared an offensive war against swindling and oppression." The author proclaimed that the "tyrants" they opposed "tremble before the fierce commotion of an angered and outraged people. They see the approaching storm and read their doom in its foreshadowed results."[36] An 1855 article in an Indiana paper, on "the reign of free whisky," decried partisan judges who had overturned prohibition laws and warned, "They have shown the corruption which is within them, and they will yet feel the anger of an outraged people."[37] An 1860 article celebrating Abraham Lincoln's growing popularity noted that "the rising enthusiasm of the country in favor of HONEST OLD ABE sends a chill of despair to the hearts of the leaders in the pro-Slavery camp" because the supporters of slavery would not be able to stop "the loud roar of the millions of angry men" who supported Lincoln.[38] When, in 1877, Rutherford P. Hayes was installed as president after losing the popular vote, editorialists noted that it was only appropriate for voters to be upset, "for a nation that Submits to wrong without protest, and bears outrage without anger and the determination to remedy the evils, is in the way of its decline."[39] Through such articles, newspapers fanned the flames of political anger and did so self-consciously, with the hope of selling more papers.[40]

Americans did not perceive political anger to be a problem, for many believed the feeling had a legitimate role in political reform. Across the nation, citizens routinely held what they called "indignation meetings." Historian Michael Woods traced their history across the nineteenth century, and he estimates that between 1830 and 1900, there was perhaps one indignation meeting every four and a half days in America.[41] At such events, citizens vented their outrage at the latest political developments. Woods finds they were particularly widespread as the sectional divide between North and South grew and as debates about slavery became more heated. The meetings, he argues, helped to create a shared political sensibility based on shared sentiments. Nineteenth-century Americans believed "in indignation's power to knit citizens into morally elevated and thoroughly united communities."[42]

It was not just slavery that sparked indignation, however, for Americans were willing to express anger for any number of causes at indignation meetings: in 1839, Welsh citizens in Columbus, Ohio, held an indignation meeting to protest the fact that the governor's message was not translated into Welsh, while another was held in New York, after a toast to Queen Victoria at a New York banquet received more "rapturous cheers" than did a toast to the president. Residents of Dearborn County, Indiana, held an indignation meeting in 1845 to decry the acquittal of an accused murderer. From causes large to small, Americans throughout the nineteenth century, and continuing into the twentieth, gave public voice to their anger and discontent at indignation meetings, and they did so with little worry that these political passions might be deemed unseemly or irrational.[43] On such occasions, righteous anger was seen as a force for virtue, essential for the maintenance of a moral order and civic unity.

Indignation meetings were not the only opportunity to express anger. Throughout the nineteenth century, white men, in particular, found many a place where their anger was tolerated, even encouraged. Anger was not uncommon in the halls of Congress, where disputes often devolved into fisticuffs or worse. While the caning of Charles Sumner on the Senate floor in 1856 was the most famous example of outraged anger leading to violence in Congress, it was by no means the only time such outbursts occurred. An 1852 paper, for instance, carried a headline "Mr. Kennedy's Face, with Senator Borland's Fist in It," and described the fight that had broken out between Kennedy, the superintendent of the Census, and Senator Borland of Arkansas.[44]

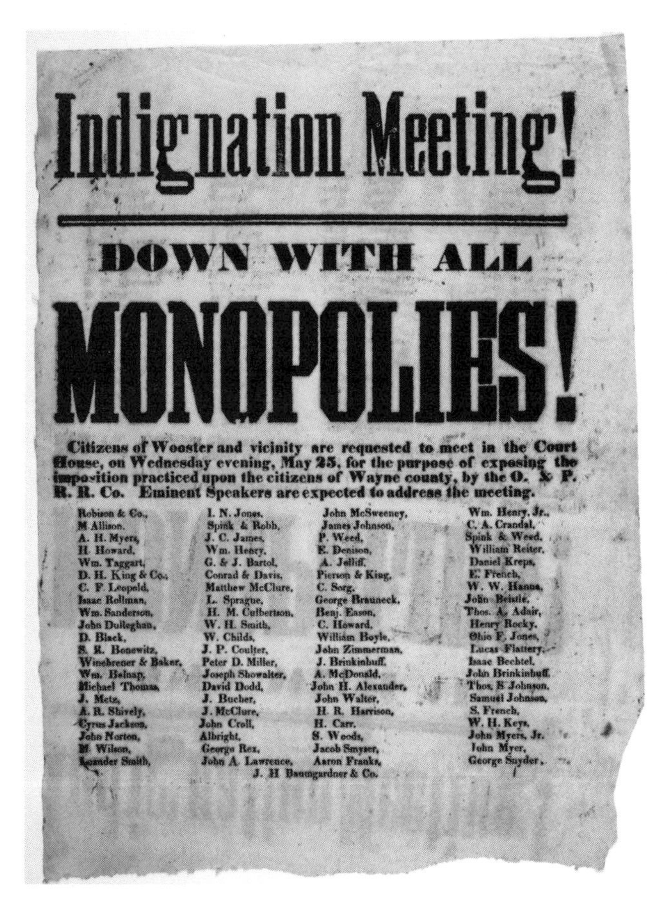

FIGURE 6.3 Indignation meetings were common throughout the nineteenth century and offered a way to express collective anger at social injustices. In May 1853, citizens of Wooster, Ohio, held an indignation meeting to protest the policies of the Ohio and Pennsylvania Railroad Company. *Courtesy of American Antiquarian Society*

Stuart Bryan, a page boy at the Oklahoma Constitutional Convention, recalled that during the deliberations over the state constitution, fist-fights often broke out. The sergeant-at-arms always quickly intervened, ended the brawls, and made the disputants shake hands before they returned to legislating.[45] The violent attacks were frequently condemned as excessive, yet they continued.

FIGURE 6.4 Outraged citizens held an indignation meeting in April 1865 at the McLean County courthouse in Bloomington, Illinois, after the assassination of Abraham Lincoln. *Courtesy of the McLean County Museum of History, Bloomington, Illinois*

They could also be found at work sites and on factory floors, where laborers and their bosses frequently clashed. In 1834, workers in Mansfield, Massachusetts, beat up their boss when he refused to increase pay and supply them with more "grog." In Pennsylvania, angry railway workers burned a bridge to protest a wage reduction.[46] It was not always laborers versus their bosses, however. Sometimes workers fought each other as well. In 1873, two men employed by the gasworks in Wheeling, West Virginia, argued, and one of them, James Fitzpatrick, "in a fit of anger . . . seized a heavy wrench and struck Patrick Brown three times." After the attack, Brown was unconscious and Fitzpatrick fled to Canada.[47] Such angry and often deadly quarrels were common into the early twentieth century. In 1919, for instance, Charles Brown, a "marble quarryman" from Fairhaven, Vermont, was at work when he argued with George Burso, an eighteen-year-old coworker, over a shovel. Burso hit Brown with his fist; Brown retaliated by "hitting the boy a smashing blow with the shovel."[48]

While Americans frequently decried the violence that might result from anger, many observers remained largely unworried about the feeling itself. As one editorialist noted in 1897, "Love and anger are the right and left hands of the soul. Love caresses the lovable, anger smites the hateful, and in a rightly balanced character they work together for the advancement of truth and righteousness. . . . Anger may be and often is a sin, but true anger, anger which arises from a proper cause, is always a virtue."[49]

G. Stanley Hall, the pioneering psychologist, found such attitudes in his survey research. He sent out questionnaires to more than two thousand people about their angry feelings and published the results in 1899. A twenty-nine-year-old male told him, "I believe in causes of offence; it is better to have the matter out, for a good rage freely vented gives an easement like the 'peace which passeth all understanding.'" A thirty-year-old male agreed, writing, "I plead for more anger in school. There is a point where patience ceases to be a moral or pedagogical virtue, but is mere flabbiness." Even women, who felt greater pressure to curb the feeling, occasionally praised it. A twenty-two-year-old woman reported, "Now rage is a sort of intoxication. I am exhilarated with a sort of unnatural happiness."[50]

Hall himself wrote approvingly of the feeling, confiding, sometimes "anger should be worked off by legitimate and regulated fighting . . . to stand up squarely in a give and take conflict, whether with fists or with straight out piece of one's mind to each other, teaches a wholesome sense of responsibility and also gives a hearty man-making type of courage." Watching others fight out their anger also could be invigorating. Hall himself felt "a quite unparalleled tingling of fibre and a peculiar mental inebriation. . . . The common experiences of life seem dull, there is a zest of heroic achievement, of staking all for the chance of victory. . . . They give a sense . . . that life is warfare, that the struggle for survival never intermits."[51] Hall's endorsement of anger reflected Social Darwinian principles popular during the era. Anger was a useful (if primitive) emotion, essential to survival, or, as he put it elsewhere, "Life is a battle, . . . anger is the acme of the manifestation of Schopenhauer's will to live, achieve and excel."[52]

Yet he and his interviewees acknowledged that, when taken to extremes, anger could bring condemnation upon those who displayed it too easily. Some of Hall's respondents reported that while the feeling initially felt good, after their rage subsided they were beset by other feelings. A thirty-one-year-old male reported, "A violent outbreak leaves me worn out in body and mind. I am strong and healthy, but after my last could hardly stand." Women generally reported more abject feelings of regret. A twenty-six-year-old told Hall, "My feeling afterward is a misery too great to speak of or even write. I know it a most dreadful sin, and remorse is deeper in proportion as the object was dearly loved." A twenty-eight-year-old lamented that after angry outbursts, "I feel humiliated in my own eyes because I have failed in what I have most desired, namely, control."[53]

What many worried about when they described their angry outbursts was how they might appear to others, for while anger was acceptable for some to display, one needed to know where and when. To indulge in rage in front of onlookers was to risk condemnation, or at the very least profound self-consciousness and self-doubt. Hall reported that "to have made an exhibition of temper before a stranger is so mortifying as to usually reinforce all the instincts of control. Some confess to having a very ugly or even dangerous temper, but declare that no person has ever seen its malignity." Women worried perhaps more than men did about how their anger made them appear to observers. A twenty-year-old woman reported, "The more strangers are about, the less my irritability troubles me. Their presence is the best control. I am far worse at home." Some used mirrors to gauge how their anger made them appear to others—a nineteen-year-old did so and did not like the way the feeling transformed her face. "Once I chanced to look in the glass when I was angry, and I did look so perversely ugly, that I now think twice before letting go. My face gets broad, heavy, babyish, the corners of my mouth go down, and I frown awfully."

If some were reluctant to show their anger to strangers, others were embarrassed when they displayed it even to their relatives. This was particularly true for women. A twenty-seven-year-old woman told Hall that when she was a child, "Once my favorite uncle dropped into the nursery and found me on the floor kicking and screaming. He was shocked and said I looked more like a beast than a little girl. I was so ashamed that it cured me entirely."[54] Women and girls continued to face substantial pressure to stifle their anger in order to conform to rules of propriety and femininity. Their supposed lack of anger was interpreted as a sign of their contentment with their subordinate role and place in life.

Such expectations also governed African Americans' expressions of anger, for even after Emancipation they faced pressing demands to limit displays of the feeling. Whites often regarded black anger as "impudence," a sign of shameless disrespect and discontent, and considered displays of the emotion a challenge to white supremacy. Antilynching activist Ida B. Wells recorded that, in the late nineteenth century, African American expressions of anger, such as "quarreling with white men" and "making threats," were often regarded as "capital offenses" and punished by lynching.[55] Even when they did not face death, African Americans who were publicly angry faced peril. "Impudent Negro. Aaron Franklin Given Severe Thrashing and Shot At," reported the *Paducah Sun*. According to the reporter, Franklin, an "impudent and sassy" man, had "cursed and abused" Tom Davidson, a white baggageman at the Illinois Central train depot. In return, he was hit with a club and then shot. The reporter declared, "All who witnessed the affray say Davidson was justifiable."[56] In 1905, in a café at the Saint Louis train station, an African American customer, again described as an "impudent negro," "demanded to be served before white people." The waitress "threw a cup of hot coffee with the cup and saucer" in his face, and received "congratulatory letters from all over the country."[57]

Indeed, when African Americans made virtually any call for equality, no matter how tempered, they might be labeled as angry and their claims dismissed, while the angry responses of whites to their demands were celebrated. This was the case in 1914, when a delegation of black federal employees met with President Woodrow Wilson to protest his segregation of the Civil Service. The meeting inflamed tempers on both sides. "Negroes Anger Wilson" ran the headline, and the reporter explained that the president had objected to the manner of the delegation and particularly its leader, W. M. Trotter. "The President told the delegation Trotter had spoiled his cause by his offensive manner . . . and that Trotter was the only man who had ever addressed the president in such an angry tone and with such a background of passion."[58] In white eyes, African American anger was a sign of insubordination and revealed a lack of deference, for the feeling challenged racial hierarchies. It was acceptable for Wilson to be angered by the delegation, but not the reverse.

Up through the early twentieth century, then, women's and African Americans' anger was condemned, but white men enjoyed great latitude for anger—it was seen as a core part of their masculine identity. Writers continued to celebrate the emotion into the new century. One claimed that, "properly directed, Anger's influence for good is immeasurable. Controlled and held in leash, Anger is like the steam pent up in the mechanism of a steam engine—a source of energy, power and usefulness. . . . Anger is responsible for almost every step of progress." He counseled his readers, "Use your anger to drive you in the direction of the worth while. Hitch anger to your ambition and to your efforts. . . . Harness anger . . . and you will think, work, and plan harder, longer and more productively."[59]

Twentieth-Century Anger

This view was gradually supplanted by a very different vision of the feeling. In the twentieth century, writers, psychologists, business leaders, and moral reformers came to condemn anger, even white male anger, as a new emotional order began to take shape within the American economy. While writers such as G. Stanley Hall celebrated the feeling as a primordial force and source of power that could drive men to excellence, starting in the late 1910s some came to describe it as a symptom of inadequacy, a result of one's limitations. Frank Crane, a Presbyterian minister, declared, "Anger is always a sign of weakness. It is our irritation at our own impotence. . . . Anger is always against what is or seems stronger than ourselves."[60]

Not only was it a sign of one's own individual weakness, but many social commentators suggested that those who expressed their anger would never advance. They compromised their chances for success when they became angry, for they became inefficient. In 1917, an editorialist pointed out, "Anger never has helped you to finish a job sooner or do it better. It has never added one cent to your earnings. . . . [A]n angry man is frequently a careless man. Some men get angry when a tool doesn't work right. When a man loses his temper over a little thing of that kind he stands a mighty good chance to lose something else along with it—a finger or hand or his head. . . . Anger burns up bodily energy. . . . Burns it up without accomplishing anything but harm."[61] Another writer labeled the feeling a "poisonous mental reptile" and argued that "Anger stands between a man and wealth, for the acquiring of wealth . . . requires good judgement. The angry man is a self-indulgent man. . . . Anger puts a man at a disadvantage."[62] Reverend Dutton, a Methodist Episcopalian minister in Kansas, chimed in from the pulpit, noting, "A bad temper in business is an exhibition of the poorest kind of judgement; a man who cannot control his temper is a failure in business." Those who displayed anger were "unsafe and undependable."[63]

While anger had once been seen as a sign of courage, masculinity, and strength, it was coming to be regarded as a liability in the twentieth century. Key to this shift was the growth of large, bureaucratically governed corporations, and the emergence of an organizational society more generally. The new imperative in large-scale businesses and in urban society was to get along with others and to work together as part of larger institutions—be they corporate offices, large factories, military units, or universities. A number of historians and sociologists have suggested that a new set of personality traits came to be valued in such circumstances, and a new type of labor—emotional labor—was required. Historian Angel Kwolek-Folland observes, for instance, that men at work in corporations in late nineteenth-century America no longer were their own bosses. As a result, they "faced a new work experience that denigrated self-seeking ambition, demanded subservience, and undercut men's ability to master their own destiny. Rationalized management and job specialization shattered nineteenth-century ideals of manhood, undermining the notions of personal will and independence. . . . Managerial capitalism's demand for subservience thwarted personal autonomy and threatened to feminize labor."[64]

In the corporate workplace, which relied on a smoothly running bureaucracy, many came to believe that anger should have no role, and rules about its expression proliferated and tightened over the course of the century. As Peter Stearns has shown, starting in the 1920s and 1930s, government and big businesses created personnel departments, which were charged with recruiting suitable employees who could work together efficiently and harmoniously. Roughly a third of large corporations had created such departments by the 1930s, and their influence grew in both the private and public sectors during World War II.[65] They were staffed by industrial psychologists, human relations experts, and counselors who dedicated themselves to reducing strife, discontent, and anger in the workplace. During the war, psychologist H. Meltzer explained that "workers, as human beings, experience frustration in industrial settings." This was a widespread issue and it represented a real problem, for it harmed worker efficiency: according to one study that Meltzer cited "negative emotions decreased production (6.8%) more than did positive emotions."[66] Personnel departments were there to weed out potentially angry job applicants as well as to help the existing workforce adjust to frustrations on the job.

As workplace culture changed, employees themselves made an effort to suppress their anger. According to Stearns, mid-level managers began to tout their ability to manage their tempers as an admirable asset. Those who worked for them, even those involved in the labor movement, also sought to control the feeling in themselves. As the

United Auto Workers union instructed its members, "A lost temper usually means a lost argument."[67] Significantly, despite this new pressure on both blue- and white-collar employees to eliminate friction and displays of temper, executives and CEOs were often allowed to still express anger, for it was a sign of forceful leadership.[68]

As a result, by the mid-twentieth century, workers frequently felt anger, but they had internalized workplace rules that required them to repress the feeling. Studs Terkel interviewed employees in a range of jobs during the 1960s and early 1970s who told him that they choked down their anger at work only to express it after hours. Mike LeFevre, who worked in a steel mill, told Terkel that rather than go home and rest, he always felt, "I gotta get it out. I want to be able to turn around to somebody and say, 'Hey, fuck you.' You know? . . .'Cause all day I wanted to tell my foreman to go fuck himself, but I can't. So I find a guy in a tavern. Tell him that. And he tells me too. I've been in brawls. He's punching me, and I'm punching him, because we actually want to punch somebody else. The most that'll happen is the bartender will bar us from the tavern. But at work, you'll lose your job."[69]

Workers in the expanding service sector faced much the same dilemma. In her pioneering work on emotional labor, sociologist Arlie Russell Hochschild sat with flight attendants as they were trained to suppress or redirect their anger. The trainer asked, "What do you do when you get angry?" The attendants' responses included "Cuss. Want to hit a passenger. Yell in a bucket. Cry. Eat. Smoke a cigarette. Talk to myself." The instructor counseled, "If you think about the *other* person and why they're so upset, you've taken attention off of yourself and your own frustration. And you won't feel so angry." Many of the flight attendants were more cynical about these attempts to muffle their anger. One confided, "If they call me 'honey' or 'sweetheart' or 'little lady' in a certain tone of voice, I feel demeaned. . . . But when I get called 'bitch' or 'slut,' I get angry." That feeling, however, was not allowed in a flight attendant's emotional repertoire. "Now the company wants to say, look, that's too bad, that's not nice, but it's all in the line of public-contact work. I had a woman throw hot coffee at me, and do you think the company would back me up? . . . No. They say don't get angry at that; it's a tough job, and part of the job is to take this abuse in stride."[70] Anger must be swallowed.

Over the course of the twentieth century, American businesses continued to police anger, though their techniques changed over time. By the end of the century, some corporations were hiring clinical psychologists to teach "anger management" classes to their employees. In 1998, Phil D'Agostino, a therapist who gave one hundred presentations a year on anger, told a reporter, "No one has a right to be angry, and it's delusional to think that anger can be productive."[71] There was also new alarm about the feeling on the job in the late 1990s, because of a rash of shootings carried out by disgruntled employees, leading some psychologists to create a new category of "workplace rage," and to develop screening tests designed to identify potentially violent job candidates.[72]

Twentieth-Century Technologies and Anger off the Job

Yet while anti-anger programs proliferated during the twentieth century, simultaneously many found new places to vent their frustrations and fury. These emotions, taboo at work, often found expression in private spaces during leisure hours.[73] For instance, in the 1920s and 1930s, many tuned in to radio broadcasts. Some had initially prophesied that radio would make people sociable and mend old divisions, for it "was . . . making the world so small." By hearing the same broadcasts, "Nations twelve thousand miles apart were neighbors. . . . It was all tending so to unite them, to lead them toward that cooperation that had been the tendency of community always."[74] Yet others quickly concluded that radio's effects might be more divisive than this. Academics of the era feared that the radio exerted great power over listeners and had the ability to create new types of angry mobs, separated by distance but motivated by the same broadcasts. As two scholars of the radio noted in 1935, "Now, as never before, crowd mentality may be created and sustained without the contagion of personal contact." These crowds "are less violent and less dangerous" than actual mobs of people, for "the radio can create racial hatred but not itself achieve a lynching." Yet the authors nevertheless lamented the fact that "to a degree the fostering of the mob spirit must be counted as one of the by-products of radio."[75]

Listeners generally did not act on their anger through group violence, but they could act on it through other means, for they saw the radio as a highly interactive, two-way form of communication. They talked—and wrote—back to radio, treating broadcasts as part of a shared conversation. They filled the mails with letters responding to particular broadcasts they had heard. As historians Lawrence and Cornelia Levine point out, broadcasters "literally trained their listeners to write letters by constantly urging them to send in their opinions and responses to programs. . . . NBC received 383,000 letters in 1926 . . . 1 million in 1929 . . . 7 million in 1931. CBS was more successful, receiving 12.6 million letters from its listeners in 1931 alone."[76]

Sometimes listeners wrote to radio stations, other times to public officials, and often they wrote with anger. In 1932, the *New York Times* reported on listeners' responses to the radio broadcast of the Democratic Convention. "It seems that the only relief a radio listener gets when stirred to a pitch of anger or inspired to the heights of joy is to shut off the set, tune to another station, or write a letter. Listening-in for a decade has created a habit of letter-writing. The letter gives vent to the feelings, whether injured or appreciative. The letter is one way the unseen audience has of 'getting back at the radio.'"[77]

During the 1930s, radio listeners posted thousands of letters detailing their emotional responses to the broadcasts they heard. Franklin Delano Roosevelt's Fireside Chats, for instance, sparked a range of responses. His early broadcasts earned him widespread praise from those who heard them, while later chats often brought more critical responses. After a 1937 radio broadcast in which he set forth his proposal to increase the number of Supreme Court Justices, he received a deluge of angry letters. One correspondent wrote him, "You . . . and the whole ring of vampires at Washington should be tarred, feathered and set on fire. . . . The dungeon of Hell will never be filled until you and your diabolical gang are dangling on pitchforks over the flames. . . .

These are the sentiments of every decent American today, which you are not."[78]

Fanning the flames of this anger were FDR's many radio critics. Among the most popular was Father Charles Coughlin, whom some have dubbed the "father of hate radio." He had an avid listening audience of thirty to forty-five million listeners each week, and appealed to them with what a biographer termed "angry tones of condemnation." In one oration, he described FDR in outraged and incendiary terms as "the great betrayer and liar, Franklin D. Roosevelt, he who promised to drive the money changers from the temple and succeeded in driving the farmers from their homesteads and the citizens from their homes in the cities."[79] In a 1937 talk, he proclaimed that "somebody must be blamed," and increasingly declared it was the Jews, whom he accused of controlling the New Deal as well as all international finance.[80]

Many who were angered and offended by his inflammatory broadcasts wrote to the Federal Communications Commission (FCC), asking the agency to ban Coughlin's programs, decrying the great power he was exerting over his listeners. As historian David Goodman writes, "A New York businessman sent a telegram to remind the President of 'the uncanny power of the radio as a medium for influencing mob psychology.' A Wisconsin man wrote to FDR to ask whether the president realized that 'this man, with his rabble-rousing voice, with his uncanny way of making . . . the most destructive falsehood appear to the uneducated in the guise of a profound truth . . . reaches, personally and effectively, millions of Americans every Sunday afternoon?'" In 1939, Dorothy Omens, a Jewish resident of Chicago, described how Coughlin's broadcasts had altered her perceptions of the radio. "I've sat for three days now in my home listening to these broadcasts and staring into my radio. Can one hate a piece of wood? Yet my lovely modern radio looks despicable to me. My favorite programs have lost their significance and sound like mockery." A New York City listener wrote that Coughlin's words were "inciting to riot and violence and provoking race hatred . . . with attendant disorders and hooliganism." Seventeen-year-old Marilyn Grunin, from the Bronx, believed Coughlin was spreading "terror and fright among right-thinking Americans."

In 1939, a Baltimore listener, appalled by Coughlin's attacks on the president, wrote Roosevelt that he found the broadcast "vile in the extreme": it "left me nauseated and filled with resentment to think that such unjust and untrue statements may be made in private, and when it is broadcast to the whole world by the medium."[81] Father Coughlin's broadcasts were eventually shut down during World War II, but in subsequent decades, other broadcasters found that radio worked well for the transmission of anger.

Starting in the late 1970s, talk radio came into its own. Historian Susan Douglas estimates that radio stations that broadcast no music, only talk, or news and talk, grew dramatically in the 1980s and 1990s—from around 200 stations to 850 by 1994.[82] The number of stations continued to grow in the twenty-first century, and as of 2014, there were almost two thousand.[83] One key factor in their growth was a change in FCC policy in 1987, which allowed stations to broadcast only one perspective, a departure from the earlier "Fairness Doctrine," which had insisted that stations airing politicized editorials allow equal time for the opposing side to respond. Under Ronald Reagan this policy was abandoned. As stations broadcast one-sided programming, many hosts used anger strategically, as a way to woo listeners, most of them male. Douglas explains:

> There was a new dynamic here, one that had been developing since at least the late 1960s, in which certain radio shows sought to rile up their audiences, following the notion that fury equals—and begets—attention, and thus profits. Unlike TV in the 1950s and early 1960s, which sought to avoid controversy so as not to alienate its audiences, talk radio pursued controversy and, again in total contradiction to the earlier years, used this as a selling point to advertisers looking for loyal, large, engaged audiences.[84]

Political scientist Murray Levin, who explored the meaning of talk radio in the 1980s, concludes that it was "a particularly sensitive barometer of alienation because the hosts promote controversy and urge their constituencies to reveal the petty and grand humiliations dealt by the state, big business, and authority." He regards the radio call-in show as "the prime conduit for proletarian despair." The discussions, he says, were "full of anger towards the politicians and parties and towards big business and labor."[85]

These shows' audiences were largely male, and the programming fanned a particular type of masculine anger. The most famous hosts were Rush Limbaugh, Don Imus, and Howard Stern who offered—and continue to offer—a counterpoint to some of the dominant emotional codes that had banished anger from the workplace and the public square. Their broadcasts were angry and aggressive, and, as Douglas points out, they challenged "buttoned-down, upper-middle-class, corporate versions of masculinity."[86]

By some accounts, the rise of talk radio was fueled by a growing conservative electorate, frustrated that their opinions were not being voiced by what they derisively called "lamestream media." However social scientists Jeffrey Berry and Sarah Sobieraj posit instead that talk radio actually first emerged because of Reagan's deregulatory initiatives and then expanded because the profitability of music radio had been eroded by internet streaming services like Pandora. In other words, profit and the "peddling [of] political outrage" were the main engines driving talk radio's ascendancy. While an anger fueled by capitalism might not be unprecedented, it has not existed "on the scale of what is available today." For Berry and Sobieraj, there is cause for real concern because much of modern radio commentary "is designed to make us as angry and fearful as possible."[87]

If radio could provoke and encourage outrage, so too could TV, although its role in the history of American anger and social friction is somewhat more ambiguous. On the one hand, there have been, of late, many nostalgic glances backward to an era when TV was dominated by the "Big Three"—NBC, CBS, and ABC. Supposedly this limited array of programming created national unity of opinion. According to proponents of this position, because Americans consumed the same programs and newscasts, they shared a worldview and were more bonded to each other. Yet some have also credited this era of mass media with creating discontent and anger—feelings that might prompt social protest and a fight for justice. TV might bring disturbing images into one's living room and galvanize political sentiment. As James Reston reported in the *New York Times* after civil rights protesters in Selma were met with violence: "It is the almost instantaneous television reporting of the struggle in the streets of Selma, Ala., that has transformed what would have been mainly a local event a generation ago into a national issue overnight. Even the segregationists who have been attacking the photographers and spraying black paint on their TV lenses understand the point."[88]

And as historian Julian Zelizer points out, the Big Three networks did not themselves shy away from broadcasting divisive debates. For instance, in 1968, ABC hosted a series of debates between Gore Vidal and William F. Buckley Jr. According to Zelizer, "During the most infamous moment from the debates that occurred on August 28, Vidal accused his adversary of being a 'pro- or crypto-Nazi.' Buckley, barely able to contain his anger, started in on Vidal by saying 'Now listen, you queer,' . . . before leaning his body over in menacing fashion and saying he would 'sock' the writer in the 'goddamn face.'"[89]

Even when it was not covering protests and debates, mainstream TV could also stoke discontent merely by broadcasting an image of life that was supposed to be the right of all, the life of suburban bounty portrayed in popular sitcoms. Those who saw these images but were unable to achieve that lifestyle often felt marginalized. As a result, in 1974, S. I. Hayakawa suggested that "the civil rights revolution of the past fifteen years was triggered by television." He maintained that African Americans watching TV saw shows and advertisements that glorified the bounties of consumer culture and a just and open society; such programs told viewers "You are an American. You are entitled to eat, drink and wear what other Americans eat, drink, and wear. You are a member of our national community." But Hayakawa suggested that because of poverty, many viewers were "denied . . . the chance to share in all the beautiful things advertised. Would you not be frustrated and angry?" This was a common reaction to television, he claimed, for it "creates wants—and nurtures discontent."[90] By glorifying a supposedly shared culture, TV allowed viewers to see with new clarity what they were lacking. Their own exclusion from the American Dream might become manifest on the TV screen.

Though TV, when it was dominated by just a few broadcasters, already had divisive effects that could nurture anger and discontent, many suggested that this tendency grew with the rise of cable television. In 1977, 90 percent of the TV programming that Americans viewed appeared on ABC, CBS, or NBC. By 2003, that figure had fallen to 29 percent.[91] As this transformation was under way, a number of political scientists, sociologists, and media scholars began to worry that this decline spelled the end of comity and a shared public sphere. Political scientist Elihu Katz argues that TV had once offered "citizens the experience of shared contemplation of matters of state that demand attention." Such viewing "rejuvenated" the polity. Alas, he warns, "The waning of television augurs ill not only for participatory democracy but also for the nation itself. The media had a lot to do with shaping the nation and holding it together—not just politically but also economically, socially, and culturally."[92] Communications scholar James Webster suggests that cable TV companies were in fact fracturing the viewing public intentionally. Channels needed to be able to attract distinct audiences, and to do so, they specialized their offerings so that they might better attract advertising revenues, "an attempt to produce a particular audience demographic that can be sold to advertisers."[93]

Not all scholars agree that the rise of cable is the chief cause of political polarization. While audiences have indeed become fractured, many viewers, rather than watching either Fox or MSNBC, are watching HBO or the Hallmark Channel. Some scholars suggest that this means more moderate viewers tune out the political clamor and vitriol altogether and disengage from political life.[94] Such arguments support Robert Putnam's contention that as Americans came to watch more and more television, their level of civic and social engagement declined accordingly. Isolated in their own houses, they might rail against politicians or turn away from politics and tune in to a sitcom, but in either case, they did so alone.[95]

TV and radio, which provided Americans the opportunity to vent in their leisure time if not at work, were joined by another technology that also offered a private space for anger—cars. While Americans needed to suppress anger at work, they did not have to do so on the commute to and from it. Americans had discussed angry or aggressive driving since the early twentieth century. The emergence of the word "road rage" in 1988 signaled a new concern with the problem, as alarming (though much debated) statistics began to circulate.[96] The US Department of Transportation also drew attention to the issue of anger behind the wheel in 1996 when it suggested that two-thirds of the 41,000 auto deaths that year were a result of aggressive driving, while a 1997 Department of Transportation report found a 51 percent increase in such driving during the 1990s. More than two-thirds of Americans surveyed in the 1990s perceived a decline in the courtesy of their fellow drivers.[97] Researchers identified a number of causes for the uptick, including an increase in the number of miles that Americans were driving and a growing number of vehicles on the road.[98]

Perhaps the auto became a site for anger because the car represented a space in which one could enjoy what one scholar termed "relative anonymity" and "ease of escape."[99] While anger was prohibited in American workplaces and in face-to-face encounters, cars offered some a way to express the feeling while still hiding their identity.

With their TVs, their talk radio, their cars, Americans sat alone, separated from others. And it was in private that they felt most at ease expressing anger. Having internalized the workplace dictates about anger management, feeling acutely the limits that had been placed on their emotions at work, they felt comfortable being outraged when they could not be observed, when there were seemingly no consequences or constraints.

Cyberspace: The New Home of Mind

With the birth of the internet, many found a new place for such anger—online. This, however, was not precisely the outcome many of the web's early promoters had hoped for. For a good part of the 1990s, Americans had naïvely believed, as they had during the emergence of many prior communication technologies, that the internet would promulgate utopian forms of freedom and democracy. For instance, the late John Perry Barlow enunciated such hopes in *A Declaration of the Independence of Cyberspace* in 1996. Barlow, a former Wyoming rancher and lyricist for the Grateful Dead, penned, among other famous songs, "Mexicali Blues." In addition to his musical talents, Barlow was also an essayist and the founder of the Electronic Frontier Foundation, an organization created in 1990 and renowned for promoting freedom of expression on the internet. Barlow thought that the 1996 Telecommunications Reform Act, which regulated some forms of online communication, jeopardized these freedoms; its passage spurred Barlow to pen his declaration.[100]

In a jeremiad reminiscent of both Jefferson's Declaration of Independence and Marx's *Communist Manifesto*, Barlow accused US legislators of betraying America's founding values: "In the United States, you have today created a law, the Telecommunications Reform Act, which repudiates your own Constitution and insults the dreams of Jefferson, Washington, Mill, Madison, DeToqueville, and Brandeis." Barlow proclaimed that these dreams needed to be "born anew" in cyberspace, writing, "I declare the global social space we are building to be naturally independent of the tyrannies you seek to impose on us. You have no moral right to rule us."[101]

Barlow and the thousands of Web users who reposted his declaration on their own websites idealized the internet as a space that was free from the "tyrannies" of the physical world.[102] In cyberspace, a full range of human sentiments could be expressed, liberated from any control. He observed, "You entrust your bureaucracies with the parental responsibilities you are too cowardly to confront yourselves. In our world, all the sentiments and expressions of humanity, from the debasing to the angelic, are parts of a seamless whole, the global conversation of bits."[103]

In spite of this freedom, Barlow believed that cyberspace would have its own "social contract" and would be governed by what he obliquely referred to as the "golden rule." Furthermore, the prejudices and discord that were so closely attached to embodied selves would, in Barlow's view, also be transcended when people left their bodies and related to each other digitally. He predicted: "We are creating a world that all may enter without privilege or prejudice accorded by race, economic power, military force, or station of birth. We are creating a world where anyone, anywhere may express his or her beliefs, no matter how singular, without fear of being coerced into silence or conformity."[104]

The internet, as Barlow described it, offered a new and possibly more constructive outlet in which to communicate. One could take off one's emotional shackles, but because one was not bound by the biases and prejudices of the physical self, those feelings were less prone to run riot. As he put it in his concluding paragraph, "We will create a civilization of the Mind in Cyberspace. May it be more humane and fair than the world your governments have made before."[105] It is no wonder that Barlow drew acolytes, enthralled by these possibilities of an unconstrained self in harmony with others.

His utopian hopes were only imperfectly realized. If the internet has offered greater freedom of expression, that expression has often devolved into caustic anger, flaming, and trolling, rather than harmony. Among our interviewees there was a general perception that anger and incivility are on the rise in the United States, visible in all parts of life, but especially in politics, and that the emotion was most apparent online. These perceptions are backed up by various polls. In quantitative surveys conducted by the public relations firm Weber Shandwick and Powell Tate, data suggest that 69 percent of Americans believe that "the U.S. has a major civility problem," and 75 percent of Americans agree that "incivility in America has risen to crisis levels"—a 5 percent increase since 2016.[106] These findings are also supported by two 2017 Pew studies that focused on partisanship. The authors noted that there is increasing division in the American electorate and that fewer and fewer Americans embrace a mix of liberal and conservative views.[107] Meanwhile, the political scientists Alan Abramowitz and Steven Webster note that there is increasing "negative partisanship" in America, which they define as the tendency to regard the other party with antipathy.[108] Such antipathies often play out on the Web. A 2017 Pew study, "The Future of Free Speech, Trolls, Anonymity and Fake News Online," polled an array of media experts, most of whom worried about the uncivil behavior taking place online, and predicted that such behavior "will persist" and perhaps even "get worse."[109]

The popular press has also noted the increasingly angry tone of online conversations. A 2013 NPR report noted, "It's no secret that angry online ranters have become a real scourge of the Internet. The equation goes something like this: Normal person + anonymity + audience = terrible person."[110] The problem seems only to have worsened in the intervening years—in 2017, *Time Magazine* declared that the nation was suffering from "a rage flu," which was "contagious." As the journalist noted, "All this anger and derision in which we're marinating isn't healthy. Not for us, not for our kids and certainly not for the country. But as a nation, we can't seem to quit. We're so primed to be mad about something every morning, it's almost disappointing when there isn't an infuriating tweet to share or a bit of our moral turf to defend waiting in our phones."[111]

Anger, then, has become newly visible again in American life, and particularly so online. Today, the feeling has been democratized. In the colonial era, some believed only God could be legitimately angry, while others contended the passion was acceptable for social elites as well. In the nineteenth century, white men of all classes were given free rein to express the feeling, but women and African Americans often felt pressure to repress the feeling. In the twentieth century, save for company executives, workers of all races and sexes encountered new restraints on their anger. Today, as a result of social media and the internet, men and women from all backgrounds have found greater opportunity to express the feeling, though not all avail themselves of such options.

As a result, anger today is far more visible and public than it once was. Some, like their nineteenth-century forebears, see the feeling as useful, a prompt for social and moral change, and they work to use their own anger as an impetus for action and to motivate others to take action as well. They are angry, but with an eye to the consequences of the emotion, and try to express it carefully. Others feel less constrained and limited and seem to have few worries about the consequences of anger—expressing it freely and publicly. Still others have transferred the modern rules against anger into their online activities; many have internalized the emotional style valued in twentieth-century corporate America, described by Hochschild as the "managed heart," by David Riesman as the "other-directed type," and by Christopher Lasch as the narcissist. Today, these are the workers who must commodify their feelings, who make every effort to mask anger and to keep it in check—at least at work or in the marketplace.[112] For them, the internet, rather than providing a place of remove from the managed heart and its demand for emotional conformity and equanimity, has simply imported such demands online, thereby expanding the space in which they are manifest. And then there are individuals who move between emotional styles, changing their modes of expression according to their circumstances, sometimes abandoning one mode for another.

These varieties of online anger differ from nineteenth-century forms, for the people expressing it are physically removed from each other. That social and geographical distance can affect both the tone and the power of the emotion, for it can make anger flare more intensely but it also may burn out more quickly, accomplishing less. If one is merely letting off steam that may be of little concern; however, if one is fighting for lasting social change, that may prove demoralizing and disillusioning.

Constructive Anger in Cyberspace

When we talked with our interviewees, they sometimes spoke approvingly about online anger and regarded it as a legitimate response to injustices festering in America at large. To these people, the expression of anger was a virtue, because it helped redress inequalities. Many saw current online movements as offering new and long-overdue opportunities for the disempowered and disenfranchised to give voice to their indignation.

For example, when we spoke with Anna Renkosiak about anger, she mentioned that the #MeToo movement was finally allowing women's anger at workforce victimization to be "heard."[113] Women had long been told that anger was a forbidden emotion; now they were publicly expressing it and getting results. Leslie Jamison, for instance, noted this transformation in herself, confessing in the *New York Times* in 2018, "I Used to Insist I Didn't Get Angry; Not Anymore."[114] And *Time Magazine* believed this new emotional style was being embraced by a broad swath of American women:

> Women have had it with bosses and co-workers who not only cross boundaries but don't even seem to know that boundaries exist. They've had it with the fear of retaliation, of being blackballed, of being fired from a job they can't afford to lose. They've had it with the code of going along to get along. . . . These silence breakers have started a revolution of refusal, gathering strength by the day, and in the past two months alone, their collective anger has spurred immediate and shocking results: nearly every day, CEOs have been fired, moguls toppled, icons disgraced. In some cases, criminal charges have been brought.[115]

Others saw this same kind of righteous indignation as the root emotion of #BlackLivesMatter. White supremacists had long tried to contain and dismiss African American anger and the demands for justice that accompanied it, but the feeling was gaining new visibility and legitimacy in the public sphere. As Yolanda Pierce, the dean of the Howard University Divinity School, notes,

> The righteous anger that animates an anti-racist movement like Black Lives Matter calls on individuals to discern if they are paying attention to the systems of injustice surrounding them, the varieties of oppression that exist like the very air we breathe. Why are you not angry about the death of a 12-year-old black child playing in the park, at the hands of law enforcement? How are you not enraged about the myriad ways the modern prison industrial complex perpetuates chattel bondage?[116]

And many we interviewed welcomed that anger, seeing it as a legitimate reaction to injustice and violence. Katie Schall, for instance, saw

#BlackLivesMatter as a positive development: "In the last 3 years there's like Black Lives Matter . . . people are feeling more disenfranchised and they have an avenue to tell people about it . . . online. And I think there is something good about talking about the negative things happening in society."[117]

Jennifer, a forty-one-year-old nurse from Iowa, likewise endorsed such emotion as a useful political tool. Reflecting on #BlackLives-Matter, #MeToo, and the Left's opposition to Trump policies, she noted, "I think anger can be constructive if you react to it wisely. . . . [It can be] a stepping stone to do something or change something. . . . I think anger can be a helpful emotion." Jennifer credited her own uptick in political activism to her anger—and to her access to online media. She noted that before the election she had rarely communicated with her legislators; now, thanks both to her anger and to the availability of programs and apps like Resistbot and 5 Calls (which helped her send messages and which provided information for those messages), she sent significantly more texts and emails to her representatives to protest Republican policies than in earlier years.[118]

Such activism, born of anger, was something many political leaders welcomed and even encouraged. Senator Jim Dabakis, a Democratic Utah state legislator whom we interviewed in the fall of 2017, was willing to speak charitably about anger—in this case about the anger his constituents were feeling in the wake of Trump's election. As Dabakis saw it, the challenge was to find a way to channel anger toward positive ends: "A lot of people . . . just don't care about politics [but] suddenly it's in their life for the first time. . . . They're angry. They're upset. They're looking for a place to act on that. And so it's been certainly a target for me to find those people and to organize them and to try to focus them into political action. There's no coyness about it."[119]

To do this, he used Twitter, Facebook, and an online newsletter. On social media, he worked to strike the right emotional tone, sometimes mixing happy or comedic posts with more indignant ones, in order to keep his communications from becoming too dark. He believed that in some cases, however, deep-seated anger needed to be acknowledged;

the challenge was finding a way to do so that did not alienate the larger constituencies with whom one was trying to work. To this point, Dabakis spoke about Utah's Grand Staircase Escalante National Monument, which was dear to the people he was representing, and whose acreage Trump had recently halved.

> When you are going to tear up Grand Escalante after all these years and turn the most dinosaur precious part into the world's largest coal mine, I mean that deserves complete, utter, massive foaming-at-the-mouth outrage. It really does. But how do you affect the long term policy? Keep your base that are foaming at the mouth like they have a voice somewhere and still be rational enough so that you don't lose your credibility with the other public that know something is going on but they're not quite sure.

In managing his constituency's passions, Dabakis generally tried to avoid inflaming anger "to the point you become a lunatic." And he allowed that in some situations, "I don't know where you go the day after anger." Anger deserved to be acknowledged. At the same time, it needed to be expressed in ways that recruited, rather than alienated, less impassioned citizens.[120]

In the interest of maintaining partisan balance, we also interviewed Senator Todd Weiler, a Republican senator in the Utah Legislature. At the start of our conversation, Weiler described his "sarcastic" yet "transparent" personality, and how he used Twitter to project who he was:

> I've been on Twitter for 10 years. . . . I'm kind of sarcastic and Twitter to me is kind of sarcastic. So I have a long history of tweeting. . . . You know I am who I am, I think I'm very open and transparent. . . . Most people don't want a politician to be this whitewashed, vanilla-flavored everything. . . . I'm not going to pretend like I'm someone. . . . I'm a real person, I have real opinions, most of them are my own, [and] I think they're right.

Weiler stated that "I'm not an angry person by nature," yet like Dabakis, he recognized that anger could serve a political purpose. Only, for Weiler it was right-wing anger that was worth highlighting:

> I think there is a lot of pent-up frustration. And maybe anger at the left-wing media. The MSNBCs of the world. And others. But I think MSNBC is the most egregious. . . . There's some bias there. So there's elements in me, "Yeah, you go tell them." . . . And that's what I think people are pushing back on. . . . I think a lot of people, like me, have felt like they've had their finger on the scale for decades now. And Trump is at least calling them out for it. I think the fact that he's doing it, I think a lot of people are like "you go get 'em."[121]

Both Dabakis and Weiler treated anger with caution. They accorded anger respect and described it as a legitimate feeling often triggered by abiding social injustice. At the same time, the legislators acknowledged that the feeling might not always get expressed in ways that were strategically helpful. But when it was directed wisely, they contended that it could be used to political advantage.

Although the senators only wanted to go so far in intellectualizing their concerns about anger and the internet, they nonetheless touch on challenges that sociologist Zeynep Tufekci raises in her 2017 book, *Twitter and Tear Gas: The Power and Fragility of Networked Protest.* As a longtime student of social media and social movements, Tufekci spent many years studying how social media was used to mobilize dissent in Turkey, Egypt, and the United States. She acknowledges that many marginalized issues like "economic inequality to racial injustice to trade to police misconduct, were brought to the forefront through the force of social media engagement." But at the same time, she sees "movement after movement falter because of a lack of organizational depth and experience. . . . [T]he capabilities that fueled their organizing prowess sometimes also set the stage for what later tripped them up, especially when they were unable to engage in the tactical and decision-making maneuvers all movements must master to survive."[122] Whereas nineteenth-century indignation meetings physically and emotionally united people, galvanizing them to organize and fight jointly for change, the effects of Twitter anger seemed less binding.

While the senators were not working with constituencies that were as inflamed as most of Tufekci's, they nonetheless were facing similar challenges. Social media could be used to mobilize angry constituencies who might effect short-term political change. But it was unclear whether those changes would prevail when playing the long game. The internet was, from some perspectives, very angry.[123] But to prognosticate whether that anger could be used for long-term advantage was to look through a glass darkly.

Snarky on Twitter, Nicer in Person, or Trolls and Tribalism

While some maintained that this less binding form of anger was still useful, many also spoke to the ways that online media were offering unprecedented opportunities to express it, in ways that were sometimes destructive and cruel. The remove of social media, the fact that one did not have to say things face-to-face, that one could be anonymous, seemed to heighten people's willingness to express anger more vigorously than they might in person, to loosen their sense of constraints. Senator Weiler, for example, ruminated on the good and the bad of social media and how it had reshaped anger:

> Politicians and policy makers are a lot more accessible and I think that's good. But . . . because I'm opinionated, because I'm active on social media, I have basically volunteered to be a punching bag. So I have people—not just all over the state of Utah, but all over the country—when something happens, "Oh, I'm going to log on to Twitter and punch Senator Weiler." . . . That can be fun but it can also wear on you a little bit. . . . They know where to kick me. . . . They can do it in ten seconds and move on with their lives.

This newfound accountability and transparency was not something to dismiss. It pained him to be a punching bag, but Weiler was not complaining; he had volunteered for it and was willing to take some hits even as he threw some of his own. That was just part of his job.

At the same time, Weiler allowed that people were "going to do things on social media that they would never do if they had face-to-face contact." When we asked him why, he explained, "I just think you don't have to look someone in the eyes." To underscore the point, the senator offered the following analogy:

> If I'm at a buffet, I would probably never cut in front of you in line to get the Jello. . . . But if I'm merging on to the freeway, I probably would cut you off if I saw a chance. And so social media is the freeway. . . . If I cut you off on the freeway . . . I don't know who you are. I'm probably never going to see you again. Chances are, 99 percent chance, you don't know who I am. . . . And you know I'm not a mean driver but, yeah, if I can get around you, I'm probably going to shoot that gap. And that's what people are doing to me on Twitter. And that's what I'm doing to some people on Twitter.

Just as cars had offered a new opportunity for men and women to express anger and road rage they would normally contain, so too did social media. Whether one posted anonymously, or whether one was simply at a geographical remove, the end result for Weiler, and for many others, was that one could use a looser rhetoric. These were the reasons Weiler's Twitter byline is "Snarky on Twitter, Nicer in Person."

In the scheme of behaviors, "snarky" does not seem too bad, and Weiler was willing to forgive, and even celebrate, these small expressions of animosity and sarcasm, even as he suffered abuse for them. But he did not believe such expressions of anger should be taken too far. When we asked whether he excused similar behavior in Trump's tweeting, he took care to delineate differences:

> One of the differences between me as a state senator and Donald Trump is formulations. Donald Trump, this week, he retweeted a tweet from a radical group in Great Britain. He retweeted a video of a refugee beating up somebody. And that has had international ramifications. . . . If I had to choose Obama's kind of boring personality versus Trump's flamboyant . . . fly off the handle, I'd [choose] . . . something closer to Obama. . . . So I do wish Trump had a filter or General Kelly would take away his Twitter account.

If Trump's behavior was degrading international relations, Weiler also worried that it was reshaping America's emotional norms, changing how people expressed themselves. "I do think that Trump is part of the problem. Because people who would otherwise have been dignified have said, 'Well, OK, if Trump is doing that . . . now I'm going to tell the world what I really think.' . . . He's so crass and so unfiltered. So Trump has certainly made it worse."[124]

Many Americans, on both the Left and the Right, share this sense that the president was popularizing a new emotional style (or perhaps resuscitating an older one—dating back to the nineteenth century, when anger was more visible in public life). Trump and Twitter, and a failure to think about repercussions, were raising the limits for allowable anger. Senator Dabakis, for example, maintained: "I don't think we've seen a more skilled politician vis-à-vis anger than President Trump. . . . he's willing to use that anger regardless of the broader consequences. . . . To me that's a slap in the face to the idea that we have a country that ought to be and must be . . . respectful, not just tolerant, but respectful of a wide variety and range of issues."[125] Anna made similar connections between Trump, Twitter, and anger. She believed his emotional style was reshaping American discourse: "Look at Donald Trump. . . . Almost every . . . single thing on his Twitter is bitter or angry. . . . That's now the norm. . . . Anger and people who support that anger and then they feel like they can be just as angry."[126]

Americans had been angry online before the election, but the people we interviewed feared it was becoming even more common now, and that many of its new manifestations were destructive rather than productive. Jennifer believed there were some people who just enjoyed being provocative and mean merely because they now could. Social media gave them a new power. She knew people who had been "trolled" online, and said of the trolls, "I feel like there are people sitting in their underwear in their mom's basement that are just looking for fights to make."[127]

There was a sense that creating online fights and drama was emotionally satisfying for some. As Katie observed, social media—and the anger it spawned—had turned much of online exchange into a kind of spectacle, albeit of a sordid variety.

> It just like degrades. I see it all the time currently. . . . You post something about a kitten and then . . . ten comments down, there's someone . . . who is like pro-Trump or anti-Trump and then it just devolves. We were talking about a kitten! . . . I know there's like people who like to instigate. So I think there's two kinds of people. People who genuinely have a passion about an opinion and then people who would just like to see it turn into a shit show, 'cause it's a fun trainwreck to watch.[128]

The anger spawned by "instigators" was in turn compounded by what our interviewees variously referred to as "segmentation," "cocooning," "bubbles," "echo chambers," or "amplification chambers." This sense that the internet was dividing America into separate camps, and that people were congregating online with others who shared their views, is an observation that intellectuals like Cass Sunstein began to note in the late 1990s (and that others from Eli Pariser to danah boyd have discussed since then).[129] Now, however, that sense has become widely acknowledged. As forty-two-year-old Colin Parker, a contractor for the government, recently retired from the Air Force, and a self-identified conservative with a strong libertarian bent, observed, "We kind of encapsulate ourselves with our thoughts now":

> I know for me personally I've pretty much given up on social media. Just because it is so polarizing. Take Facebook. . . . You

> get everything fed to you. It all depends on what you like and what you don't like. So eventually you only get to hear what you want to hear. . . . And then all of a sudden somebody else that you know has a different viewpoint . . . and they're completely sure that . . . they're right in their viewpoint. 'Cause they're fed the same but opposite amount of just crap. And you get to the point where it's like everybody's so sure of what they feel, it must be fact now.[130]

Greg described this phenomenon as "amplifying chambers." He explained, "A lot of people say 'echo chambers' . . . but I think what really happens is you hear your own voice amplified over and over. You hear everyone else saying the same things and so you buy into certain mindsets more earnestly, even if they aren't necessarily as realistic."[131] Social media grouped like-minded people, reinforced their existing beliefs, and made them less willing to hear opposing viewpoints.

As people talked about filter bubbles, they often raised questions about the principles and underlying design of social media platforms, suggesting that it was also to blame. Colin observed that if Trump was resetting anger norms, so too was Twitter: "I look at Twitter and maybe it's just because of President Trump, I don't know, but I look at Twitter . . . and it's just like 'that's a bad system.' . . . I'm not for shutting down any portion of the First Amendment, but that is a system designed for . . . a verbal meme. . . . You have to get straight to your point, and nine times out of ten somebody can't do that and they just come out with 'F You.'" For Colin, the telegraphic or truncated quality of modern social media was making the public more predisposed to anger.[132]

There was also an emerging sense that technology companies had an interest in fomenting the feeling, regardless of the subject being discussed. For Katie, anger seemed to serve the business needs built into the Facebook algorithm:

> I think the algorithms in Facebook want to send you the biggest bang for your buck. . . . If they are not getting any emotion,

> whatever that emotion is, then people aren't going to come back to Facebook . . . to see ads that people are buying. So there is like a profit motive in making sure that when you go on Facebook you are seeing things that make you react. . . . Whether that means inciting anger or inciting great feels—love, things like that—then that's what they care about.[133]

Politically active and very astute, Phil, a twenty-three-year-old college student from Iowa, was equally cynical about social media's motives. He suggested that to maximize profits, social media companies grouped people, feeding them ads and sometimes fake news that would keep them focused on a site. "Anger is very profitable, just like war is profitable. . . . So what you are doing, you're just further polarizing the market for monetary gain. . . . If you want the answer, follow the money."[134]

Giving people only what they liked to see, hear, and read in their filter bubbles had powerful effects, reinforcing their own opinions, affirming their stance, validating their sense of worth and rightness. Some therefore saw a connection between the rise of anger and the rise of narcissism and a larger sense of self. Erica, a thirty-one-year-old nanny from Florida, believed that her generation had grown up with a sense that they were "entitled to a conflict-free life." These expectations were reinforced by the insulating effect of social media bubbles and the constant affirmation that social media "likes" offered. In the short term, those bubbles might keep anger at bay. But when people eventually came into contact with others who had different opinions, the result was anger, for they had lost the ability to tolerate difference, or to understand that their own perspectives were not the only ones extant in the world. As Erica put it:

> If you feel entitled to no conflict you are going to be hypersensitive to conflict around you. So you are then going to interpret your environment as very angry and very confrontational. . . . It . . . plays into narcissism. Any little disagreement is a slight, an attack on your character. . . . You feel like you are entitled to no

> conflict, there's no room for someone to be a good person and also disagree with you. . . . You . . . get this hypersensitivity to anger, therefore you would perceive more anger, and also increase your own defensiveness, . . . which . . . increase[s] other people's perception of anger.

Feeling entitled to a conflict-free life did not mute anger. In fact, it exacerbated it, as Erica explained: "It means you get angry all the time. Because people are being so mean to you by not agreeing with you. . . . It's a common manifestation of narcissism too. You can't believe that so many people would be so mean to you. Or have a different opinion. . . . It makes you very hypersensitive."[135] Erica's insight was in line with the findings of contemporary psychologists, who argue that those afflicted with narcissistic personality disorder "have difficulty tolerating criticism or defeat, and may be left feeling humiliated or empty when they experience an 'injury' in the form of criticism or rejection."[136] Americans' concern with their self, their love of affirmation, was heightening not just their sense of self-importance but also their anger.

Angry outbursts on social media were not just a response to slights against oneself, but also offered ways to enhance and publicize one's sense of self. If one fought and battled, one could be noticed. Anna, for instance, suggested that some started online battles for "the attention."[137] Colin agreed, noting that people "really want to be the voice in the center of the room. The one that everyone's got their eyes on. . . . People just want to be noticed and we live in this world where everybody wants their fifteen minutes of fame." Often, he said, such a drive for attention led people to post about how offended they were by news items, other people's posts, politicians' comments. "If you are offended you are going to become angry. And then your anger will make your fingers type a little bit faster. The keyboard's making a little bit more noise and then all of a sudden you're typing in all caps and everything else. And then you've alienated everyone you know because of something silly you posted on Facebook."[138]

As people typed out their anger, they could easily forget that other, real people with feelings would read their posts. Staring at one's phone or keyboard could be an insulating experience, just as being behind the wheel of a car could be. It put one at a distance and turned people into abstractions. Scholars had first noticed that people were more likely to engage in uninhibited "flaming" with the rise of email in the 1990s, but the tendency to do so seemed to be magnified on social media.[139] Time and again, interviewees talked about the fact that not having to look at the people they were addressing changed the tenor of messages. Barry Gomberg, the Ogden, Utah, lawyer, noted, "People, in part sheltered by the anonymity that the internet allows, are willing to abandon civility and express thoughts in ways that are so much more aggressive. And seemingly [more] hurtful than they used to." Thinking about the distance the internet created between people, the way it allowed individuals to insult freely, Barry said, "Just got this strange image of the bag we put over a condemned person's head. Before we shoot him before a firing squad or before we electrocute them."[140] For Barry, the internet was like a bag or a mask; it enabled people to be more callous then they otherwise might be. The metaphor of the mask was repeated in other interviews. Cooper Ferrario observed, "It's easier, it's totally easier, to say things that you wouldn't say if you were looking at someone in the face. . . . It's a mask, it's a barrier and it's protective."[141] When one masked one's own identity by being anonymous, the effect might be heightened.[142] As other people became abstractions and one's sense of self grew, the particular design of social media allowed users to feel confirmed in their own beliefs, more sure of their own positions. Their sense of their own rightness was fortified.

No one we interviewed in 2017 and 2018 seemed terribly sanguine about this state of affairs. But it was JulieAnn Winward, the writer and poet, who had become truly disenchanted and had decided (as she had done more than once in the past) to disconnect from social media because of the way that the segmentation encouraged what she called tribalism:

> Honestly, I can't even look at Facebook anymore. My friend pool is made up consistently of people who have the same ideology as I do and we do. I can't even handle that anymore. I hate them

> all. I hate liberals. I hate progressives. I hate conservatives. I hate the libertarians. I hate them all. I hate them all because they have all become so completely rigid and so completely fundamentalist that there is absolutely no room for dialogue because everybody has lost the ability to have a mind. And they have all devolved into tribal mentality. And social media has facilitated it—which is so ironic. That something of the technological age, meant to connect, has actually formed once again warring tribes that do not get along, and are not altruistic and not interested in working together for group survival, but are simply interested in being right within their group. And so it's toxic and I don't do it. I can't afford to do it.[143]

The effort to have the right opinions, to burnish one's identity through "virtue signaling," in order to receive likes and retweets, might pay off for people individually, but for JulieAnn, it seemed to have larger civic costs.

The Branded Self, the Managed Heart

If there was a group that tried to use anger to achieve extrinsic political goals, and another that has succumbed to trolling and tribalism, there was a third group that was more circumspect and did not feel free to let loose, despite all of the outrage that they were exposed to on the internet. When they got angry anyplace, it was at home, face-to-face with family members—the one place they now felt was safe to express the taboo emotion. Mores had changed from the nineteenth century. Whereas once home had been idealized as an anger-free sanctuary, these twenty-first-century Americans believed that it now had become the place one could be angry. In contrast, the online public sphere was a more risky and consequential venue. Phil from Iowa captured this outlook: "Where can you express anger? You can only express anger in your personal life because if you express anger in your public life, it's a mark against your character and it's hard enough in this world without anything else. . . . Everyone has to be perfect because the race to the top gets harder and harder."[144]

This belief was fairly widespread among middle-class professionals and many young people aspiring to that status. Susan Theiling, a fifty-two-year-old insurance consultant, worked hard to avoid anger online: "I kind of follow the rule . . . that if you don't have something good to say or something that could go on the front page of the *New York Times,* you don't need to be sharing." Her husband, Chuck, an ecologist for the Army Corps of Engineers, agreed, noting, "Even going back to the stone age of email, I was always really careful not to put too much in. . . . It's the place to be somewhat civil, somewhat professional, and I think I still am old enough to look at the internet a little bit like that; it's not my living room, I can't disagree with everybody; it's more public." While he was not comfortable arguing online, he was happy to do it in person with those close to him: "I rant at certain members of my family," a statement his wife confirmed. He continued, "I love to rant. . . . Ninety percent of the time I'm pretty mellow and calm and 10 percent of the time I just go crazy." But, significantly, he did so offline, not online.[145]

Like Chuck, those who did not express anger online often still felt the emotion; they just believed it was imprudent and perhaps unkind to express it publicly. D. M., a forty-year-old college counselor who had recently returned to Utah after many years in Washington, D.C., told us about her income anxieties and her worries about the level of inequality in the United States:

> I'm angry about politics, I'm angry that it's . . . gotten so bad. . . . I feel frustrated for me 'cause I'm trying to build up a retirement. . . . One day I'd like to own a home and that seems so unattainable these days. . . . It makes me upset; it makes me want to bawl; it makes me want to scream. . . . I'm trying to find the most healthy way to deal with that. . . . Big companies . . . it's always just a race to the bottom . . . they're always finding ways to pay their employees less and less.

In spite of this hurt, and this perception that she was part of an underclass that was being exploited, D. M. had reservations about expressing such sentiments online, because the digital world, in her opinion, did not foster empathy. Instead, she merely "lurked" and observed others:

> I feel like online . . . because you're not face-to-face with someone, you can't see the hurt in their face when you say something really terrible to them, and people just are out to . . . make the best joke and try to get a comment to go viral. . . . Some of them are super funny but some of them are just mean. . . . I don't want to be a part of that. . . . I try and keep my online presence to a minimum. . . . I only lurk. . . . I lurk and I read.

D. M. was looking for a "healthy way" to deal with her anger. She looked for places where it might be contained and rendered harmless. Sometimes she would get mad when she was driving and yell in her car, but she often felt guilty about it afterward. The solution she thought worked best was to write her feelings down on paper—she was keeping an "anger journal" in order to avoid becoming angry in public. She explained, "When I get really upset I go and I write in a journal, I call it my anger journal . . . and I just write out all this stuff that's making me really, really upset and then I burn it. . . . I burn the pages. . . . It helps me vent; it helps me get it out; and then I'm not hurting people's feelings, because I would feel terrible if I hurt somebody's feelings . . . and I don't want to be an online troll."[146]

No one else we interviewed kept an "anger journal" or seemed to go to the formal lengths that D. M. did to keep a lid on her anger. But we certainly met others who, though they might feel angry, worked to contain the emotion. Eddie Baxter told us, "I'm half black, half white and we have a lot of the issues going on with police shootings." Online, some of his white relatives were dismissive of groups like Black Lives Matter, with whom Eddie was affiliated, and on occasion they made racially insensitive comments. Sometimes he unfriended them rather than engage. On other occasions, when he encountered posts that made him angry, he was deliberate in how he responded: "I take my time. I try to make sure that my facts are straight. . . . I have my wife proofread it multiple times." Even after editing and proofreading, he still often held back: "Sometimes I'll have a whole post ready to go and I'll just save and leave it and wake up the next morning to decide whether or not I want to post it—only when it's on controversial stuff." He exercised such care, he explained, because "what we're posting online . . . a lot more people are seeing that. . . . A lot of folks are seeing what you're posting, so I want to stand up for the right thing. I want to be perceived in a positive light."[147] Will, the Grinnell student from Delaware, shared this perspective. He confided that he expressed strong, sometimes provocative, opinions online, "I post stuff that are like pretty charged politically sometimes one way or another," but he tried to do so with care. "Just be a decent human being on Facebook that's my motto."[148]

Some might be provoked by what they saw online but chose not to respond whatsoever. Eulogio Alejandre, a school principal, avoided online fights, for he worried that the internet might divide communities and families. "Somebody puts down that 'I agree with the President; Mexicans are rapists,' . . . and they're my friends from high school. . . . If I allow myself to get into the conversation, the division is going to be wider, and not only with me and that person, but his friends and my friends are now creating tribes. . . . Whereas if I don't answer, . . . I [can] still say 'Hi, how are you?,' we [can still] see each other at the high school reunions." In this manner, Eulogio worked to preserve comity.[149] Barry took a similar approach: "I studiously avoid it [fights online] because it looks so unproductive. So needlessly hostile. . . . It both surprises and disappoints in how quickly so many conversations degenerate into unproductive, hurtful bile. It's largely why I avoid participating in social media. It just seems like a fight club."[150]

People not only worked to limit their anger but also endeavored to avoid being labeled as angry, for many saw it as a pejorative term. This was especially true for Jeff Hellrung, the truck driver and retired Air Force mechanic, who had always voted Republican and had voted for Trump in the 2016 election. Although many of our interviewees explained Trump's win as an expression of right-wing anger, Jeff objected to that: "I don't like that term 'angry' . . . OK? . . . The liberals and the mainstream media . . . want to use that as an excuse on why Trump got elected . . . and I don't think that's why he got elected. . . . The reason Trump got elected and a lot of people voted for him is because the alternative was worse. Now why do you have to be angry to make that choice?" To punctuate this point, Jeff noted, "There's many reasons why I and many Americans voted for Trump in the general election. Anger wasn't necessarily one of them. I vote Republican because my views are conservative, not because I'm angry."[151]

Jeff's views reflect larger Republican sentiments that the charge of anger was an effort to delegitimate right-wing politics, to make them seem irrational. For example, D. C. McAllister, a writer for the conservative publication *The Federalist,* objects to being called angry by the Left. She maintains that her position and that of her compatriots is a moral stance: "Voting for someone . . . is a moral statement. . . . [T]o rip that from people and to say that they are just peasants and they're angry and they have no kind of authority is to denigrate people and this is how we got Donald Trump in the first place."[152]

People do not want to be labeled as angry because the emotion is freighted with negative connotations, regarded as a sign of immaturity, of irrationality. As Susan Theiling put it, "When I am sending an email at work or putting together a Facebook post. I swear to god, I go over it three or four times. . . . Is my punctuation correct? . . . We all have a responsibility to some sort of a civil level of communication." She saw angry tweets, such as those of the president, as symptomatic of "someone who is perpetually stuck at thirteen years old; I mean he is a seventh-grade boy."[153]

Greg also viewed anger in a negative light—albeit from a different perspective. For Greg, anger was a "secondary emotion," best avoided. As Greg put it, "You look into the psychology of emotion and anger is a secondary emotion. It is usually representative of other emotions like disappointment or fear or things like that, but we are unable to express them adequately and unable to deal with them, and so the way it manifests itself is this frustrated anger anxiety." During our conversation, we raised the possibility that anger might be virtuous, that it might actually be triggered by the realization that there are injustices in the world, and that it was an emotional mode for redressing those injustices. Greg, however, resisted this way of talking about anger. To him, angry people were those who had not examined their feelings closely enough:

> I actually don't think anger is a realization. I think anger is an "I haven't realized yet what I'm feeling." I feel like anger is an outpouring of primary emotions like disappointment or shame. . . . I feel like anger is sort of a coping mechanism while we try to figure out what we are actually feeling. . . . Sometimes our inclination [is] to go online and be angry. . . . This isn't a thinking brain that is trying to sort through things. This is an emotional brain that's saying I don't know what to do. I recognize pain both in myself and other people. I don't know what to do with these emotions. And so I'm going to throw it out there—in a really raw format.[154]

From Greg's perspective, anger was a suspect emotion, and if people indulged in it online, they probably were not handling or managing their feelings very wisely.

There were also those who contained their anger, not just because it was a mark of immaturity or a sign of not knowing themselves, but because it could have damaging effects on their futures. Phil had worked on the Bernie Sanders campaign. He was very concerned about economic inequality and felt frustrated that this issue was too often eclipsed by what he termed "social politics." However, he felt that he was not "allowed" to be angry because of the identity politics of the progressive Left with whom he was allied:

> You're talking about identity politics, . . . you're talking about gay rights. . . . You're talking about trans rights and dealing with minorities. . . . I'm part of the camp that says we have very real economic issues that affect everyone . . . everyone has anger but when you're dealing with the progressive Left, you're dealing

> with people who are also kind of disenfranchising people from their anger, because they're talking about their anger as if it's something only they're allowed to have. . . . You can't have the anger because you're life's so much more fortunate . . . because you're white, because you're male, because you're cis. . . . There is a backlash against people saying "I am angry; I have a right to be angry; why can't I be angry?"

Phil had thought about how best to express his frustration, and had decided he could not do so online. He worried a great deal about showing the feeling, fearful that he would offend people and then be attacked online. When we asked him whether he felt constrained in his ability to express anger, he replied emphatically: "Absolutely. . . . I can't express anger." And when we asked him whether the relative anonymity of the internet might loosen some of these constraints, he demurred. The problem was that the internet, in contrast to Barlow's fevered hopes that it would be a place apart, was not at a remove from the physical world. For Phil, the distinctions between the physical and the virtual were so slight that it made little difference to think that the tyrannies he was feeling offline would somehow not also be present online: "If you want to talk about the curated self on social media you have to talk about the curated self in real life, because social media is the same as real life . . . especially . . . Facebook, which is my primary . . . social media. . . . My name is attached to that and any time your name is attached to something you're not talking about [an] anonymous online entity; you're talking about you as a person. . . . It's an expression of your self and subject to the same constraints."

Phil was corroborating a concern shared by many others who, at least since the revelations of Edward Snowden, worry about increased online and offline surveillance. For these people, many parts of the internet are not that anonymous. The surveillance apparatuses of states and corporations are growing in power, effectively extending the reach and tyrannies of the physical workplace to the virtual world as well. This reach curbs people's emotional expression and makes them conform to the emotional style approved by employers. This was certainly true of Phil:

> If you express anger, you have to do something that I don't want to do, which is put your name on it; that's why we're having this conversation without my last name. If you want to talk about something, you better be ready to fight for it, and if you're ready to fight for it you have to be worried about offending someone, 'cause if you offend . . . it ends very badly for you. Tell me of a job where you can't get fired because of workplace discrimination or because you said something wrong. If you take a stand and say we're all equal or something, if you take a stand and say that I don't care about your particular social issue or I don't think it's important, well then you're in danger of alienating just about everyone in the intelligentsia—which happens to be the group I'd like to belong to.[155]

Nurturing aspirations for the future and fearing consequences, he preferred to swallow his anger rather than express it publicly.

For many like Phil, the internet was a place where one needed to build a personal brand and also a place where that brand was being surveilled. To indulge in anger in these settings might not have the same repercussions as indulging in it around the company water cooler, but it was similar. The emotional norms might be more attenuated, yet they still applied, unless one wanted to venture on to some dark niche of the Web that is not yet monitored. As Colin noted, online there is often "a lack of fear of repercussion. If you said some of that stuff to someone face-to-face, you might walk away with a bloody nose. And if you say it online, you are anonymous. Nobody can ever track you down." However, such a sense of freedom was illusory, as Colin observed, "Every now and then you see that there are repercussions. Somebody will post something just really stupid, racist, or sexist or something like that, and then you find out they were doing it on company time, the company's Twitter page, or maybe they were elected president, oh wait, that doesn't happen. But no, and then the next thing you know they are out of a job."[156]

In the interest of preserving a reputation and developing a personal brand, of selling oneself online, many aspiring Americans edit out their anger from their online personae. While advocates of self-branding often make gestures to the ideal of being one's authentic self, in practice their marketing tips counsel the projection of a more edited self, so as to increase one's personal allure and the sparkle of one's brand to potential consumers or employers. MSNBC anchor Chris Matthews, or Eminem, or Donald Trump might find their brand burnished when colored by anger, but for the average professional creating a persona on LinkedIn, or for the entrepreneur developing a start-up, the twentieth-century strictures on anger still apply. More importantly those strictures have been extended by the growth of the internet. It is no sphere of emotional liberty, no free psychic space for people trying to hang on to or enhance their professional reputations.

It might be tempting to believe that the imperatives to manage one's heart have been lifted. But the reach of the modern economy has in fact been extended online and offline. To prevail in it, often the passions must still be contained, even when one is physically far from the company water cooler.

The Fluid Self: From Limitedness to Limitlessness and Back Again

Although these distinct patterns of anger were visible online, some people moved between these emotional styles, at times repressing the feeling and other times, and in other venues, expressing it.[157] Uma Thurman, as she took part in the #MeToo movement, illustrates how one can deploy one style and then another. In October 2017, a week or two after the Harvey Weinstein scandal reached a crescendo, a reporter asked Thurman how she felt about it. She replied with restraint: "I don't have a tidy soundbite for you . . . because . . . I've learned that when I have spoken in anger, I usually regret the way that I have expressed myself. So I've been waiting to feel less angry. And when I'm ready, I'll say what I have to say."[158] But then, a month later, she posted an image of herself on Instagram with the following caption:

> H A P P Y T H A N K S G I V I N G I am grateful today, to be alive, for all those I love, and for all those who have the courage to stand up for others. I said I was angry recently, and I have a few reasons, #metoo, in case you couldn't tell by the look on my face. I feel it's important to take your time, be fair, be exact, so . . . Happy Thanksgiving Everyone! (Except you Harvey, and all your wicked conspirators—I'm glad it's going slowly—you don't deserve a bullet)—stay tuned Uma Thurman.[159]

The categories that separate the reserved, from the ones who try to express their anger productively, to the more vitriolic and impulsive, are fluid. For Americans, there is no settled consensus on where and when to express anger. To one person, anger can be a display of righteous indignation at a perceived injustice, as many who use the hashtag #MeToo have professed. For those who are less understanding, it can be described as a manifestation of what Matt Damon called "outrage culture."[160] But often times that consensus does not even reside easily within the self. These vacillations are on display in Thurman's responses and they were also manifest when we talked with E., who worked as a mental health counselor in Utah. On the one hand, she acknowledged that anger had positive social potential and that it was a virtue: "I don't see anger as a negative. Anger to me comes up when . . . something needs to shift . . . something needs to be different. . . . That's the only reason other people get loud, like I'm not being heard. . . . It can be an act of engagement."

Shortly thereafter, she qualified this, explaining that she was not sure the anger she was feeling was productive or leading to social change, because the political climate seemed so polarized and because no one seemed to be listening:

> I'm really scared about where mine's at right now. . . . I get angry and that kind of motivates me to engage . . . so corrective things

> can be done. . . . [But now] there's no uptake. . . . I've had this experience over and over where it's not going to be taken up and used to influence next steps. . . . It doesn't feel like anger anymore. It just feels like numbness[,] . . . resignation, which feels worse. . . . It has moved to resignation . . . because the people [politicians] I need to connect with aren't willing to connect.[161]

E. regarded anger as many modern Americans do, as something that had multivalent potential. It could effect good or bad ends and it was a constant challenge to determine when it was something that should be expressed or repressed. As Erica put it, "Understanding where your anger is appropriate and when it's inappropriate" was a "skill of adulthood."[162] It certainly was part of the modern emotional style so many had been raised on.

Often, therefore, people were selective about when and to whom they expressed their anger. Many said they moderated their tone when they had face-to-face relationships with those with whom they were conversing online. Katie noted, for instance, that the tone on Nextdoor.com, the social media site that connects people in the same neighborhood, was more tempered than many other places, perhaps because users realized they actually had to live with one another offline. Maintaining good relations counted for more, and if one failed to do so, there could be real consequences to what one said. As a result, the tenor was more restrained. "I just think it's the personal connection. They know they are going to see the other person at some point."[163]

Such attention to consequences often served to limit anger. For example, Senator Dabakis hoped that when anger was channeled by people with the right intentions, it could be directed toward good outcomes. At the same time he worried about consequences. As he explained, "I don't know where you go the day after anger. . . . It felt good for a moment, but where do we go now? . . . Look, play this through. You [are angry] . . . on Friday. What happens Monday?" The consideration of consequences placed limits on the amount of anger he was willing to indulge in, but for people unworried about repercussions, such limits did not exist.[164]

As a result, for those who felt unbounded, unconstrained, and unconcerned about consequences, anger was an emotion they could indulge with little reservation. Susan noted that people often felt liberated to post whatever they wanted: "It seems like it has allowed people to feel like there's a freedom—that I can throw it out there and nobody's going to know or I can just do whatever I want."[165] Gregory Noel lamented the fact that, online, people "get a chance to just say whatever they want without the consequences of having to face that individual." Such freedom made them more "brazen" online.[166] In this narrow sense, the expression of anger goes hand in hand with the feeling of limitlessness, giving people the ability to feel powerful and influential, if only momentarily. Jennifer, for instance, speculated that online trolls' comments were a way for them to "feel like they're as good as the people that they feel . . . are tearing them down . . . or economic opportunities they've lost. . . . For some people it's out of frustration."[167]

While anger might give one a momentary sense of unbounded, limitless power, it often sprang from a sense of limits and sometimes reinforced that sense. Being angry was often an attempt to overcome constraints, limits, the lack of power born of inequity. Greg Bayles had noticed this, observing that "some of the anger that we feel is that same realization that we are a lot smaller than we thought we were and there are a lot bigger powers influencing us than we previously supposed."[168] Psychologist Leon Seltzer notes, "I'm convinced that anger is employed universally to bolster a diminished sense of personal power."[169] Yet when one expresses anger online, one's sense of self is not bolstered for long. A 2013 study found that while ranting on the internet initially gave people a sense of relief, it did not persist; posting angry rants might make one feel even angrier in the long term.[170] Angry outbursts, born of a desire to feel powerful and a belief that online there were not limits to free expression, could also lead to feelings of frustration and inadequacy.

Many Americans, Left and Right, today feel this oppression, powerlessness, and anger. Raised on the American ideology of self-reliance, which suggested one's fate was under one's own control, encouraged by technology companies to believe they had new powers, and taught by psychologists to believe their emotional life was their own to shape, they found themselves nevertheless limited, constrained, oppressed. On the one hand, they had hopes of infinite freedom and reach; on the other hand, they had strong evidence of the many constraints that shaped and limited their identities. The anger so visible online today speaks to the fact that the awe and beneficial effects that technology promised—and that technology evangelists like Barlow also promoted—have either not actually been realized or have been unevenly distributed. Technology's power to provide uplift is, unfortunately, not limitless, and anger is, at least in part, a realization of these limits.

That this anger today seems so raw and ubiquitous, that its appearance surprised so many, points to an important transformation, for the feeling, suppressed for a century by psychologists and business leaders, has come roaring back. Few would have been surprised by its presence in public life in earlier centuries. In the nineteenth century, many believed that anger had great social utility, and might be used to fight sin and injustice. Nineteenth-century Americans recognized the power of indignation, though they did not believe all were entitled to express it. White men in particular found their anger affirmed, for it was a part of masculine identity, a way of asserting oneself, one's commitments, and one's causes. They often proudly and publicly displayed their angry passions in parades, protests, moral reform movements, street fights, and newspapers, while women and African Americans had to contain their discontents. In the twentieth century, as corporate capitalism came to structure work life and market relations, men and women, whites and blacks alike, found little space to express their anger in public (though their bosses still might). Instead, they used the technologies of leisure—cars, radios, televisions, phones—to express the feeling, but also to contain it. In the private spaces of the home or automobile, men and women could shout and rant, giving voice to the feelings they might suppress at work. While anger persisted in their private lives, it was far less visible in public, and many believed that it was possible to vanquish the feeling entirely.[171]

This illusion, that anger could be eliminated, was shattered with the birth of online media, for Americans have found new opportunities to display publicly the ire they once kept private. Some have used these opportunities to enunciate cries for long-overdue reforms, for an end to injustice, racism, sexism, and inequality. Such online movements recall the indignation meetings of the nineteenth century, which, in their struggles against slavery, helped bring about positive change.

Yet contemporary technology does more than just provide a new venue for long-standing anger. In some cases, it has given it new shape. While indignation movements of the nineteenth century joined protesters together physically, allowing them to meet, to establish trust, and to organize, American anger today is often expressed in more isolated spaces—projected into the public sphere, but via a phone or a computer in one's own house. Physically isolated from one another, many feel at once empowered, with few restraints on their words, and also powerless, because they are ultimately sitting on their own, apart from others. Anger, in recent years, has become more absent from face-to-face interactions, but has gained an unprecedented foothold in the digital world. Expressed in isolation, on Twitter or Facebook, anger may be more apparent, yet less powerful as an agent of change than it was in earlier years.

Conclusion

We live in Ogden, Utah, a town that sits on the side of the Wasatch Mountains, overlooking the so-called Silicon Slopes. Every morning we take advantage of that slope and look out over a developing metropolis and, beyond that, to the Great Salt Lake. When we began this routine in 2002, we listened to the radio and took in the view. Then we started checking on the news—in the beginning in the paper, then, as the years passed, more frequently on a phone. We thought we could take it all in: the view, the news, the radio, our conversation. Our attention seemed limitless. But at some point, we realized we were forgetting to look out the window and not quite catching all the news, or what one of us was saying to the other.

We were drawn to all the internet had to offer, but found that sometimes it offered too much, or not the right thing. It was changing our emotions, our expectations, our behaviors. Our metamorphosis is not unique. Like most Americans, we have experienced the new feelings of the silicon-driven economy—the dopamine fix when we first look at our phones in the morning, the hope that our posts might go viral, the sociability of digital connections, the incessant distractions and siren call of our screens, the freedom to vent, and often the regret that followed, the nagging sense that we were missing out on something by looking down at our phones rather than up at the world before us.

This book is an attempt to make sense of those new feelings. Married to each other, and to academic traditions that encourage us to scrutinize and historicize daily life, we began to turn our attention to the world we saw about us, and how it was being transformed—and how we were too. Often we worried that we were overintellectualizing our daily habits and emotions. And perhaps, on some occasions, we were. At the same time, as we worked on this book, we discovered that our research was helping us to better navigate our feelings and make more informed choices about when to be narcissistic, lonely, bored, inattentive, awed, and angry. Studying this transformation has provided a compass that enhances our ability to decide when to see life as limited or limitless. Perhaps its most valuable benefit has been that it has revealed how our individual feelings are shaped by the American experience, past and present, and reminded us that our emotions are not natural, inevitable, or fixed; they change and are changeable.

In some ways, the new American emotional style we have been experiencing and observing is a sign of emotional liberation and democracy. For instance, anger, a feeling that seventeenth-century colonists believed to be the sole purview of God, eighteenth-century Americans saw as the prerogative of social elites, and nineteenth-century observers believed was legitimately expressed only by white men, today has become the right of all. Similarly, the ability to present and celebrate oneself to the world through words and pictures, long the privilege of just the rich who could read, write, and sit for portraits, has become available to everyone who uses the internet. Self-presentation, self-adoration, and aggrandizement are now rights we all share. The problem of how to fill empty moments, a problem born of too much leisure, is a condition earlier generations dreamed of. Today, however, empty moments and the emotional dilemma of how to fill them have become widespread in the United States.

The idea that sociability is in the power of all, and that human connection is a right, also repudiates older notions that loneliness was inevitable. Then too the belief that men and women of all races are suited for intellectual work and not just stultifying drudgery is another sign of this change, as is the belief that all of us have still-untapped mental powers if we can only figure out how to access them. Finally, our sense of awe at our own powers speaks to a rising sense of our own significance and a loss of awe at the powers of the universe. In the documentary *The Singularity,* Ray Kurzweil gives voice to this new sense of limitless possibility, seemingly now the right of all in the high-tech world.

> This is what it is to be human. To go beyond our limitations. We didn't stay on the ground. We didn't stay on the planet. We didn't stay with the limitations or our biology. Human life expectancy was twenty-six, ten thousand years ago. We are the species that goes beyond its limitations and we're going to continue to do this in an exponential fashion so we will become much more intelligent. We will be able to expand our horizons . . . beyond the earth ultimately. . . . Even if we change the nature of our biology, we will still be human.[1]

In Kurzweil's view, transcending limits is the new human condition.

These new emotional patterns—from our right to unfettered anger and unconstrained self-promotion to our awe at our own transcendence—have helped create in many a sense of an unbounded self, unconstrained by tradition or by mortality. And this emerging, seemingly limitless emotional style of the twenty-first century is in many ways far more egalitarian than those that prevailed in earlier years.

That is on the plus side. However, we also see that the technologies that have helped facilitate this change have led us to sometimes forget that we are flesh and blood and that those we correspond with, tweet to and about, are as well.

Earlier generations of Americans did not find themselves cast upon the stage as often as we are. Their sense of self was smaller and that self was not reflected back at them as often as it is now. The compulsion to see oneself through the eyes of others was not triggered by technology as much as it is today. Likewise, Americans in the eighteenth, nineteenth, and early twentieth centuries experienced loneliness differently than we do. Many saw it as a condition that they had no choice but to endure. Moralists of the time, who celebrated the virtues of solitude, provided frameworks for regarding aloneness in a positive light. Those messages, of course, have not disappeared entirely from

American discourse. However, today we experience loneliness differently than our ancestors did. In the twenty-first century, ministers and philosophers have been increasingly displaced by psychologists. Instead of cultivating and celebrating solitariness, psychologists encourage us to regard it as an affliction that should be avoided, portraying it as a deadly epidemic with real physical consequences. Technology companies join the chorus of advice, telling consumers that this frightening condition can be cured with a simple swipe or click.

Earlier generations also experienced drudgery differently than do modern Americans. Because the word "boredom" only gradually came into use in the mid-nineteenth century, Americans categorized, evaluated, and experienced tedium differently than they do today. To be sure, past generations were aware of dull moments and dull days, and they perceived that time sometimes passed slowly, that certain activities engaged their attention more fully than others. However, they did not react to these conditions with as much distaste and revulsion as we do today. After all, they did not have smartphones at their disposal to banish every dreary interlude. Today, Americans regard constant stimulation as something that borders on an entitlement, as essential to mental health.

Complementing this shift is a related story about focus and distraction. If eighteenth- and nineteenth-century Americans did not feel entitled to constant entertainment, they also had humbler expectations about their ability to concentrate. Today, the phrase "pay attention" enjoys much more currency than it did a century ago, and that change is emblematic of a shift in how mental focus is regarded. Nineteenth-century Americans envisioned the mind as limited, and they worried about the dangers of overtaxing their brains, for they believed they possessed finite cognitive capacities. In today's world, however, many of us imagine that our powers to pay attention can be amplified by taking Adderall or caffeine, and we believe that we can pay attention to more things simply by opening more tabs in our browsers. The deep irony is that today, while we consider our intellects to be boundless, attention is an increasingly scarce and valuable resource. Companies like Google and Facebook have built their businesses on their ability to capture and sell our interest and time; to them, both the value of attention and its scarcity are self-evident. However, for many who consume their services, these market realities have not yet become fully apparent. We continue to squander our attention, to give it away for a few "free" services, as if it were bounteous and unlimited.

Some hoped that the new technologies and the new social media platforms might liberate Americans' emotional lives, might allow them to give voice to their sentiments, including their discontents. And when they express their anger online, many today do feel, at least fleetingly, newly powerful, liberated from traditional rules governing American emotional life. Nineteenth-century Americans believed they should curb their anger in some circumstances but also acknowledged that the feeling had the potential to bring about change. Anger therefore had a legitimate role in public discourse, at least for some. In the twentieth century, the feeling was banished from the workplace, as fears of labor unrest grew, and as the service economy came to demand cheerful employees. Consequently, many Americans learned to express the feeling only in private, as they sat in their houses and talked back to the TV or the radio, or as they drove in traffic and yelled at other drivers, safe behind rolled-up windows. With the rise of online media, however, millions of Americans have found a new space to express anger, and consequently, the emotion in the twenty-first century is in many ways far more visible than it was in the twentieth, though not always expressed in particularly productive ways.

Taken together, these intertangled histories depict a change in how many Americans see themselves. Rather than being awed by the power of the universe and regarding themselves as limited, as fated to experience loneliness, boredom, and powerlessness, many have come to believe there need be no limits on their experience, no inherent unhappiness. This new vision is at once filled with hope and fraught with disappointment. Some of the hopes may be partially realized; we may end up knowing more people than we once did, thanks to social media, but we will not make aloneness disappear. Now that rules on vanity have been relaxed, we may be able to present ourselves more perfectly and freely to others; however, such new freedom to self-promote will not necessarily make us popular, and when we offer burnished self-portraits, we will not be able to present ourselves in all the rich detail that truly makes us human. We may develop a grander sense of self but overlook the grandeur of the larger world. We may find ever-changing news and games online as merchants vie for our attention and offer cures for boredom, but we will never be able to apprehend it all, no matter how fast we read or how many links we click. Before our eyes, new rules are appearing about these emotions, and new expectations, born of new, seemingly unbounded opportunities. But every so often, even with the latest iPhone, we bump up against the reality that we are finite and imperfect.

The environmentalist and author Bill McKibben explains the appeal, as well as the downside, of such a limitless self.

> Clearly there's a part of us that's all about more. And that's the part that finds all this appealing. There's nobody that looks at the prospect of immortality, superhuman powers [without thinking] "Oh, that would be kind of cool." . . . [But] the much more profound part of the human being is the part of us that is able to look at it a second time and say "No, that's not where I want to go. No, I don't want to be King Midas who everything I touch turns to gold. Sounds good for the first forty-five minutes, and after that it kind of sucks. No, I don't need to be Prometheus. No, I don't need to be Superman. In fact, I'm most human when I'm not."[2]

McKibben reminds us that our increasingly powerful tools will not inevitably lead us toward utopia and transcendence. We cannot expect that technology will lead us to the good life on its own accord. Only our own wisdom and good judgment—and a self-consciousness about the emotional and social implications of our devices—can take us there. To this same concern, Langdon Winner, in *The Whale and the Reactor*, observed, "In an age in which the inexhaustible power of scientific technology makes all things possible, it remains to be seen where we will draw the line, where we will be able to say, here are possibilities that wisdom suggest we avoid. I am convinced that any philosophy of technology worth its salt must eventually ask, How can we limit modern technology to match our best sense of who we are and the kind of world we would like to build?"[3] Winner's observation reminds us that we, as humans, are not just distinguished by our tools, we are distinguished by our capacity and responsibility to design and choose how to use them.

It is not just an issue of designing and choosing our tools carefully, but also of thinking about our feelings. The history of emotions reminds us that our feelings are not inherent or natural, and do not belong to us alone, but instead reflect widely shared mores and conventions. This book illustrates that those mores do change and that those changes affect our emotional lives. Contemporary Americans often try to reshape their feelings on their own—through self-help. But lasting change is more difficult. Such transformations happen on a societal level and also must occur collectively, for our feelings are not just private and personal, but social and political. They have implications for the world we live in and the world we want to live in.

To that end, as we envision new emotional norms or recover old ones, we should also consider which technologies help us realize them. Those norms should acknowledge the importance of solitude and the dangers of selfishness, and should foster emotional habits that connect us to others more profoundly. We should be deliberate as we design and choose our tools, making sure that they forward rather than frustrate these goals.

Taking such injunctions to heart as we look out over the Silicon Slopes, we realize that we are living literally on a boundary. And that boundary is not just the mountains with wooded pines on one side of us and the booming, digitally driven metropolis on the other side. It is also that our feelings are being shaped by carbon and by silicon—by our embodied carbon selves and by the selves that have been captured in silicon. A future self, still part carbon, but perhaps more silicon, beckons. It is incumbent on us to define its limits and to recognize that we might be most human when those limits are acknowledged.

NOTES

ACKNOWLEDGMENTS

INDEX

Notes

INTRODUCTION

1. Eli Gay, Berkeley, California, January 31, 2016, interviewed by Luke Fernandez and Susan Matt.

2. Nicholas Carr, "Is Google Making Us Stupid?," *Atlantic,* July / August 2008, http://www.theatlantic.com/magazine/archive/2008/07/is-google-making-us-stupid/306868/; Stephen Marche, "Is Facebook Making Us Lonely?," *Atlantic,* May 2012, http://www.theatlantic.com/magazine/archive/2012/05/is-facebook-making-us-lonely/308930/; "Is Social Media to Blame for the Rise in Narcissism?," *Psychology Today,* http://www.psychologytoday.com/blog/compassion-matters/201211/is-social-media-blame-the-rise-in-narcissism, all accessed June 9, 2018.

3. Caroline Clark Richards, April 5, 1857, in Caroline Cowles Richards, *Village Life in America, 1852–1872, Including the Period of the American Civil War as Told in the Diary of a School-Girl* (New York: Henry Holt, 1913), 80.

4. John Perry Barlow, "A Declaration of the Independence of Cyberspace," February 8, 1996, Electronic Frontier Foundation, https://www.eff.org/cyberspace-independence, accessed January 13, 2018.

5. Franklin Foer, *World without Mind: The Existential Threat of Big Tech* (New York: Penguin, 2017), 1, 45–49, 51–52.

6. Noam Cohen, "Facebook Doesn't Like What It Sees When It Looks in the Mirror," *New York Times,* January 16, 2018, sec. Opinion, https://www.nytimes.com/2018/01/16/opinion/facebook-zuckerberg-public-content.html; Mark Zuckerberg, "A Letter to Our Daughter," December 1, 2015, https://www.facebook.com/notes/mark-zuckerberg/a-letter-to-our-daughter/10153375081581634/.

7. Daniel Horowitz, *Happier? The History of a Cultural Movement That Aspired to Transform America* (New York: Oxford University Press, 2018), 5.

8. Noteworthy among those who discuss such an idea are David Brooks, whose work we discuss in Chapter 1, and Dacher Keltner, Paul Piff, and colleagues, whose works we discuss in Chapter 5. See, for instance, Brooks, *The Road to Character* (New York: Random House, 2015); Michelle Shiota, Dacher Keltner, and John Oliver, "Positive Emotion Dispositions Differentially Associated with Big Five Personality and Attachment Style," *Journal of Positive Psychology* 1, no. 2 (April 2006): 61–71; Paul K. Piff et al., "Awe, the Small Self, and Prosocial Behavior," *Journal of Personality and Social Psychology* 108, no. 6 (2015): 883–99, https://doi.org/10.1037/pspi0000018.

9. Erik Brynjolfsson and Andrew McAfee, *The Second Machine Age: Work, Progress, and Prosperity in a Time of Brilliant Technologies* (New York: W. W. Norton, 2016); Thomas L. Friedman, *The World Is Flat: A Brief History of the Twenty-First Century* (New York: Farrar, Straus and Giroux, 2005); Kevin Kelly, *What Technology Wants* (New York: Viking, 2010).

10. Jonathan Taplin, *Move Fast and Break Things: How Facebook, Google, and Amazon Cornered Culture and Undermined Democracy* (Boston: Little, Brown, 2017); Ronald J. Deibert, *Black Code: Surveillance, Privacy, and the Dark Side of the Internet* (Toronto: Signal, 2013); Laura Poitras, *Citizenfour* (2014), documentary, http://www.imdb.com/title/tt4044364/; Douglas Rushkoff, *Throwing Rocks at the Google Bus: How Growth Became the Enemy of Prosperity* (New York: Portfolio, 2017); Jaron Lanier, *Who Owns the Future?* (New York: Simon and Schuster, 2014).

11. Vauhini Vara, "How Utah Became the Next Silicon Valley," *New Yorker,* February 3, 2015, https://www.newyorker.com/business/currency/utah-became-next-silicon-valley; Barry Leiner et al., "Brief History of the Internet," *Internet Society* (blog), https://www.internetsociety.org/internet/history-internet/brief-history-internet/, accessed May 5, 2018; Lee Davidson, "Urbanites: Nine of 10 Utahns Live on 1 Percent of State's Land," *Salt Lake Tribune,* March 27, 2012, http://archive.sltrib.com/article.php?id=53794385&itype=cmsid.

12. "Declining Majority of Online Adults Say the Internet Has Been Good for Society," Pew Research Center, Internet and Technology, April 30, 2018, http://www.pewinternet.org/2018/04/30/declining-majority-of-online-adults-say-the-internet-has-been-good-for-society/, accessed May 20, 2018.

13. According to a recent Pew study, 75 percent of African Americans, 77 percent of Hispanics, and 77 percent of whites own them. "Mobile Fact Sheet," February 5, 2018, Pew Research Center, Internet and Technology, http://www.pewinternet.org/fact-sheet/mobile/, accessed May 14, 2018.

14. Eighty-two percent of those earning between $30,000 and $49,000 have one; also 83 percent of those earning $50,000 and $74,999, and 93 percent of those with incomes above $75,000 own them; see "Mobile Fact Sheet."

15. "Internet / Broadband Fact Sheet," February 5, 2018, Pew Research Center, Internet and Technology, http://www.pewinternet.org/fact-sheet/internet-broadband/; "Mobile Fact Sheet."

16. "Declining Majority of Online Adults Say the Internet Has Been Good for Society."

17. Class identity varies, depending on how it is measured. For instance, 70 percent of the nation claims membership in the middle class, though only around 49 percent meet the technical income definition. "Are You in the American Middle Class?," Pew Research Center, May 11, 2016, http://www.pewresearch.org/fact-tank/2016/05/11/are-you-in-the-american-middle-class/, accessed May 4, 2018; "70% of Americans Consider Themselves Middle Class—but Only 50% Are," CNBC, June 30, 2017, https://www.cnbc.com/2017/06/30/70-percent-of-americans-consider-themselves-middle-class-but-only-50-percent-are.html, accessed May 4, 2018.

18. Julia Glum, "Millennials Selfies: Young Adults Will Take More Than 25,000 Pictures of Themselves during Their Lifetimes: Report," *International Business Times*, September 22, 2015, http://www.ibtimes.com/millennials-selfies-young-adults-will-take-more-25000-pictures-themselves-during-2108417.

19. Alta Martin, Ogden, Utah, October 28, 2015, interviewed by Luke Fernandez and Susan Matt.

20. Camree Larkin, Ogden, Utah, October 30, 2015, interviewed by Luke Fernandez and Susan Matt.

21. Lucy Larcom, Diary, September 12, 1861, in *Lucy Larcom: Life, Letters, and Diary*, ed. Daniel Dulany Addison (Boston: Houghton, Mifflin, 1895), 104. See also Paul C. Helmreich, "Lucy Larcom at Wheaton," *New England Quarterly* 63, no. 1 (March 1990): 109–20.

22. Ben Hur Wilson, "Telegraph Pioneering," *Palimpsest* (November 1925): 381–82; Alvin F. Harlow, *Old Wires and New Waves* (1936; New York: Arno and the *New York Times*, 1971), 137, 140, 141, 157. More episodes are described in Chapters 2 and 5.

23. John Griffin, "Mark Zuckerberg's Big Idea: The 'Next 5 Billion' People," CNNMoney, August 20, 2013, http://money.cnn.com/2013/08/20/technology/social/facebook-zuckerberg-5-billion/index.html. For discussions about the separation anxiety that many feel when away from their phones, see, for instance, Russell B. Clayton, Glenn Leshner, and Anthony Almond, "The Extended iSelf: The Impact of iPhone Separation on Cognition, Emotion, and Physiology," *Journal of Computer-Mediated Communication* (January 8, 2015).

24. William C. Gannett, *Blessed Be Drudgery: And Other Papers* (Glasgow: David Bryce and Sons, 1890), 16. See also Daniel T. Rodgers, *The Work Ethic in Industrial America, 1850–1920* (Chicago: University of Chicago Press, 1979), 14.

25. Adam Kaslikowski, Salt Lake City, Utah, April 13, 2016, interviewed by Luke Fernandez and Susan Matt.

26. Friedrich Nietzsche, *The Gay Science: With a Prelude in Rhymes and an Appendix of Songs* (New York: Knopf Doubleday, 2010), 108; Patricia Spacks, *Boredom: The Literary History of a State of Mind* (Chicago: University of Chicago Press, 1996), 2; Ralph Waldo Emerson, "Self-Reliance," in *The American Intellectual Tradition, Vol. 1, 1630–1865,* ed. David A. Hollinger and Charles Capper (New York: Oxford University Press, 2001), 354–68.

27. Alyson Gausby, "Attention Spans," Microsoft Canada Consumer Insights Report, 2015, Scribd, https://www.scribd.com/document/265348695/Microsoft-Attention-Spans-Research-Report, accessed January 28, 2018.

28. The science fiction writer William Gibson coined this phrase, arguing that "the future is already here—it's just unevenly distributed." See "Broadband Blues," *Economist,* June 21, 2001, http://www.economist.com/node/666610, accessed January 30, 2018.

29. The most prominent technology critic of late to seriously entertain technological determinism is Nicholas Carr. See Nicholas Carr, *The Shallows: What the Internet Is Doing to Our Brains* (New York: W. W. Norton, 2011); and Nicholas Carr, *The Glass Cage: How Our Computers Are Changing Us* (New York: W. W. Norton, 2015). Others of prominent note include Langdon Winner, *The Whale and the Reactor: A Search for Limits in an Age of High Technology* (Chicago: University of Chicago Press, 1989); and Kevin Kelly, *What Technology Wants* (New York: Viking, 2010).

30. danah boyd, *It's Complicated: The Social Life of Networked Teens* (New Haven, CT: Yale University Press, 2014), 212.

31. See G. Stanley Hall, *Adolescence: Its Psychology and Its Relations to Physiology, Anthropology, Sociology, Sex, Crime, Religion, and Education* (New York: D. Appleton, 1904); Kelly Schrum, *Some Wore Bobby Sox: The Emergence of Teenage Girls' Culture, 1920–1945* (New York: Palgrave Macmillan, 2004); John Demos and Virginia Demos, "Adolescence in Historical Perspective," *Journal of Marriage and Family* 31, no. 4 (November 1969): 632–38.

32. For an introduction to the history of emotions, see, for example, the many works of Peter N. Stearns, including *American Cool: Constructing a Twentieth-Century Emotional Style* (New York: New York University Press, 1994); *Shame: A Brief History* (Urbana: University of Illinois Press, 2017); Peter N. Stearns and Carol Z. Stearns, "Emotionology: Clarifying the History of Emotions and

Emotional Standards," *American Historical Review* 90, no. 4 (October 1985): 813–36; and Susan J. Matt and Peter Stearns, *Doing Emotions History* (Urbana: University of Illinois Press, 2013). See also William Reddy, *The Navigation of Feeling: A Framework for the History of Emotions* (Cambridge: Cambridge University Press, 2001); Barbara H. Rosenwein, "Worrying about Emotions in History," *American Historical Review* 107 (June 2002): 821–45; Barbara H. Rosenwein and Riccardo Cristiani, *What Is the History of Emotions?* (Cambridge, UK: Polity Press, 2018); Rob Boddice, *The History of the Emotions* (Manchester, UK: Manchester University Press, 2018); Jan Plamper, *The History of Emotions: An Introduction* (New York: Oxford University Press, 2015); Ute Frevert, *Emotions in History Lost and Found* (Budapest: Central European University Press, 2011); David Konstan, *The Emotions of the Ancient Greeks: Studies in Aristotle and Classical Literature* (Toronto: University of Toronto Press, 2006); Susan J. Matt, "Emotions in American History," *American Historian* (August 2016); and Susan J. Matt, "Current Emotion Research in History: Or, Doing History from the Inside Out," *Emotion Review* 3 (January 2011): 117–24.

33. On basic emotions, see Paul Ekman, "The Universal Smile: Face Muscles Talk Every Language," *Psychology Today* (1975): 35–39; Paul Ekman, "Expression and the Nature of Emotion," in *Approaches to Emotion,* ed. K. R. Scherer and P. Ekman (Hillsdale, NJ: Psychology Press, 1984), 319–43; and Paul Ekman, "The Argument and Evidence about Universals in Facial Expressions of Emotion," in *Handbook of Social Psychophysiology,* ed. H. Wagner and A. Manstead (Chichester, UK: John Wiley and Sons, 1989), 143–64.

34. Lisa Feldman Barrett, "Solving the Emotion Paradox: Categorization and the Experience of Emotion," *Personality and Social Psychology Review* 10, no. 1 (2006): 20–46; Lisa Feldman Barrett, *How Emotions Are Made: The Secret Life of the Brain* (New York: Houghton Mifflin Harcourt, 2017); James R. Averill, "The Future of Social Constructionism: Introduction to a Special Section of *Emotion Review*," *Emotion Review* 4, no. 3 (July 2012): 215–20; Rosenwein and Cristiani, *What Is the History of Emotions?,* 6–25; William M. Reddy, "Humanists and the Experimental Study of Emotion," in *Science and Emotions after 1945: A Transatlantic Perspective*, ed. Frank Beiss and Daniel M. Gross (Chicago: University of Chicago Press, 2014), 41–66.

35. J. A. Russell, "Is There Universal Recognition of Emotion from Facial Expression? A Review of the Cross Cultural Studies," *Psychological Bulletin* (1994): 102–41; J. A. Russell, "Reading Emotions from and into Faces: Resurrecting a Dimensional-Contextual Perspective," in *The Psychology of Facial Expression,* ed. J. A. Russell and J. M. Fernández-Dols (Cambridge: Cambridge University Press, 1998), 295–320; Catherine A. Lutz, *Unnatural Emotions: Everyday Sentiments on a Micronesian Atoll and Their Challenge to Western Theory*

(Chicago: University of Chicago Press, 1988); Anna Wierzbicka, *Emotions across Languages and Cultures: Diversity and Universals* (Cambridge: Cambridge University Press, 1999); Frevert, *Emotions in History.*

36. Thomas Dixon, "'Emotion': The History of a Keyword in Crisis," *Emotion Review* 4 (2012): 338–44.

37. Reddy, "Humanists and the Experimental Study of Emotion," 41–66; Luiz Pessoa, "Beyond Brain Regions: Network Perspective of Cognition-Emotion Interactions," *Behavioral and Brain Sciences* 35, no. 3 (2012): 158–59.

38. Reddy, "Humanists and the Experimental Study of Emotion," 41–66; Reddy, *Navigation of Feeling,* 105.

39. Barbara Rosenwein's work on emotional communities is a useful reminder of the variability within cultures. See Rosenwein, "Worrying about Emotions." For a discussion about the perils and promise of describing the American self, see Peter N. Stearns, "American Selfie: Studying the National Character," *Journal of Social History* 51, no. 3 (February 2018): 500–525, https://doi.org/10.1093/jsh/shx099.

1. FROM VANITY TO NARCISSISM

1. Jean M. Twenge and W. Keith Campbell, *The Narcissism Epidemic: Living in the Age of Entitlement* (New York: Atria, 2010); David Brooks, *The Road to Character* (New York: Random House, 2015); Michael Maccoby, *The Productive Narcissist: The Promise and Peril of Visionary Leadership* (New York: Broadway, 2003). See also Jeffrey Kluger, *The Narcissist Next Door: Understanding the Monster in Your Family, in Your Office, in Your Bed, in Your World* (New York: Riverhead, 2014); Drew Pinsky and S. Mark Young, *The Mirror Effect: How Celebrity Narcissism Is Seducing America* (New York: Harper, 2009).

2. Jesse F. Battan, "The 'New Narcissism' in 20th-Century America: The Shadow and Substance of Social Change," *Journal of Social History* 17, no. 2 (December 1, 1983): 205.

3. Randi Kreger, "An End to Narcissistic Personality Disorder in DSM-5?," *Psychology Today,* http://www.psychologytoday.com/blog/stop-walking-eggshells/201012/end-narcissistic-personality-disorder-in-dsm-5, accessed July 19, 2015; Charles Zanor, "Narcissistic Disorder to Be Eliminated in Diagnostic Manual," *New York Times,* November 29, 2010, http://www.nytimes.com/2010/11/30/health/views/30mind.html. Despite the prediction that it would be dropped, NPD was kept in *DSM* 5. For a comprehensive survey of the ambiguities and disagreements about when narcissism should be used, see Battan, "'New Narcissism' in 20th-Century America." As Battan puts it, "As a

controversial category of clinical pathology as well as a general symbol of selfishness and alienation, the concept of narcissism merely clouds our understanding of . . . complex social and cultural issues, and neither measures nor explains them in an effective way," 211.

4. Elizabeth Lunbeck also suggests that contemporary American narcissism is different from what it once was, because succeeding generations of psychoanalysts and cultural commentators have redefined it. In short, Freud's narcissism is not the same as Christopher Lasch's. One difference she identifies is that the modern American narcissist needs outside affirmation whereas the narcissist that Freud described did not. Elizabeth Lunbeck, *The Americanization of Narcissism* (Cambridge, MA: Harvard University Press, 2014).

5. Lunbeck, *Americanization of Narcissism,* 83, 84.

6. Thomas Dixon, *From Passions to Emotions: The Creation of a Secular Psychological Category* (Cambridge: Cambridge University Press, 2003).

7. "Vanity," *Oxford English Dictionary,* http://www.oed.com/view/Entry/221396?redirectedFrom=vanity#eid.

8. "Vanitas (Latin, 'emptiness', 'Worthlessness'.)," in *The Bloomsbury Guide to Art,* ed. Shearer West (London: Bloomsbury, 1996), http://search.credoreference.com/content/entry/bga/vanitas_latin_emptiness_worthlessness/0.

9. John Bunyan, *The Pilgrim's Progress* (Boston: Ginn and Company, 1917), 71–72.

10. Early Christian theologians initially categorized pride and vanity as separate sins, but eventually came to consider them part of the same moral problem and grouped them together, making vanity a subset of pride. Henry Fairlie, "Pride or Superbia," *New Republic* (November 12, 1977): 19–23.

11. Andrew Jones, *Morbus Satanicus. The Devil's Disease; OR The Sin of Pride Arraigned and Condemned,* 14th ed. (Edinburgh, 1699).

12. Solomon Williams, *The Vanity of Human Life. A Sermon Preach'd at Coventry, Septemb. 17th 1752* (Boston: S. Kneeland, 1754), 5, 11.

13. Timothy Dwight, *Theology, Explained and Defended, in A SERIES OF SERMONS,* vol. IV (Middletown, CT: William Baynes, 1819), 508, 298, 307, 417, 465–66.

14. Delia Locke, April 22, 1879, in Delia Locke Diary, 1875–1879, Locke-Hammond Family Papers, MSS 110, Holt-Atherton Special Collections, University of the Pacific, http://www.pacific.edu/university-libraries/find/holt-atherton-special-collections/delia-locke-diaries.html.

15. *The New Instructive History of Miss Patty Proud, or, The Downfall of Vanity: With the Reward of Good Nature* (Windsor, VT: Farnsworth and Churchill, 1810);

Arnaud Berquin, *The History of Caroline: Or, a Lesson to Cure VANITY* (New Haven, CT: Sidney's Press, 1818), 4, 5, 11–14.

16. Stephen Greenblatt, *Renaissance Self-Fashioning: From More to Shakespeare* (Chicago: University of Chicago Press, 1980), 2–3.

17. John F. Kasson, *Rudeness and Civility: Manners in Nineteenth-Century Urban America* (New York: Hill and Wang, 1990), 6–7, 13, 18.

18. *The Ladies' Vase, or Polite Manual for Young Ladies. Original and Selected. By an American Lady,* 8th ed. (Hartford, CT: H. S. Parsons, 1849), 37, 42.

19. *A Manual of Politeness: Comprising the Principles of Etiquette, and Rules of Behaviour in Genteel Society for Persons of Both Sexes* (Philadelphia: J. B. Lippincott, 1859), 80.

20. M. G. Sundell, "Introduction," in *Twentieth Century Interpretations of Vanity Fair: A Collection of Critical Essays*, ed. M. G. Sundell (Englewood Cliffs, NJ: Prentice-Hall, 1969), 5–7; Catherine Peters, *Thackeray: A Writer's Life* (Phoenix Mill, UK: Sutton, 1999), 144–51.

21. Louisa M. Alcott, *Little Women, Or, Meg, Jo, Beth and Amy* (Boston: Roberts Brothers, 1868), 138, 141.

22. Louis P. Masur, "'Age of the First Person Singular': The Vocabulary of the Self in New England, 1780–1850," *Journal of American Studies* 25, no. 2 (August 1991): 189–211.

23. "Likes and Dislikes," *Ripley (MS) Transcript*, January 11, 1838.

24. "Small Things Are Great to Little Minds," *Baton-Rouge (LA) Gazette,* June 1, 1844.

25. Richard R. John, *Spreading the News: The American Postal System from Franklin to Morse* (Cambridge, MA: Harvard University Press, 1995), 21–22, 157–59.

26. David M. Henkin, *The Postal Age: The Emergence of Modern Communications in Nineteenth-Century America* (Chicago: University of Chicago Press, 2006), 23–25, 1, 2, 33–34. For further discussion of African American letter writing, see Christopher Hager, *Word by Word: Emancipation and the Act of Writing* (Cambridge, MA: Harvard University Press, 2013).

27. *Memoir of Asahel Grant, Missionary to the Nestorians* (New York: M. W. Dodd, 1847), 113. This was a widespread convention. See Henkin, *Postal Age*, 109–18; William Merrill Decker, *Epistolary Practices: Letter Writing in America before Telecommunications* (Chapel Hill: University of North Carolina Press, 1998), 95–96.

28. Clay quoted in Decker, *Epistolary Practices,* 96. For an example of an enslaved African American using this form, see Hager, *Word by Word,* 84.

29. *The Letter Writer's Own Book or, the Art of Polite Correspondence, Containing a Variety of Plain and Elegant Letters, on Business, Love, Courtship, Marriage, Relationship, Friendship, &c. with Forms of Complimentary Cards, and Directions for Letter Writing. To Which Are Added Forms of Mortgages, Deeds, Bonds, Powers of Attorney, &C.* (Philadelphia: John B. Perry, 1850), 5.

30. Deidre M. Mahoney, "'More Than an Accomplishment': Advice on Letter Writing for Nineteenth-Century American Women," *Huntington Library Quarterly* 66, no. 3 / 4, 412–19; Wm. A. Alcott, *Gift Book for Young Ladies or Familiar Letters on Their Acquaintances, Male and Female, Employments, Friendships, &c.* (Auburn, NY: Derby and Miller, 1853), 129–30.

31. Sarah Josepha Hale, *Manners or, Happy Homes and Good Society all the Year Round* (1868; New York: Arno, 1972), 138.

32. "Egotism," *Bossier Banner* (Bellevue, Bossier Parish, LA), August 12, 1909.

33. "Letter Writing and Letter Writers," *Littell's Living Age* 52, no. 661 (January 24, 1857): 199. Another guide queried, "Have you ever observed that the letters of friends are filled with egotism? For my part I think very suspiciously of every letter that is not." See *The Art of Letter-Writing, Illustrated by Examples from the Best Authors* (London: T. Nelson and Sons, 1858), 9.

34. Jenny MacNeven to her mother, May 20, 1840, Box 1, Folder 42, William James MacNeven Collection, 1802–1848, MS 2009-02, Burns Library, Boston College, Boston, MA.

35. Allen White to Mary Hatten, March 19, 1867, in William Allen White, *The Autobiography of William Allen White* (New York: Macmillan, 1946), 17–18.

36. Frederick Douglass, "Lecture on Pictures," in *Picturing Frederick Douglass: An Illustrated Biography of the Nineteenth-Century's Most Photographed American*, ed. John Stauffer, Zoe Trodd, and Celeste-Marie Bernier (New York: W. W. Norton, 2015), 127.

37. Robert Taft, *Photography and the American Scene: A Social History, 1839–1889* (New York: Dover, 1964), 60–61, 76; Richard Rudisill, *Mirror Image: The Influence of the Daguerreotype on American Society* (Albuquerque: University of New Mexico Press, 1971), 197–98.

38. Robert Lewis, "Photographing the California Gold Rush," *History Today* 52, no. 3 (2002) 11–17; Henkin, *The Postal Age*, 57–60.

39. Taft, *Photography and the American Scene,* 78–81.

40. Quoted in Rudisill, *Mirror Image,* 50.

41. Ibid., 53.

42. "The Daguerreotype," *Knickerbocker or New-York Monthly Magazine* 14 (1839): 561.

43. "Daguerreotypes," *Littell's Living Age* 9, no. 110 (June 20, 1846): 552.

44. Marcus Aurelius Root, *The Camera and the Pencil or The Heliographic Art* (1864; Pawlet, VT: Helios, 1971), 44, 46–48.

45. Geoffrey Batchen, "Individualism and Conformity: Photographic Portraiture in the Nineteenth Century," *New York Journal of American History* 66, no. 3 (2006): 19; Robert Hirsch, *Seizing the Light: A Social History of Photography,* 2nd ed. (Boston: McGraw Hill, 2009), 25–26.

46. Quoted in John Adams-Graf, "In Rags for Riches: A Daguerrian Survey of Forty-Niners' Clothing," *Dress* 22 (November 1995): 60.

47. T. S. Arthur, "American Characteristics. No. V.—The Daguerreotypist," *Godey's Lady's Book,* May 1849, 355.

48. "Daguerreotypes," 551–52; Rudisill, *Mirror Image,* 206–7.

49. Deborah Willis, *Reflections in Black: A History of Black Photographers, 1840 to the Present* (New York: W. W. Norton, 2000), 10.

50. "New Wrinkle in Photography," *Evening Statesman* (Walla Walla, WA), April 7, 1905.

51. "A Shrewd Photographer," *Omaha (NE) Daily Bee,* March 6, 1892.

52. "Wiles of Photographers, How They Manage to Get Pictures of Celebrated People," *Somerset (OH) Press,* February 20, 1879.

53. "What Next?," *Nashville (TN) Patriot,* March 12, 1860.

54. "Postage-Stamp Photographs the Latest Contribution of Science to Vanity," *Iola (KS) Register,* June 5, 1885.

55. "Vanity, Saith the Preacher," *Minneapolis (MN) Journal,* December 12, 1905.

56. Stauffer, Trodd, and Bernier, discovered Douglass was the most photographed person; see *Picturing Frederick Douglass;* Ginger Hill, "'Rightly Viewed': Theorizations of the Self in Frederick Douglass' Lectures on Pictures," in *Pictures and Progress: Early Photography and the Making of African American Identity,* ed. Maurice O. Wallace and Shawn Michelle Smith (Durham, NC: Duke University Press, 2012), 56; Teresa Zackodnik, "The 'Green-Backs of Civilization': Sojourner Truth and Portrait Photography," *American Studies* 46, no. 2 (Summer 2005): 117–43.

57. Frederick Douglass, "Lecture on Pictures (1861)," in Stauffer, Trodd, and Bernier, *Picturing Frederick Douglass,* 128–29. See also pp. xii–xiii, xv in the same book.

58. Ibid., 127.

59. Frederick Douglass, "Negro Portraits," *Liberator* (April 20, 1849): 2, http://fair-use.org/the-liberator/1849/04/20/the-liberator-19-16.pdf, accessed May 15, 2018; Laura Wexler, "'A More Perfect Likeness': Frederick

Douglass and the Image of the Nation," in Wallace and Smith, *Pictures and Progress,* 21.

60. Stauffer, Trodd, and Bernier, *Picturing Frederick Douglass,* ix–xviii.

61. Nell Irvin Painter, *Sojourner Truth: A Life, a Symbol* (New York: W. W. Norton, 1996), 186, 187, 198–99. See also Zackodnik, "The 'Green-Backs of Civilization.'"

62. For more on the history of African Americans' relationship to early photography, see Willis, *Reflections in Black;* and Jackie Napolean Wilson, *Hidden Witness: African-American Images from the Dawn of Photography to the Civil War* (New York: St. Martin's, 1999).

63. Root, *Camera and the Pencil,* 26–27.

64. Batchen, "Individualism and Conformity"; Masur, "'Age of the First Person Singular,'" 208–11.

65. Henkin, *Postal Age,* 56–60.

66. Susan Annette Newberry, "Commerce and Ritual in American Daguerrian Portraitures, 1839–1859" (Ph.D. diss., Cornell University, 1999), 44.

67. Jay Ruby, *Secure the Shadow: Death and Photography in America* (Cambridge, MA: MIT Press, 1995).

68. Nancy Martha West, *Kodak and the Lens of Nostalgia* (Charlottesville: University Press of Virginia, 2000), 139. West quotes Ruby on this point. See also Gary S. Cross and Robert N. Proctor, *Packaged Pleasures: How Technology and Marketing Revolutionized Desire* (Chicago: University of Chicago Press, 2015), 187.

69. Delia Locke, March 5, 1859, Delia Locke Diary, 1858–1861, Locke-Hammond Family Papers, MSS 110, Holt-Atherton Special Collections, University of the Pacific.

70. Delia Locke, February 10, 1878, Delia Locke Diary, 1875–1879, in ibid.

71. Quoted in Rudisill, *Mirror Image,* 215; Samuel C. Busey, *Records of the Columbia Historical Society, Washington, DC* 3 (1900): 87.

72. Newberry, "Commerce and Ritual in American Daguerrian Portraitures"; Andrew H. Baldwin, Broadside, ca. 1870, reprinted in "Inside the Collections; Acquisition Enhances Study of Mourning," *Inkwell: A Newsletter for Strong Museum Associates* (Summer 1990): 179–80.

73. Batchen, "Individualism and Conformity," 16.

74. Patricia Kelly Hall and Steven Ruggles, "'Restless in the Midst of Their Prosperity': New Evidence on the Internal Migration of Americans, 1850–2000," *Journal of American History* 91 (December 2004): 834–35.

75. Arozina Perkins, Diary, in Polly Welts Kaufman, *Women Teachers on the Frontier* (New Haven, CT: Yale University Press, 1984), 76.

76. Abby Mansur to her sister, September 12, 1852; letter to Hanna and mother, December 9, 1854; all in Abby T. Leighton Mansur, Correspondence, 1852–1874, Manuscript Collection WA MSS S-1258 M318, Beinecke Rare Books and Manuscript Collection, Yale University, New Haven, CT.

77. "Heyn Elite Studio Portraits of Quality Ad," *Great Falls (MT) Daily Tribune,* October 5, 1919.

78. *Evening Herald* (Klamath Falls, OR), November 24, 1920.

79. Todd Gustavson, *500 Cameras: 170 Years of Photographic Innovation* (New York: Sterling Signature, 2011), 99.

80. West, *Kodak and the Lens of Nostalgia,* 1.

81. Christina Kotchemidova, "From Good Cheer to 'Drive-by Smiling': A Social History of Cheerfulness," *Journal of Social History* 39, no. 1 (Autumn 2005): 5–37.

82. Christina Kotchemidova, "Why We Say 'Cheese': Producing the Smile in Snapshot Photography," *Critical Studies in Media Communication* 22, no. 1 (March 1, 2005): 2, doi:10.1080/0739318042000331853.

83. West, *Kodak and the Lens of Nostalgia.*

84. Kasson, *Rudeness and Civility,* 166. Rebecca Kathleen Shrum, in "Mirroring Others / Fashioning Selves: A History of the Looking Glass in America" (Ph.D. diss., University of South Carolina, 2007), 27, 41, suggests that looking glasses were more widely diffused a bit earlier than Kasson does.

85. Mark Pendergast, *Mirror, Mirror: A History of the Human Love Affair with Reflection* (New York: Basic Books, 2003), 249.

86. "Popular Image Worship," *Roanoke (VA) Times,* January 2, 1891.

87. "Minister to Modern Vanity: The Looking-glass and How It Looks to Men and Women," *Columbus (NE) Journal,* January 31, 1906.

88. Shrum, "Mirroring Others / Fashioning Selves," 164–66. Shrum offers alternative explanations for the practice.

89. "All about Looking-Glasses by Mrs. F. F. Victor," *True Northwest* (Portland, OR), August 25, 1871.

90. "Men as Vain as Women—Watch Them as They Pass a Mirror and See for Yourself," *Lafayette (LA) Advertiser,* January 27, 1894; "Mirrors and Grace," *Evening Statesman* (Walla Walla, WA), February 26, 1906.

91. Mary Paik Lee, *Quiet Odyssey: A Pioneer Korean Woman in America,* ed. Sucheng Chan (Seattle: University of Washington Press, 1990), 46.

92. See for instance, *Evening World* (New York, NY), July 7, 1905, 12.

93. "Vanity," http://www.oed.com/view/Entry/221396, accessed May 9, 2016.

94. "Vanity, Thy Name Is Man!" *Omaha (NE) Daily Bee,* September 13, 1913.

95. Eva S. Moskowitz, *In Therapy We Trust: America's Obsession with Self-Fulfillment* (Baltimore, MD: Johns Hopkins University Press, 2001), 20.

96. John M. Reisman, *A History of Clinical Psychology,* 2nd ed. (New York: Brunner-Routledge, 1991), 154; Richard M. Huber, *The American Idea of Success* (New York: McGraw-Hill, 1971), 177–85.

97. Lunbeck, *Americanization of Narcissism,* 83.

98. Ibid., 33.

99. Ibid., 83–84.

100. *Day Book* (Chicago, IL), August 15, 1912, emphasis added.

101. Susan J. Matt, *Keeping Up with the Joneses: Envy in American Consumer Society, 1890–1930* (Philadelphia: University of Pennsylvania Press, 2003), 81–95; Huber, *American Idea of Success,* 226–50.

102. Dale Carnegie, *How to Win Friends and Influence People* (New York: Simon and Schuster, 1936), 103, 120.

103. Ad for Ingram's Velveola Souveraine Face Powder, *Sun* (New York, NY), March 9, 1913.

104. Ad for Styleplus clothes, *Pensacola (FL) Journal,* April 26, 1914.

105. West, *Kodak and the Lens of Nostalgia,* 109.

106. "A Virtue of Conceit" by Mrs. Humphrey ("MADGE"), *El Paso (TX) Herald,* May 27, 1911.

107. "Vanity Is Sanity," *Delineator,* September 1936, 30–31.

108. Annie Maude Dee Porter, Diary, March 20, 1931, Special Collections, Stewart Library, Weber State University, Ogden, Utah, https://cdm.weber.edu/digital/collection/ANN/search.

109. Annie Maude Dee Porter, Diary, August 10, 1938, in ibid.

110. Letter from Simeon Styles, "The Cultivation of Vanity," *Christian Century,* January 31, 1951, 137.

111. Peter Stearns, "Self Esteem," in *Encyclopedia of Children and Childhood in History and Society*, Vol. 3, ed. Paula S. Fass (New York: Macmillan, 2004), 736–38.

112. Moskowitz, *In Therapy We Trust,* 225–27.

113. Lunbeck, *Americanization of Narcissism,* 37–58.

114. David Riesman, Nathan Glazer, and Reuel Denney, *The Lonely Crowd, Revised Edition: A Study of the Changing American Character* (New Haven, CT: Yale University Press, 2001).

115. On this point, see also Vivian Gornick, "In Defense of Narcissism," *Boston Review,* May 5, 2014, http://bostonreview.net/books-ideas/vivian-gornick-defense-narcissism-elizabeth-lunbeck-christopher-lasch-feminism, accessed September 6, 2017; and Brooks, *Road to Character,* 247–49.

116. Tom Wolfe, "The 'Me' Decade and the Third Great Awakening," *New York Magazine,* August 23, 1976, http://nymag.com/news/features/45938/.

117. Christopher Lasch, *The Culture of Narcissism: American Life in an Age of Diminishing Expectations,* rev. ed. (New York: W. W. Norton, 1979).

118. Ibid., 10.

119. Lunbeck, *Americanization of Narcissism,* 37–80.

120. Robert Raskin and Howard Terry, "A Principal-Components Analysis of the Narcissistic Personality Inventory and Further Evidence of Its Construct Validity," *Journal of Personality and Social Psychology* 54, no. 5 (1988): 890–902.

121. Constance Rosenblum, "Narcissism on Upswing, According to Experts," *Lima (OH) News,* July 21, 1978.

122. Moskowitz, *In Therapy We Trust,* 1.

123. Ibid., 1, 4.

124. Ibid., 270.

125. Gwendolyn Seidman, "Is Facebook Really Turning Us into Narcissists?" *Psychology Today,* August 11, 2014, https://www.psychologytoday.com/blog/close-encounters/201408/is-facebook-really-turning-us-narcissists, accessed September 4, 2015; Julie Beck, "How to Spot a Narcissist Online," *Atlantic,* January 16, 2014, http://www.theatlantic.com/health/archive/2014/01/how-to-spot-a-narcissist-online/283099/; Tara Parker-Pope, "Does Facebook Turn People into Narcissists?," *New York Times* (blog), May 17, 2012, http://well.blogs.nytimes.com/2012/05/17/does-facebook-turn-people-into-narcissists/, accessed July 28, 2015.

126. In *The Americanization of Narcissism,* Lunbeck points out that the definition of narcissism varied among these observers, and also differed from the Freudian description.

127. Beck, "How to Spot a Narcissist Online." Also see Brooks, *Road to Character,* chap. 10.

128. Adam Kaslikowski, Salt Lake City, Utah, April 13, 2016, interviewed by Luke Fernandez and Susan Matt.

129. Katherine J. Schall, Ogden, Utah, October 18, 2014, interviewed by Luke Fernandez and Susan Matt.

130. Michelle Braeden, Ogden, Utah, October 17, 2014, interviewed by Luke Fernandez and Susan Matt.

131. Cooper Ferrario, Ogden, Utah, June 3, 2015, interviewed by Luke Fernandez and Susan Matt.

132. Anna Renkosiak, Bettendorf, Iowa, August 12, 2015, interviewed by Susan Matt.

133. Kaslikowski interview.

134. Schall interview, October 18, 2014

135. William Speigle, Ogden, Utah, June 3, 2015, interviewed by Luke Fernandez and Susan Matt.

136. Emily Meredith, Oakland, California, January 30, 2016, interviewed by Luke Fernandez and Susan Matt.

137. Leah Bourne, "New Infographic Shows That Over a Million Selfies Are Taken Every Day," *StyleCaster,* http://stylecaster.com/beauty-high/selfies-infographic/, accessed September 4, 2015.

138. "Generations and Selfies," Millennials in Adulthood, March 5, 2014, http://www.pewsocialtrends.org/2014/03/07/millennials-in-adulthood/sdt-next-america-03-07-2014-0-04/, accessed May 15, 2018.

139. JulieAnn Carter-Winward and Kent Winward, Ogden, Utah, August 24, 2015, interviewed by Luke Fernandez and Susan Matt.

140. Christine Fullmer, Ogden, Utah, November 7, 2014, interviewed by Luke Fernandez and Susan Matt.

141. Camree Larkin, Ogden, Utah, October 30, 2015, interviewed by Luke Fernandez and Susan Matt.

142. Skyler Coombs, Ogden, Utah, October 26, 2015, interviewed by Luke Fernandez and Susan Matt.

143. K., Ogden, Utah, October 27, 2015, interviewed by Luke Fernandez and Susan Matt.

144. Ibid.

145. Eddie Baxter, Ogden, Utah, May 24, 2018, interviewed by Luke Fernandez and Susan Matt.

146. Larkin interview.

147. Alta Martin, Ogden, Utah, October 28, 2015, interviewed by Luke Fernandez and Susan Matt.

148. Schall interview, October 18, 2014,.

149. Meredith interview.

150. Gregory Noel, Ogden, Utah, May 17, 2018, interviewed by Luke Fernandez and Susan Matt.

151. Baxter interview.

152. Renkosiak interview, August 12, 2015.

153. Hailee Money, Ogden, Utah, October 30, 2015, interviewed by Luke Fernandez and Susan Matt.

154. L., Ogden, Utah, October 19, 2015, interviewed by Luke Fernandez and Susan Matt.

155. Meredith interview.

156. Diana Lopez, Ogden, Utah, May 22, 2018, interviewed by Luke Fernandez and Susan Matt.

157. Larkin interview.

158. Meredith interview.

159. David Nemer and Guo Freeman, "Selfies | Empowering the Marginalized: Rethinking Selfies in the Slums of Brazil," *International Journal of Communication* 9 (May 15, 2015): 1832–47.

160. Erin Ryan, "Selfies Aren't Empowering. They're a Cry for Help," Jezebel, http://jezebel.com/selfies-arent-empowering-theyre-a-cry-for-help-1468965365, accessed September 5, 2015.

161. Minh-Ha T. Pham, "'I Click and Post and Breathe, Waiting for Others to See What I See': On #FeministSelfies, Outfit Photos, and Networked Vanity," *Fashion Theory* 19, no. 2 (February 1, 2015): 221–41, doi:10.2752/175174115X14168357992436.

162. Ibid., 224.

163. Soraya Nadia McDonald, "After Michael Brown's killing, #IfTheyGunnedMeDown Shows How Selfies Shape History," *Washington Post,* August 11, 2014.

164. See http://www.jeffbullas.com/2012/08/02/blogging-statistics-facts-and-figures-in-2012-infographic/.

165. Scott Rosenberg, *Say Everything: How Blogging Began, What It's Becoming, and Why It Matters* (New York: Broadway, 2010), 333.

166. Claire Tanner, "Digital Narcissism: Social Networking, Blogging and the Tethered Self," in *Vanity: 21st Century Selves,* ed. Suzanne Fraser, Jane-Maree Maher, and Claire Tanner (Basingstoke, UK: Palgrave Macmillan, 2013).

167. JulieAnn currently blogs at http://jacarterwinward.com/. Winwards interview, August 24, 2015.

168. Ibid.

169. http://kikiandtea.com/2014/08/bloggers-narcissistic/, accessed January 16, 2016.

170. http://therichlifeonabudget.com/2012/04/why-i-blog.html, accessed January 16, 2016.

171. Rosenberg, *Say Everything,* 356.

172. Winwards interview, August 24, 2015.

173. As Walter Ong puts it in *Orality and Literacy,* "Like the oracle or the prophet, the book relays an utterance from a source, the one who really 'said' or wrote the book. The author might be challenged if only he or she could be reached, but the author cannot be reached in any book. There is no way directly to refute a text. After absolutely total and devastating refutation, it says exactly the same thing as before." Walter Ong, *Orality and Literacy: 30th Anniversary Edition,* 3rd ed. (London: Routledge, 2012), 78.

174. Tony Danova et al., "Just 3.3 Million Fitness Trackers Were Sold in the US in the Past Year," *Business Insider,* May 5, 2014, http://www.businessinsider.com/33-million-fitness-trackers-were-sold-in-the-us-in-the-past-year-2014-5, accessed September 4, 2015.

175. "Fitbit Community Grows to More Than 25 Million Active Users in 2017," https://investor.fitbit.com/press/press-releases/press-release-details/2018/Fitbit-Community-Grows-to-More-Than-25-Million-Active-Users-in-2017/default.aspx, and https://www.fitbit.com/home, both accessed October 21, 2018. See also "An Insurance Company Wants You to Hand over Your Fitbit Data So It Can Make More Money. Should you?" *Washington Post,* September 25, 2018, https://www.washingtonpost.com/business/2018/09/25/an-insurance-company-wants-you-hand-over-your-fitbit-data-so-they-can-make-more-money-should-you/?utm_term=.460525a00c9e, accessed October 21, 2018.

176. Evgeny Morozov, *To Save Everything, Click Here: The Folly of Technological Solutionism,* (New York: PublicAffairs, 2014). See especially chaps. 7 and 8.

177. Gary Wolf, "The Data-Driven Life," *New York Times,* April 28, 2010, https://www.nytimes.com/2010/05/02/magazine/02self-measurement-t.html. See also Morozov, *To Save Everything,* 314, 317.

178. Gordon Bell, *Total Recall: How the E-Memory Revolution Will Change Everything* (New York: Dutton Adult, 2009), 3. See also Morozov, *To Save Everything,* 268–76.

179. Lindsey Rainwater, "Are We Becoming Too Self Absorbed? #TotallyTrendyTuesday," Lindsey RainH2o, http://www.lindseyrainh2o.com/blog/2015/8

/18/internally-driven-by-external-validation-self-absorbed-or-self-aware-have-we-gone-too-far-totallytrendytuesday, accessed May 6, 2016.

180. As Nicholas Carr has speculated, the shift to what he calls "technologies of the self" might be symptomatic of a nation that has already built the technologies that cater to more basic needs. Citing Abraham Maslow's hierarchy of needs, Carr believes that we've developed the technologies that meet needs lower on the Maslow pyramid, like having all of the physiological necessities as well as physical security. Although there's still an underclass for whom these have not been realized, over the course of the nineteenth and twentieth centuries, a large portion of Americans have gradually acquired access to better shelter, transportation, and health care. With the technological infrastructure already in place to satisfy these requirements, we have now turned our attention to the satisfaction of needs that sit highest on the Maslow hierarchy—namely the development of the inner self or, as Maslow calls it, "self-actualization." Nicholas Carr, "Does Innovation Arc toward Decadence?" *ROUGH TYPE,* http://www.roughtype.com/?p=5452, accessed July 19, 2015.

181. In *The American Technological Sublime,* David Nye argues that technologies like the first transcontinental railroad, the Golden Gate Bridge, the Hoover Dam, or the Apollo space program made individual American lives more comfortable. But in addition, they also induced Americans to marvel at the power of the nation's engineering achievements and to celebrate their common participation in those achievements. Nye called this feeling "the technological sublime," and in his view, it served to subsume individual selves into the nation. As Nye puts it, "When experienced by large groups the sublime can weld society together. In moments of sublimity, human beings temporarily disregard divisions among elements of the community. . . . Since the early nineteenth century the technological sublime has been one of America's central 'ideas about itself'—a defining ideal helping to bind together a multicultural society. . . . [T]he sublime has served as an element of social cohesion."

While Americans are still building and maintaining infrastructure that evokes the technological sublime, in the late twentieth and early twenty-first centuries they are spending increasing amounts of time building and celebrating technologies that cater to the self. This progression, from the technological sublime to a technology for the cultivation, expression, measurement, and hyper-documentation of the inner self, is, of course, not absolutely linear. They still make technologies that prompt civic pride and that inspire them to consider their association with the nation at large. See David E. Nye, *American Technological Sublime* (Cambridge, MA: MIT Press, 1996), xiii–xiv.

182. David Rose asserts that good design should incorporate "enchantment" and that "enchanted objects" should include five design elements, including the capability of "socialization" in which the object helps add "connections to friends, loved ones, and colleagues." See David Rose, *Enchanted Objects: Innovation, Design, and the Future of Technology* (New York: Scribner, 2014), 194.

183. Most of the time people try to maintain some semblance between their offline and online selves in the name of authenticity. As the sociologist Nathan Jurgenson argues, there may be times when one does not want to honor this virtue. See Nathan Jurgenson, "The Liquid Self," http://blog.snapchat.com/post/61770468323/the-liquid-self, accessed July 28, 2015.

184. Peter Steiner, Cartoon, "On the Internet, Nobody Knows You're a Dog," *New Yorker*, July 5, 1993.

185. Warren Trezevant observed, "I just assume everything I write is read so you know I always make sure I communicate from that viewpoint." Warren and Harriet Trezevant, Oakland, California, January 31, 2016, interviewed by Luke Fernandez and Susan Matt.

186. Kaslikowski interview.

187. Q., Ogden, Utah, October 19, 2015, interviewed by Luke Fernandez and Susan Matt.

188. Trezevants interview.

2. THE LONELY CLOUD

1. Stephen Marche, "Is Facebook Making Us Lonely?," *Atlantic*, May 2012, http://www.theatlantic.com/magazine/archive/2012/05/is-facebook-making-us-lonely/308930/, accessed June 11, 2018.

2. Eric Klinenberg, *Going Solo: The Extraordinary Rise and Surprising Appeal of Living Alone* (New York: Penguin, 2012); Eric Klinenberg, "Facebook Isn't Making Us Lonely," *Slate*, April 2012, http://www.slate.com/articles/life/culturebox/2012/04/is_facebook_making_us_lonely_no_the_atlantic_cover_story_is_wrong_.html; Claude S. Fischer, *Still Connected: Family and Friends in America since 1970* (New York: Russell Sage Foundation, 2011); Claude Fischer, "The Loneliness Scare," *Boston Review*, April 2012, http://bostonreview.net/Fischer-loneliness-modern-culture_n, 2011; Zeynep Tufekci, "Social Media's Small, Positive Role in Human Relationships," *Atlantic*, http://www.theatlantic.com/technology/archive/2012/04/social-medias-small-positive-role-in-human-relationships/256346/, all accessed June 11, 2018.

3. Jean M. Twenge, *iGen: Why Today's Super-Connected Kids Are Growing Up Less Rebellious, More Tolerant, Less Happy—and Completely Unprepared for Adulthood—and What That Means for the Rest of Us* (New York: Atria, 2017).

4. Lynn Okura, "How One School Is Battling the Loneliness Epidemic," *Huffington Post,* September 3, 2015, http://www.huffingtonpost.com/entry/loneliness-epidemic-in-schools_55e76101e4b0b7a9633b8ddb, accessed July 2, 2016; Carolyn Gregoire, "Why Loneliness Is a Growing Public Health Concern—and What We Can Do about It," *Huffington Post,* March 21, 2015, http://www.huffingtonpost.com/2015/03/21/science-loneliness_n_6864066.html, accessed July 2, 2016.

5. One frequently cited estimate comes from John Cacioppo and William Patrick's book *Loneliness,* which claims that one in five Americans suffer from loneliness. However, some of the data they use for this date to the 1980s, while other studies they cite looked only at senior citizens. John T. Cacioppo and William Patrick, *Loneliness: Human Nature and the Need for Social Connection* (New York: W. W. Norton, 2008), 5. AARP commissioned a study and found that 35 percent of Americans forty-five and over whom they surveyed reported being lonely. "Loneliness among Older Adults: A National Survey of Adults 45+," http://www.aarp.org/research/topics/life/info-2014/loneliness_2010.html, accessed June 24, 2016. Claude Fischer, in *Still Connected,* suggests the figures are much lower.

6. "Just Say Hello, a message from Skype and *O, The Oprah Magazine,*" https://www.youtube.com/watch?v=xn6kmhNdcYg, accessed April 24, 2014.

7. Julianne Holt-Lunstad et al., "Loneliness and Social Isolation as Risk Factors for Mortality: A Meta-Analytic Review," *Perspectives on Psychological Science* 10, no. 2 (2015): 227–37; Cacioppo and Patrick, *Loneliness;* Olga Khazan, "How Loneliness Begets Loneliness," *Atlantic,* April 6, 2017, https://www.theatlantic.com/health/archive/2017/04/how-loneliness-begets-loneliness/521841/; John Cacioppo and Stephanie Cacioppo, "Loneliness Is an Epidemic in Need of a Treatment," *New Scientist,* December 30, 2014, https://www.newscientist.com/article/dn26739-loneliness-is-a-modern-epidemic-in-need-of-treatment/.

8. "Loneliness Grows from Individual Ache to Public Health Hazard," *Washington Post,* January 31, 2016.

9. Ibid.

10. "lonely, adj."; "loneliness, n."; "lonesome, adj.": all on Oxford English Dictionary Online, June 2016, Oxford University Press, http://www.oed.com/view/Entry/109974?redirectedFrom=lonesome, accessed July 2, 2016.

11. Genesis 2:18; James T. Johnson, "The Covenant Idea and the Puritan View of Marriage," *Journal of the History of Ideas* 32, no. 1 (1971): 107.

12. Max Weber, *The Protestant Ethic and the Spirit of Capitalism* (New York: Charles Scribner's Sons, 1958), 104.

13. David Brainerd to his brother, April 30, 1743, in Jonathan Edwards, *An Account of the Life of the Late Reverend Mr. David Brainerd, Minister of the Gospel, Missionary to the Indians, from the Honourable Society in Scotland, for the Propagation of Christian Knowledge, and Pastor of a Church of Christian Indians in New Jersey* (Edinburgh: John Gray and Gavin Alston, 1765), 264, 265.

14. Rebecca Dickinson, Diary, September 2, 1787; and September 12, 1787, all quoted in Marla Miller, "'My Part Alone': The World of Rebecca Dickinson, 1787-1802," *New England Quarterly* 71, no. 3 (September 1998): 361, 351, 352.

15. Alexis de Tocqueville, *Democracy in America,* ed. J. P. Mayer, trans. George Lawrence (New York: Harper Perennial, 1988), 508.

16. For a discussion of an emotion related to loneliness—homesickness—which was also widespread in the nineteenth century, see Susan J. Matt, *Homesickness: An American History* (New York: Oxford University Press, 2011).

17. Carin Rubenstein and Phillip Shaver, "The Experience of Loneliness," in *Loneliness: A Sourcebook of Current Theory, Research, and Therapy,* ed. Letitia Anne Peplau and Daniel Perlman (New York: John Wiley and Sons, 1982), 207–8.

18. Patricia Kelly Hall and Steven Ruggles, "'Restless in the Midst of Their Prosperity': New Evidence on the Internal Migration of Americans, 1850-2000," *Journal of American History* 91 (December 2004): 834–35.

19. *Experiences and Personal Narrative of Uncle Tom Jones: Who Was for Forty Years a Slave. Also The Surprising Adventures of Wild Tom of the Island Retreat, a Fugitive Negro from South Carolina* (New York: George C. Holbrook, 1854), 24–25.

20. *Memoir of Old Elizabeth, a Coloured Woman* (Philadelphia: Collins, Printer 1863), in *Six Women's Slave Narratives,* introd. William L. Andrews (New York: Oxford University Press, 1988), 3–7.

21. *Aunt Sally: or, The Cross the Way to Freedom: A Narrative of the Slave-Life and Purchase of the Mother of Rev. Isaac Williams, of Detroit, Michigan* (Cincinnati, OH: Western Tract and Book Reform Society, 1866), 45–48.

22. For a discussion of homesickness and capitalist opportunity, see Matt, *Homesickness,* chap. 2.

23. Elisha Douglass Perkins, *Gold Rush Diary, Being the Journal of Elisha Douglass Perkins on the Overland Trail in the Spring and Summer of 1849,* ed. Thomas D. Clark (Lexington: University of Kentucky Press, 1967), 13.

24. William Rounseville Alger, *The Solitudes of Nature and of Man; Or, The Loneliness of Human Life* (Boston: Roberts Brothers, 1867), 20, 22, 23–25, 31, 34.

25. J. J. J., *Fire: Or From Loneliness to Relationship, a Book for Young Christians,* 2nd ed. (London: S. W. Partridge, n.d.), 13, 14.

26. Ibid., 17, 18.

27. "Lonesome Valley," http://maxhunter.missouristate.edu/songinformation.aspx?ID=1587; Kevin Lewis, *Lonesome: The Spiritual Meanings of American Solitude* (London: I. B. Tauris, 2009), 1, 89, 140.

28. Amory H. Bradford, *The Art of Living Alone* (New York: Dodd, Mead and Company, 1899), 47–48.

29. Ralph Waldo Emerson, "Self-Reliance," in *The American Intellectual Tradition, Vol. 1, 1630–1865,* ed. David A. Hollinger and Charles Capper (New York: Oxford University Press, 2001), 362.

30. Bradford, *Art of Living Alone,* 35, 37.

31. Henry David Thoreau, *Walden,* ed. J. Lyndon Shanley (Princeton, NJ: Princeton University Press, 1989), 133, 137.

32. Quoted in Richard Lebeaux, *Thoreau's Seasons* (Amherst: University of Massachusetts Press, 1984), 48–49; Jill Lepore, "Vast Designs: How America Came of Age," *New Yorker,* October 29, 2007; Robert Sullivan, *The Thoreau You Don't Know: What the Prophet of Environmentalism Really Meant* (New York: HarperCollins, 2009), 10.

33. Thoreau, *Walden,* 140.

34. For examples, see, for instance, Daniel Drake, *Pioneer Life in Kentucky, 1785–1800,* ed. Emmet Field Horine (New York: Henry Schuman, 1948), 118–19; and Hatty [Harriet Pelton?] to Mary Ann and Jane Conine, September 30, 1849, in Donald E. Baker and Ruth Seymour Burmester, "The Conine Family Letters, 1849–1851: Employed in Honest Business and Doing the Best We Can," *Indiana Magazine of History* 69, no. 4 (1973): 345.

35. David M. Henkin, *The Postal Age: The Emergence of Modern Communications in Nineteenth-Century America* (Chicago: University of Chicago Press, 2006).

36. Sara Ames Stebbins to Lois Stebbins, January 6, 1839, in Donald F. Carmony and Kelsey Flower, ed., "Frontier Life: Loneliness and Hope," *Indiana Magazine of History* 61 (March 1965): 53–54.

37. Christian Miller to Benjamin Kenaga, April 18, 1881, Series 1, Box 3, Folder 83, Christian Miller correspondence, 1874–82, Kenaga Family Papers, Beinecke Library, Yale University, New Haven, CT.

38. Elizabeth Emma Sullivan Stuart to her son, February 27, 1852, in *Stuart Letters of Robert and Elizabeth Sullivan Stuart and Their Children, 1819–1864,* vol. 1 (Helen Stuart Mackay-Smith Marlatt, 1961), 280.

39. Anton A. Huurdeman, *The Worldwide History of Telecommunications* (Hoboken, NJ: John Wiley and Sons, 2003), 135n.

40. Robert Luther Thompson, *Wiring a Continent: The History of the Telegraph Industry in the United States, 1832–1866* (Princeton, NJ: Princeton University Press, 1947), 369; http://www.measuringworth.com/uscompare/relativevalue.php; Tomas Nonnenmacher, "History of the US Telegraph Industry," http://eh.net/encyclopedia/history-of-the-u-s-telegraph-industry/, accessed July 3, 2016; Clarence D. Long, *Wages and Earnings in the United States, 1860–1890* (Washington, DC: National Bureau of Economic Research, 1960), 14.

41. Thoreau, *Walden,* 52.

42. Ben Hur Wilson, "Telegraph Pioneering," *Palimpsest,* November 1925, 381–82; Alvin F. Harlow, *Old Wires and New Waves* (1936; New York: Arno and the *New York Times*, 1971), 137, 140, 141, 157.

43. Harlow, *Old Wires and New Waves,* 141, 140.

44. Harlow, *Old Wires and New Waves,* 321; Phillip H. Ault, "The (Almost) Russian-American Telegraph," *American Heritage* 26 (June 1975), https://www.americanheritage.com/content/almost-russian-american-telegraph, accessed May 5, 2018; Royal BC Museum Collections, http://search-bcarchives.royalbcmuseum.bc.ca/search/advanced?f=&so0=and&sq0=hp010529+hp088710+hp088711&sf0, accessed May 5, 2018; "Hagwilget Bridge Near Hazleton, British Columbia," http://www.rdks.bc.ca/content/hagwilget-bridge-near-hazelton-british-columbia, accessed May 5, 2018.

45. *New York Times,* January 12, 1888.

46. *Evening World* (New York, NY), February 5, 1889, 4.

47. *Pittsburg Dispatch,* September 3, 1891, 3.

48. *Evening World* (New York, NY), May 20, 1893, Sporting Extra, 2.

49. *Vermont Phoenix,* August 12, 1910, 6.

50. Mark Sullivan, *The Education of an American* (New York: Doubleday, Doran, 1938), 26, 52.

51. Claude Fischer, *America Calling: A Social History of the Telephone to 1940* (Berkeley: University of California Press, 1992), 42, 46, 53.

52. "Have a Telephone," *Waterloo (IA) Semi Weekly Courier,* October 4, 1901, 6.

53. "The Farmer and the Telephone," *Telephony* 8 (December 1904): 520, quoted in Fischer, *America Calling,* 99.

54. Malcolm M. Wiley and Stuart A. Rice, *Communication Agencies and Social Life* (New York: McGraw-Hill, 1933), 153. See also United States Bureau of the Census, *Telephones: 1907* (Washington, DC: Government Printing Office, 1910),

16, 51; *Census of Electrical Industries 1927: Telephones* (Washington, DC: Government Printing Office, 1930), https://www2.census.gov/prod2/decennial/documents/14880874ch1.pdf; *Census of Electrical Industries 1927 Telegraphs*, https://www2.census.gov/prod2/decennial/documents/13472916ch1.pdf. Fischer, in *America Calling*, traces the diffusion of phones across the nation as well. See also *Historical Statistics of the United States, Colonial Times to 1970, Part 2* (Washington, DC: Bureau of the Census, 1975), https://www2.census.gov/library/publications/1975/compendia/hist_stats_colonial-1970/hist_stats_colonial-1970p2-chR.pdf.

55. C. Robert Haywood and Sandra Jarvis, eds., *"A Funnie Place, No Fences": Teenagers' Views of Kansas, 1867–1900* (Lawrence: University of Kansas Division of Continuing Education, 1992), 291.

56. Edward I. Rich, Diary, January 1–10, 1896, Dr. Edward I. and Emily Almira Cozzens Rich Diaries, Special Collections, Stewart Library, Weber State University, Ogden, UT, https://dc.weber.edu/node/23.

57. George Beard, *American Nervousness: Its Causes and Consequences* (New York: G. P. Putnam's Sons, 1881), 40–64, 96–122.

58. Stephen Kern, *The Culture of Time and Space, 1880–1918* (Cambridge, MA: Harvard University Press, 1983).

59. David G. Schuster, *Neurasthenic Nation: America's Search for Health, Happiness, and Comfort, 1869–1920* (New Brunswick, NJ: Rutgers University Press, 2011), 29–30.

60. Irwin Edman, *Human Traits and Their Social Significance* (Boston: Houghton Mifflin, 1920), 138.

61. Ibid.

62. "And Now the Lonesome Club," *Washington Herald*, September 27, 1922.

63. "Negro Hospitality," *Seattle (WA) Republican*, June 21, 1901.

64. Carey McWilliams, *Southern California: An Island on the Land* (Layton, UT: Gibbs Smith, 1973), 166; "And Now the Lonesome Club," *Washington Herald*, September 27, 1922.

65. "Less Lonely Club," *Willmar (MN) Tribune*, May 3, 1911.

66. "'Lonely Clubs' for Lonely Folks Should be Organized in Every City in the Land," *Washington Times*, February 17, 1915.

67. "Less Lonely Club."

68. Richard M. Huber, *The American Idea of Success* (New York: McGraw-Hill, 1971), 124–85.

69. Norman Vincent Peale, *The Power of Positive Thinking* (New York: Fireside, 2003), 191.

70. On the connections between advertising and emotions, see Susan J. Matt, *Keeping up with the Joneses: Envy in American Consumer Society, 1890–1930* (Philadelphia: University of Pennsylvania Press, 2003); Roland Marchand, *Advertising the American Dream: Making Way for Modernity, 1920–1940* (Berkeley: University of California Press, 1985).

71. *Alliance (NE) Herald,* April 25, 1912.

72. *Colville (WA) Examiner,* February 28, 1914.

73. "Right in Your Neighborhood," AT&T ad, 1909, Book 109, Folder 1, Box 21, Ayer AT&T Collection, N. W. Ayer Advertising Agency Records, Archives Center, National Museum of American History, Smithsonian Institution.

74. "A Highway of Communication," AT&T ad, in ibid.

75. "Finder of Men," AT&T ad, 1910, Book 109, Folder 2, in ibid.

76. *McCook (NE) Tribune,* August 4, 1905.

77. *Evening Star* (Washington, DC), April 8, 1906.

78. *Richmond (VA) Times-Dispatch,* October 3, 1918, 2.

79. *Saturday Evening Post,* November 3, 1923, 67.

80. *Saturday Evening Post,* March 16, 1935, 86.

81. Letter to Radio Announcer, September 29, 1923, Broadcasting Stations Fan Mail 1923, Folder 2B, Box 139, Series 8, George H. Clark Radioana Collection, Archives Center, National Museum of American History, Smithsonian Institution.

82. Azriel Eisenberg, *Children and Radio Programs: A Study of More Than Three Thousand Children in the New York Metropolitan Area* (New York: Columbia University Press, 1936), 155.

83. Dorothy M. Johnson, *When You and I Were Young, Whitefish* (Missoula, MT: Mountain Press, 1982), 138.

84. Elsie Robinson, "Listen, World! Learn to Be Alone," *Portsmouth (NH) Herald,* February 3, 1942.

85. Huber, *American Idea of Success,* 252.

86. Dale Carnegie, *How to Win Friends and Influence People* (New York: Simon and Schuster, 1936), 97.

87. Rollo May, *Man's Search for Himself* (New York: W. W. Norton, 1953), 28.

88. "loner, n.," Oxford English Dictionary Online, June 2016, Oxford University Press, http://www.oed.com/view/Entry/109973?redirectedFrom=loner, accessed July 4, 2016.

89. David Riesman, Nathan Glazer, and Reuel Denney, *The Lonely Crowd, Revised Edition: A Study of the Changing American Character* (New Haven, CT: Yale University Press, 2001), 307.

90. Robert Putnam, *Bowling Alone: The Collapse and Revival of American Community* (New York: Touchstone, 2001).

91. David Farber and Beth Bailey, *The Columbia Guide to America in the 1960s* (New York: Columbia University Press, 2003), 396.

92. As Hopalong Cassidy put it, "Instead of running around in all directions vainly seeking amusement, mother, father, sister and brother found it in the last place they'd suspect—right in their own living room. They also found that what they were seeking wasn't entertainment at all—it was really just being close together. Our kids have fathers and mothers again, and our fathers and mothers have their kids back with them. That, I think, is the real miracle of television." Quoted in Robert Sinclair, "Miracle of Television May Be That It Hasn't Yet Made the Family Obsolete," *Saturday Evening Post,* April 23, 1955, 12.

93. Interview with Pat Aiko (Suzuki) Amino by Mary Doi, March 30, 1998, in *Regenerations Oral History Project: Rebuilding Japanese American Families, Communities, and Civil Rights in the Resettlement Era, Chicago Region: Volume I*, ed. Japanese American National Museum, Chicago Japanese American Historical Society, Japanese American Historical Society of San Diego, and Japanese American Resource Center / Museum of San Jose, 43, http://texts.cdlib.org/view?docId=ft7n39p0cn&doc.view=entire_text.

94. Interview of Rainette Holimon, in *Oral History of Monmouth County* (Malapan, NJ: Monmouth County Library, 2001), http://www.visitmonmouth.com/oralhistory/Interview.html, accessed July 4, 2016.

95. Interview with Bill Miller by Stacy Danaher, August 19, 2000, in Center for Columbia River History, http://www.ccrh.org/comm/slough/oral/bmiller.php, accessed July 4, 2016.

96. "Hypnosis in Your Living Room," *Reader's Digest,* April 1949, 70.

97. Kathy A. Krendl and Bruce Watkins, "Understanding Television: An Exploratory Inquiry into the Reconstruction of Narrative Content," *Educational Technology Research and Development* 31, no. 4 (December 1983): 201–12; James G. Webster, "The Audience," *Journal of Broadcasting & Electronic Media* 42, no. 2 (Spring 1998).

98. Marie Winn, "The Plug-in Drug," *Saturday Evening Post*, 249 (November 1977) 40–41. The effect of television on civic life is also famously covered in Putnam, *Bowling Alone.*

99. See, for instance, "Damrosch Says Radio Will Save Family Life from Disruption by the Automobile," *New York Times,* June 13, 1930; Robert S. Lynd and Helen Merrell Lynd, *Middletown: A Study in Modern American Culture* (New York: Harcourt, Brace, 1959), 251–63.

100. Putnam, *Bowling Alone.*

101. Philip Slater, *The Pursuit of Loneliness, 20th Anniversary Edition,* 3rd ed. (Boston: Beacon, 1990), 7.

102. Ibid., 12.

103. For example, see Leon Botstein, "Children of the Lonely Crowd," *Change Magazine* 10 (May 1978). Botstein examined "The Children of *The Lonely Crowd*" and wondered whether Americans were "a more lonely crowd than our parents thirty years ago?" Botstein discussed the works of other social critics who were tackling the subject, including Christopher Lasch's *The Culture of Narcissism.* Recounting Lasch's main thesis, Botstein observed that Americans had lost interest in politics and civic associations (which often were where individuals made friends and socialized) and had instead begun to look inward as new therapeutic and self-help books advised.

104. Adele Stern, "Miniguide: A Mini Course on Loneliness and Alienation," *Scholastic Review,* October 1973, 32.

105. Ibid., 33.

106. "Chasing the Lonelies," *Seventeen,* August 1975; "I'm Lonely, You're Lonely," *American Home,* October 1976, 27–28; Carolyn See, "Christmas Can Be a Lonely Time, Too," *Today's Health,* December 1974, 52–57; Robert J. Blakely, "The Lonely Youth of Suburbia," *PTA Magazine,* April 1961, 14–16; Mary Cantwell and Amy Gross, "43 Ways to Get Unlonely," *Mademoiselle,* December 1977; Zick Rubin, "Seeking a Cure for Loneliness," *Psychology Today* 13 (October 1979): 82–90.

107. Martha Weinman Lear, "Loneliness: More Common Than the Common Cold," *Redbook,* November 1980, 33.

108. Here, for example, is one of the questions: "I am unhappy doing so many things alone." Users were then prompted to indicate whether they often, sometimes, rarely, or never felt that way. See Dan Russell, Letitia Anne Peplau, and Mary Lund Ferguson, "Developing a Measure of Loneliness," *Journal of Personality Assessment* 42, no. 3 (July 1, 1978): 290–94, doi:10.1207/s15327752jpa4203_11.

109. Letitia Anne Peplau and Daniel Perlman, *Loneliness: A Sourcebook of Current Theory, Research and Therapy* (New York: Wiley, 1982).

110. Ibid., 7, 2–3.

111. Robert Weiss, *Loneliness: The Experience of Emotional and Social Isolation* (Cambridge, MA: MIT Press, 1975).

112. Robert Weiss, "Issues in the Study of Loneliness," in Peplau and Perlman, *Loneliness: A Sourcebook,* 71–80; Letitia Anne Peplau and Daniel Perlman, "Loneliness: What Is It?," in Peplau and Perlman, *Loneliness: A Sourcebook,* 5.

113. Robert Weiss, quoted in Rubin, "Seeking a Cure for Loneliness."

114. "A Telephone Call from Out of Town Takes the Blues Out of the Night," *Life Magazine* 45, no. 5 (August 4, 1958), https://books.google.com/books?id=k1MEAAAAMBAJ&pg=PA61&lpg=PA61&dq=A+telephone+call+from+out+of+town+takes+the+blues+out+of+the+night.&source=bl&ots=bZNhNDIgnO&sig=LuoUGPl5Y-sEWnQ02XXvvj1R8yo&hl=en&sa=X&ved=0CCQQ6AEwAmoVChMI7dCMpJHcyAIVF9FjCh0XbwjZ#v=onepage&q=A%20telephone%20call%20from%20out%20of%20town%20takes%20the%20blues%20out%20of%20the%20night.&f=false.

115. "When your children grow up and go their own way . . . keep them close by Long Distance. It's the next best thing to being there," *Life Magazine* 60, no. 21 (May 27, 1966), https://books.google.com/books?id=XVYEAAAAMBAJ&pg=PA98&lpg=PA98&dq=%22when+your+children+grow+up+and+go+their+own+way%22&source=bl&ots=WnSRJK0c36&sig=EjgMAtij8NeVIGIN9cGfkcZ9BgE&hl=en&sa=X&ved=0ahUKEwjY86uvuIXMAhWKu4MKHYdiD-oQ6AEIHTAA#v=onepage&q=%22when%20your%20children%20grow%20up%20and%20go%20their%20own%20way%22&f=false, 96.

116. Confirmed in correspondence with Sheldon Hochheiser, Corporate Historian at the AT&T Archives and History Center, June 13, 2018.

117. Eric Klinenberg, *Going Solo: The Extraordinary Rise and Surprising Appeal of Living Alone* (New York: Penguin, 2013), 4–5. See also Richard Fry, "The Share of Americans Living without a Partner Has Increased, Especially among Young Adults," *Pew Research Center* (blog), October 11, 2017, http://www.pewresearch.org/fact-tank/2017/10/11/the-share-of-americans-living-without-a-partner-has-increased-especially-among-young-adults/.

118. Fischer, *Still Connected.*

119. Jacqueline Olds and Richard S. Schwartz, *The Lonely American: Drifting Apart in the Twenty-First Century* (Boston: Beacon, 2009), 4.

120. Jihan Thompson, "Just Say Hello," Skype Blogs, February 20, 2014, http://blogs.skype.com/2014/02/20/just-say-hello/, accessed April 24, 2014.

121. John Griffin, "Mark Zuckerberg's Big Idea: The 'Next 5 Billion' People," *CNN Money,* August 21, 2013, http://money.cnn.com/2013/08/20/technology/social/facebook-zuckerberg-5-billion/.

122. It is likely that corporate profits are not the only thing that drives Zuckerberg's interest in internet.org. After all, in 2011, the United Nations also declared that access to broadband should be a basic human right (cf. http://www.forbes.com/sites/randalllane/2011/11/15/the-united-nations-says-broadband-is-basic-human-right/).

123. Christian Wutz and Chris Fowles, Salt Lake City, Utah, March 25, 2016, interviewed by Luke Fernandez and Susan Matt.

124. William Powers, *Hamlet's Blackberry: Building a Good Life in the Digital Age* (New York: HarperCollins, 2010), 4.

125. Aaron Smith, "What People Like and Dislike about Facebook," Pew Research Center, February 3, 2014, http://www.pewresearch.org/fact-tank/2014/02/03/what-people-like-dislike-about-facebook/.

126. Skyler Coombs, Ogden, Utah, October 26, 2015, interviewed by Luke Fernandez and Susan Matt.

127. Cooper Ferrario, Ogden, Utah, June 3, 2015, interviewed by Luke Fernandez and Susan Matt.

128. Wutz and Fowles interview.

129. L., Ogden, Utah, October 19, 2015, interviewed by Luke Fernandez and Susan Matt; Q., Ogden, Utah, October 19, 2015, interviewed by Luke Fernandez and Susan Matt.

130. Ana Chavez, Ogden, Utah, October 19, 2015, interviewed by Luke Fernandez and Susan Matt.

131. Christine Fullmer, Ogden, Utah, November 7, 2014, interviewed by Luke Fernandez and Susan Matt.

132. Nathan Jurgenson, "Digital Dualism versus Augmented Reality," *Cyborgology, the Society Pages,* February 24, 2011, https://thesocietypages.org/cyborgology/2011/02/24/digital-dualism-versus-augmented-reality/, accessed July 4, 2016.

133. Hailee Money, Ogden, Utah, October 30, 2015, interviewed by Luke Fernandez and Susan Matt.

134. Anna Renkosiak, Bettendorf, Iowa, August 12, 2015, interviewed by Susan Matt.

135. L. interview.

136. Eli Gay, Berkeley, California, January 31, 2016, interviewed by Luke Fernandez and Susan Matt.

137. Alta Martin, Ogden, Utah, October 28, 2015, interviewed by Luke Fernandez and Susan Matt.

138. Ferrario interview.

139. Camree Larkin, Ogden, Utah, October 30, 2015, interviewed by Luke Fernandez and Susan Matt.

140. Gregory Noel, Ogden, Utah, May 17, 2018, interviewed by Luke Fernandez and Susan Matt.

141. Money interview; Martin interview; JulieAnn Carter-Winward and Kent Winward, Ogden, Utah, August 24, 2015, interviewed by Luke Fernandez and Susan Matt.

142. Warren and Harriet Trezevant, Oakland, California, January 31, 2016, interviewed by Luke Fernandez and Susan Matt.

143. Elena Ellsworth, Estes Park, Colorado, August 19, 2016, interviewed by Luke Fernandez and Susan Matt.

144. Eulogio Alejandre, Ogden, Utah, May 15, 2018, interviewed by Luke Fernandez and Susan Matt.

145. Diana Lopez, Ogden, Utah, May 22, 2018, interviewed by Luke Fernandez and Susan Matt.

146. Alejandre interview.

147. Ana Chavez, Ogden, Utah, October 19, 2015, interviewed by Luke Fernandez and Susan Matt.

148. H., Grinnell, Iowa, May 23, 2016, interviewed by Luke Fernandez and Susan Matt.

149. Renkosiak interview, August 12, 2015.

150. Winwards interview, August 24, 2015.

151. Jennifer Egan, "Love in the Time of No Time," *New York Times Magazine,* November 23, 2003, http://www.nytimes.com/2003/11/23/magazine/love-in-the-time-of-no-time.html?pagewanted=6&pagewanted=all, accessed June 11, 2018.

152. Wutz and Fowles interview.

153. N., Honolulu, Hawaii, July 9, 2015, interviewed by Luke Fernandez and Susan Matt.

154. Gay interview.

155. Debbie and William Speigle, Ogden, Utah, June 3, 2015, interviewed by Luke Fernandez and Susan Matt.

156. Mary Kaplan, Davenport, Iowa, December 28, 2014, interviewed by Susan Matt.

157. Gay interview.

158. Allen Riedy, Honolulu, Hawaii, July 10, 2015, interviewed by Luke Fernandez and Susan Matt.

159. Tufekci, "Social Media's Small, Positive Role in Human Relationships," 18.

160. Riedy interview.

161. Jeffrey L. Hellrung, Ogden, Utah, October 27, 2014, interviewed by Luke Fernandez and Susan Matt.

162. Xavier Stilson, Ogden, Utah, October 20, 2015, interviewed by Luke Fernandez and Susan Matt.

163. L. interview.

164. Coombs interview.

165. "Louis C.K. Explains Why He's Not a Fan of Smartphones," *CBS News*, http://www.cbsnews.com/news/louis-ck-explains-why-hes-not-a-fan-of-smartphones/, accessed July 4, 2016.

166. William Deresiewicz, "The End of Solitude," *Chronicle of Higher Education*, January 30, 2009, http://chronicle.com/article/The-End-of-Solitude/3708.

167. Emily Meredith, Oakland, California, January 30, 2016, interviewed by Luke Fernandez and Susan Matt.

168. "Google Ngram Viewer," https://books.google.com/ngrams/graph?content=loneliness%2Csolitude&year_start=1800&year_end=2000&corpus=15&smoothing=3&share=&direct_url=t1%3B%2Cloneliness%3B%2Cc0%3B.t1%3B%2Csolitude%3B%2Cc0, accessed May 15, 2016. The graph plots the occurrences of the word "solitude" in Google's Ngram corpus between 1800 and 2000. It should be noted that after the year 2000 occurrences of "solitude" begin to increase in Google Ngram. However, the developers of Google Ngram warn that results outside the 1800 to 2000 period may be skewed: "Before 1800, there aren't enough books to reliably quantify many of the queries that first come to mind; after 2000, the corpus composition undergoes subtle changes around the time of the inception of the Google Books project." "Google N-Gram Viewer—Culturomics," http://www.culturomics.org/Resources/A-users-guide-to-culturomics, accessed September 12, 2017. Also see Jean-Baptiste Michel et al., "Quantitative Analysis of Culture Using Millions of Digitized Books," *Science*, December 16, 2010, doi:10.1126/science.1199644; Christopher D. Bader, *Faithful Measures: New Methods in the Measurement of Religion*, ed. Roger Finke (New York: New York University Press, 2017), 290; schrisomalis, "Conservative Skewing in Google N-Gram Frequencies," *Glossographia*, July 14, 2013, https://glossographia.wordpress.com/2013/07/14/conservative-ngram-skew/.

169. Historians of the emotion argue that because emotions are in part cognitive and culturally constituted, the words one uses to describe them have a significant role in shaping emotional experience. On this point, see William M. Reddy, *The Navigation of Feeling: A Framework for the History of Emotions* (Cambridge: Cambridge University Press, 2001), 105.

170. It is intuitive to think that societies that uphold individualism would be better at fostering solitude than communitarian societies. But the evidence for this is scant. As Long and Averill argue, "Superficially, it might be assumed that individualistic societies would encourage . . . solitude more readily than would collectivist societies. We know of no evidence to suggest that such an assumption is generally true; on the contrary, anecdotal accounts suggest that collectivist societies . . . prize and encourage solitude as much as do individualistic . . . societies." See Christopher R. Long and James R. Averill, "Solitude: An Exploration of Benefits of Being Alone," *Journal for the Theory of Social Behaviour* 33, no. 1 (March 1, 2003): 21–44, doi:10.1111/1468-5914.00204.

171. William Deresiewicz, "Faux Friendship," *Chronicle of Higher Education,* December 6, 2009, http://chronicle.com/article/Faux-Friendship/49308/; Sherry Turkle, *Alone Together: Why We Expect More from Technology and Less from Each Other* (New York: Basic Books, 2012); Gary Turk, "Look Up," 2014, https://www.youtube.com/watch?v=Z7dLU6fk9QY.

172. Claude Fischer, in *Still Connected,* argues that our present concerns about the isolating effects of the internet are highly reminiscent of analogous concerns about the telephone. See Fischer, *Still Connected.* In a similar vein, Zeynep Tufekci thinks that critics are engaging in a moral panic. In Tufekci's view, the car, the isolating effects of mass media, and suburban living are much likelier causes of loneliness than social media. In fact, for Tufekci, social media's net effects probably allay and mitigate some of the isolating effects that arose from these twentieth-century technologies. And the reason why some people are discrediting it is not because it is inherently evil but because the new social media jeopardizes the status of people who have successfully capitalized off of older forms of media. Zeynep Tufekci, "The Social Internet: Frustrating, Enriching, but Not Lonely," *Public Culture* (2013), 20, https://www.academia.edu/5788740/The_Social_Internet_Frustrating_Enriching_but_Not_Lonely.

3. THE FLIGHT FROM BOREDOM

1. Edward I. Rich, 1894 Diary, May 11, 1894; May 12, 1894; May 14, 1894, May 30, 1894; June 12, 1894; June 20, 1894, all in Dr. Edward I. and Emily Almira Cozzens Rich Diaries, Stewart Library, Weber State University, Ogden, UT, https://dc.weber.edu/node/23.

2. "Americans Demand New Form of Media to Bridge Entertainment Gap while Looking from Laptop to Phone," *Onion*, July 30, 2014, http://www.theonion.com/article/americans-demand-new-form-of-media-to-bridge-enter-36579, accessed January 31, 2018.

3. Rachel Feltman, "Most Men Would Rather Shock Themselves Than Be Alone with Their Thoughts," *Washington Post*, July 3, 2014.

4. Barbara Dalle Pezze and Carlo Salzani, eds., *Essays on Boredom and Modernity* (Amsterdam: Rodopi, 2009), 10.

5. Ute Frevert, *Emotions in History: Lost and Found* (Budapest: Central European University Press, 2011), 31.

6. Siegfried Wenzel, *The Sin of Sloth: Acedia in Medieval Thought and Literature* (Chapel Hill: University of North Carolina Press), 5.

7. Frevert, *Emotions in History*, 32–33.

8. Wenzel, *Sin of Sloth*, 5, 18–19, 22; James B. Williams, "Working for Reform: *Acedia*, Benedict of Aniane and the Transformation of Working Culture in Carolingian Monasticism," in *Sin in Medieval and Early Modern Culture: The Tradition of the Seven Deadly Sins*, ed. Richard G. Newhauser and Susan J. Ridyard (York, UK: York Medieval Press, 2012), 19–42.

9. Wenzel, *Sin of Sloth*, 35, 38, 42.

10. Barbara Dalle Pezze and Carlo Salzani, "The Delicate Monster: Modernity and Boredom," in Pezze and Salzani, *Essays on Boredom and Modernity*, 9; "Ennui," Oxford English Dictionary Online, http://www.oed.com/view/Entry/62506?rskey=JF9GmN&result=1&isAdvanced=false#eid, accessed June 11, 2018.

11. Pezze and Salzani, *Essays on Boredom and Modernity*, 10. For a discussion of the use of the word "drudgery," see Noah W. Sobe, "Attention and Boredom in the 19th-Century School: The 'Drudgery' of Learning and Teaching and the Common School Reform Movement," in *Aufmerksamkeit Zur Geschichte, Theorie und Empirie eines pädagogischen Phänomens*, ed. Sabine Reh, Kathrin Berdelmann, and Jörg Dinkelaker (Wiesbaden, Germany: Springer VS, 2015), 55–70.

12. Patricia Meyer Spacks, *Boredom: The Literary History of a State of Mind* (Chicago: University of Chicago Press, 1996), 10–15.

13. Thomas Jefferson to Martha Jefferson, November 28, 1783, in *The Domestic Life of Thomas Jefferson, Compiled from Family Letters and Reminiscences*, ed. Sarah Randolph (New York: Harper and Brothers, 1871), 69.

14. Thomas Jefferson to Martha Jefferson, March 28, 1787, in ibid., 115–17.

15. Thomas Jefferson to Martha Jefferson, May 21, 1787, in ibid., 123; Roy Rivenburg, "Bias against Boredom Is Becoming an Obsession," *Los Angeles*

Times, Journal—Gazette (Fort Wayne, IN), March 1, 2003, 1D, https://hal.weber.edu/login?url=http://search.proquest.com/docview/411031855?accountid=14940; Spacks, *Boredom.*

16. John Adams to Abigail Adams, December 14, 1794, in *Letters of John Adams: Addressed to His Wife* (Boston: C. C. Little and J. Brown, 1841), 171–72.

17. John Adams to Christopher Gadsen, April 16, 1801, in *The Works of John Adams, Second President of the United States, with a Life of the Author, Notes, and Illustrations, by His Grandson, Charles Francis Adams* (Boston: Little, Brown, 1854), 585.

18. F. J. V. Broussais, *A Treatise on Physiology Applied to Pathology,* trans. John Bell and R. La Roche (Philadelphia: H. C. Carey and J. Lea, 1826), 151.

19. Andrew Combe, *The Principles of Physiology Applied to the Preservation of Health, and to the Improvement of Physical and Mental Education* (New York: Fowler and Wells, 1847), 223, 225.

20. D. S. Welling, *Information for the People, or, The Asylums of Ohio: with Miscellaneous Observations on Health, Diet and Morals, and the Causes, Symptoms and Proper Treatments of Nervous Diseases and Insanity* (Pittsburgh: Geo. Parkin, 1851), 346.

21. "Music as Medicine," *Iola (KS) Register,* July 17, 1875; "There May Be Something to This," *Arizona Republican,* December 13, 1894. See also *Indianapolis Journal,* March 16, 1902.

22. Welling, *Information for the People,* 281.

23. Louis J. Kahn, *Nervous Exhaustion, Its Cause and Cure: A series of lectures on debility and disease, with practical information on marriage, its obligations and impediments: illustrated with cases, to which is added pictures from real life, or photographic life-studies, addressed to the young, the old, the grave, the gay* (New York: L. J. Kahn, 1870), 41–42.

24. Norman Shanks Kerr, *Inebriety: Its etiology, pathology, treatment, and jurisprudence* (Philadelphia: Blakiston, 1888), 177–78.

25. *New York Tribune,* May 28, 1899.

26. "Ennui," *Pascagoula (MS) Democrat Star,* August 3, 1888.

27. "Ennui: A Misunderstood Malady from Which Americans Are Exempt, and Royalty the Greatest Sufferers," *New York Tribune,* October 15, 1893.

28. Ibid.

29. Ibid.; "Just Tired of Life So He Dies; 'Sheer Boredom' Causes Parisian to Shoot Himself before His Guests," *Daily Missoulian (MT),* March 14, 1909; "Crown Prince Dying? Says It Is Boredom," *Seattle (WA) Star,* May 24, 1921.

30. "Sunday Thoughts," *Arizona Republican,* March 20, 1892.

31. *Virginia Enterprise,* November 11, 1898.

32. "The Terrors of Ennui," *Palatka (FL) News and Advertiser,* December 9, 1910.

33. Max Weber, *The Protestant Ethic and the Spirit of Capitalism,* trans. Talcott Parsons (New York: Charles Scribner's Sons, 1958).

34. Spacks, *Boredom,* 9, 13, 15, 22. See also Pezze and Salzani, *Essays on Boredom and Modernity,* 10; Elizabeth Goodstein, *Experience without Qualities: Boredom and Modernity* (Stanford, CA: Stanford University Press, 2004), 3, 107–20; Sobe, "Attention and Boredom in the 19th-Century School."

35. Daniel Drake, *Pioneer Life in Kentucky, 1785–1800,* ed. Emmet Field Horine (New York: Henry Schuman, 1948), 57–58, 176, 179.

36. May 4, 1855, and June 14, 1855, in "The Diary of James R. Stewart, Pioneer of Osage County, April, 1855–April, 1857; May, 1858–November, 1860," *Kansas Historical Quarterly* 17 (February 1949): 8, 14.

37. Charles Ball, *Slavery in the United States: A Narrative of the Life and Adventures of Charles Ball, a Black Man . . .* (New York: John S. Taylor, 1837), 362.

38. *Narrative of James Williams, an American Slave, Who Was for Several Years a Driver on a Cotton Plantation in Alabama* (New York: American Anti-Slavery Society, 1838), 52–53.

39. *Augusta (GA) Chronicle,* November 20, 1850, quoted in Michael Woods, *Emotional and Sectional Conflict in the Antebellum United States* (New York: Cambridge University Press, 2014), 41.

40. "Foreign Immigration," *Columbia (TN) Herald,* February 3, 1871.

41. In "The Factory as Republican Community: Lowell Massachusetts," the second chapter in Kasson's *Civilizing the Machine,* there is also extensive coverage of the lives of Lowell Mills girls and the extent to which their work contributed, or detracted, from the virtue of the American republic. See John F. Kasson, *Civilizing the Machine: Technology and Republican Values in America, 1776–1900* (New York: Hill and Wang, 1999).

42. Lucy Larcom, *A New England Girlhood* (New York: Corinth, 1961), 175–82; Herbert G. Gutman, "Work, Culture, and Society in Industrializing America, 1815–1919," *American Historical Review* 78 (June 1973): 551–52.

43. M., "The Hill-side and the Fountain Rill," *Lowell (MA) Offering,* December 1840, 20; https://babel.hathitrust.org/cgi/pt?id=hvd.32044021216981;view =1up;seq=5

44. Malenda M. Edwards to Sabrina Bennett, April 4, 1839, in *Farm to Factory: Women's Letters, 1830–1860,* ed. Thomas Dublin, 2nd ed. (New York: Columbia University Press, 1993), 75.

45. F. G. A., "Susan Miller," *Lowell Offering*, August 1841, 167–71 https://babel.hathitrust.org/cgi/pt?id=hvd.32044019620947;view=1up;seq=3, ; see also David A. Zonderman, *Aspirations and Anxieties: New England Workers and the Mechanized Factory System, 1815–1850* (New York: Oxford University Press, 1992), 23–24.

46. Almira, "The Spirit of Discontent," *Lowell Offering*, April 1841, 111–14, https://babel.hathitrust.org/cgi/pt?id=hvd.32044019620947;view=1up;seq=121.

47. Ibid.

48. Tally Simpson to Mary Simpson, June 18, 1862; also Tally to Anna Tallulah Simpson, March 21, 1863, both in *"Far, Far from Home": The Wartime Letters of Dick and Tally Simpson, Third South Carolina Volunteers*, ed. Guy R. Everson and Edward H. Simpson Jr. (New York: Oxford University Press, 1994), 129, 203.

49. Larcom, *New England Girlhood*, 175–82; Gutman, "Work, Culture, and Society," 551–52.

50. Fiducia, "Fancy," *Lowell Offering*, April 1841, 117–18 https://babel.hathitrust.org/cgi/pt?id=hvd.32044019620947;view=1up;seq=128,.

51. "Pleasures of Factory Life," *Lowell Offering*, December 1840, 25, https://babel.hathitrust.org/cgi/pt?id=hvd.32044021216981;view=1up;seq=37.

52. James R. Stewart, July 27, 1855, in "Diary of James R. Stewart," 22.

53. Tally Simpson to Anna Tallulah Simpson, January 3, 1862, in *"Far, Far from Home,"* 103.

54. William C. Gannett, *Blessed Be Drudgery: And Other Papers* (Glasgow: David Bryce and Sons, 1890), 16. See also Daniel T. Rodgers, *The Work Ethic in Industrial America, 1850–1920* (Chicago: University of Chicago Press, 1979), 14.

55. Quoted in Roy Rosenzweig, *Eight Hours for What We Will: Workers & Leisure in an Industrial City, 1870–1920* (New York: Cambridge University Press, 1983), 1.

56. William Hard and Rheta Childe Dorr, "The Woman's Invasion," *Everybody's Magazine*, March 1909, 376; Gutman, "Work, Culture, and Society."

57. Keith Sward, *The Legend of Henry Ford* (New York: Rinehart, 1948), 49. Also quoted in Matthew B. Crawford, *Shop Class as Soulcraft: An Inquiry into the Value of Work* (New York: Penguin, 2010), 41–42.

58. "Henry Ford Expounds Mass Production," *New York Times*, September 19, 1926.

59. Clayton W. Fountain, *Union Guy* (New York: Viking, 1949), 17, 20–23, 29–30; Joyce Shaw Peterson, "Auto Workers and Their Work, 1900–1933," *Labor History* 22 (1981): 22.

60. Florence Lucas Sanvill, "A Woman in the Pennsylvania Silk-Mills," *Harper's Magazine,* April 1910, 651–62, excerpted in *The Human Side of American History,* ed. Richard C. Brown (Boston: Ginn and Company, 1968), 239.

61. Elton Mayo, "The Blind Spot in Scientific Management," 1, Elton Mayo Papers, Folder 9, Box 51, Harvard Business School Archives, Baker Library, Harvard University, Cambridge, MA.

62. Elton Mayo, "The Basis of Industrial Psychology," paper presented at a meeting of the Taylor Society, New York, December 5, 1924, reprinted from vol. IX, no. 6, of the Bulletin of the Taylor Society, 9, 8, in Elton Mayo Papers, Folder 8, Box 5a, Harvard Business School Archives, Baker Library.

63. For a discussion of the rest pause, see Elton Mayo, "The Human Factor in Mass-Production," Folder 49, Box 5b, Harvard Business School Archives, Baker Library; "What Do Workers Want?," a radio broadcast drafted for December 8, 1946, for WNAC Boston, 3, in Elton Mayo Papers, Folder 133, Box 5c, Harvard Business School Archives, Baker Library.

64. Rodgers, *Work Ethic in Industrial America.*

65. Fountain, *Union Guy,* 29–30; Peterson, "Auto Workers and Their Work," 22; Thomas A. Klug, "Review of *American Automobile Workers, 1900–1933,* by Joyce Shaw Peterson," *Journal of Social History* (Spring 1989): 583–85.

66. A number of historians have made this claim. See, for instance, Lawrence B. Glickman, *A Living Wage: American Workers and the Making of Consumer Society* (Ithaca, NY: Cornell University Press, 1999).

67. In a book review, historian Thomas Klug takes issue with the idea of workers consciously accepting more boring work in exchange for more money and more excitement, arguing that they had little choice in the matter. They could take the job or not but had little actual bargaining power. Klug, "Review of *American Automobile Workers, 1900–1933,*" 583–85.

68. Spacks, *Boredom,* 4–5.

69. Daniel Horowitz, *The Morality of Spending: Attitudes toward the Consumer Society in America, 1875–1940* (Baltimore, MD: Johns Hopkins University Press, 1985); William R. Leach, "Transformations in a Culture of Consumption: Women and Department Stores, 1890–1925," *Journal of American History* 71, no. 2 (September 1984): 319–42; Susan J. Matt, *Keeping Up with the Joneses: Envy in American Consumer Society, 1890–1930* (Philadelphia: University of Pennsylvania Press, 2003).

70. John F. Kasson, *Amusing the Millions: Coney Island at the Turn of the Century* (New York: Hill and Wang, 1978); Kathy Peiss, *Cheap Amusements: Working*

Women and Leisure in Turn-of-the-Century New York (Philadelphia: Temple University Press, 1986).

71. H. E. Wilkinson, *Memories of an Iowa Farm Boy* (1952; Ames: Iowa State University Press, 1994), 90–91.

72. For a discussion of telephones and their effects on social life, see Claude S. Fischer, *America Calling: A Social History of the Telephone to 1940* (Berkeley: University of California Press, 1992).

73. Fliers for the Edison Phonograph, Box 1, Folder 11, Warshaw Collection, Business Americana, Smithsonian Institution.

74. Remington Schuyler, "Static Days and Nights: The Boredom of Ranch Life Is Now Broken by Radio," *Radio Broadcast,* January 1925, 507–10.

75. Harry A. Bean to STA. WGI, September 14, 1923, in Broadcasting Stations Fan Mail 1923, Series 8, Box 139, Folder 2A, George H. Clark Radioana Collection, Archives Center, National Museum of American History, Smithsonian Institution.

76. E. E. Ricker, Lynn, Massachusetts, September 21, 1923, in ibid.

77. Abraham Meyerson, *The Foundations of Personality* (Boston: Little, Brown, 1921), 129–30, 127.

78. Robert E. Park, "Community Organization and the Romantic Temper," *Journal of Social Forces* 3, no. 4 (1925): 673–77; Stanford M. Lyman, *The Seven Deadly Sins: Society and Evil* (New York: St. Martin's, 1978), 19.

79. "Home Cure for Boredom," *Forum and Century* 84 (July 1930): 8–13, 10.

80. Spacks, *Boredom,* 4–5; Otto Fenichel, "On the Psychology of Boredom," in *The Collected Papers of Otto Fenichel, Vol. 1,* ed. Hanna Fenichel and David Rapaport (New York: W. W. Norton, 1953), 301.

81. Fountain, *Union Guy,* 38–40.

82. "Y.M.C.A. Plans Special Service for Coming Year," *Waterloo [IA] Daily Courier,* November 7, 1932.

83. Frank Thone, "Fatigue—the Enemy of War Production," *Butte Montana Standard,* April 12, 1942.

84. Richard Kurin and C. Wayne Clough, *The Smithsonian's History of America in 101 Objects* (New York: Penguin, 2013), 448.

85. "On the Old Assembly Line," http://www.lyricsmania.com/on_the_old_assembly_line_lyrics_glenn_miller.html, accessed January 31, 2018.

86. Kathryn Blood, *Negro Women War Workers* (Washington, DC: US Department of Labor, Women's Bureau, Bulletin No. 205, 1945), in *American Women*

in a World at War: Contemporary Accounts from World War II, ed. Judy Barrett Litoff and David C. Smith (Wilmington, DE: Scholarly Resources, 1997), 187.

87. "Topics of the Times," *New York Times,* February 6, 1949.

88. Hal Boyle, "TV Boredom," *Athens (OH) Messenger,* January 8, 1953, 4.

89. Harriet Van Horne, "TV Boredom Grows as Summer Wears On," *El Paso (TX) Herald Post,* July 4, 1961.

90. Woodburn Heron, "The Pathology of Boredom," *Scientific American,* 1957, 52.

91. Ibid., 56.

92. "Is Boredom Bad for You?" *McCall's* 84 (April 1957): 50.

93. V. S. Pritchett, "Temptations of Boredom," *Holiday* 37 (January 1965): 8, 16. He also wondered about the specific place that boredom occupied in the United States and believed it to be more taboo there than elsewhere. "Since Americans have made activity a virtue in itself, I doubt if boredom is respected or acknowledged as an inevitable human experience."

94. Peter Chew, "Ten Ways You Can Cheat Boredom," *Science Digest,* December 1972, 40.

95. "Bored on the Job," *Life Magazine* (September 1, 1972): 30–38.

96. Estelle Ramey, "Boredom: The Most Prevalent Disease," *Harper's Magazine,* November 1974, 18.

97. Mihaly Csikszentmihalyi, *Beyond Boredom and Anxiety. The Experience of Play in Work and Games* (San Francisco: Jossey-Bass, 1975), 35–36.

98. Marilyn Thomsen, "Mihaly Csikszentmihalyi and Flow," *Flame*, September 1, 2000, http://flame.cgu.edu/faculty/mihaly-csikszentmihalyi-flow/. Of course, many developments contributed to the popularization of flow—not least of which was that Csikszentmihalyi decided to use the word "flow" instead of the equivalent (but more academic) term "autotelic experience." But the existence of bored suburban housewives and television viewers who were looking for ways of redressing their boredom certainly heightened its popularity.

99. Arthur J. Sniden, "Boredom Is Becoming America's No. 1 Disease—Psychologists," *Lincoln (NE) Journal and Star,* May 8, 1977.

100. R. Farmer and N. Sundberg, "Boredom Proneness: The Development and Correlates of a New Scale," *Journal of Personality Assessment* 50, no. 1 (1986): 4–17. Also discussed in Teresa Belton and Esther Priyadharshini, "Boredom and Schooling: A Cross Disciplinary Exploration," *Cambridge Journal of Education* 37, no. 4 (2007): 579–95.

101. Rufus Quail, "Boredom Proneness Scale," http://www.gotoquiz.com/boredom_proness_scale, accessed May 14, 2016. See also Richard Farmer and Norman D. Sundberg, "Boredom Proneness—the Development and Correlates of a New Scale," *Journal of Personality Assessment* 50, no. 1 (March 1, 1986): 4–17, https://doi.org/10.1207/s15327752jpa5001_2.

102. Belton and Priyadharshini, "Boredom and Schooling."

103. Similar statistics that substantiate the growth of ADHD diagnoses are also cited in Chapter 4. See Gretchen LeFever, Andrea Arcona, and David Antonuccio, "ADHD among American Schoolchildren," *Scientific Review of Mental Health Practice* 2, no. 1 (Summer 2003): http://www.srmhp.org/0201/adhd.html; Matthew Smith, *Hyperactive: The Controversial History of ADHD* (London: Reaktion, 2014), 21.

104. Les Linet, "The Search for Stimulation: Understanding Attention Deficit / Hyperactivity Disorder," http://www.selfgrowth.com/articles/search_stimulation_understanding_attention_deficit_hyperactivity_disorder.html, accessed January 31, 2018, as quoted in N. Katherine Hayles, "Hyper and Deep Attention: The Generational Divide in Cognitive Modes," *Profession* 2007, no. 1 (November 26, 2007): 190, https://doi.org/10.1632/prof.2007.2007.1.187.

105. Hayles, "Hyper and Deep Attention," 190.

106. Ela Malkovsky et al., "Exploring the Relationship between Boredom and Sustained Attention," *Experimental Brain Research* 221, no. 1 (August 2012): 59–67, https://doi.org/10.1007/s00221-012-3147-z.

107. Barbara Garson, *The Electronic Sweatshop: How Computers Are Transforming the Office of the Future* (New York: Penguin, 1989), 20, 36.

108. Ibid., 146–47.

109. Paul Beaudry, David A. Green, and Benjamin M. Sand, "The Great Reversal in the Demand for Skill and Cognitive Tasks," Working Paper (National Bureau of Economic Research, March 2013), http://www.nber.org/papers/w18901; Thomas B. Edsall, "The Downward Ramp," *New York Times,* June 10, 2014, http://www.nytimes.com/2014/06/11/opinion/the-downward-ramp.html.

110. More recently, Matthew Crawford in *Shopcraft as Soulcraft*, had worked an office job after getting his doctorate but took refuge from this clerkdom by starting up his own motorcycle repair business (47). Along similar lines, in *Race against the Machine: How the Digital Revolution Is Accelerating Innovation, Driving Productivity, and Irreversibly Transforming Employment and the Economy* (Lexington, MA: Digital Frontier Press, 2012), MIT business professors Erik Brynjolfsson and Andrew McAfee believe that the internet enables a winner-take-all society that deskills nonwinners and leads to their unemployment.

There are, however, more optimistic accounts. In *The Rise of the Creative Class: And How It's Transforming Work, Leisure, Community, and Everyday Life*, unabridged ed. (Brilliance Audio, 2014), Richard Florida and Mark Boyett argue that a creative class "will continue to grow in coming decades, as more traditional economic functions are transformed into Creative Class occupations." In "Better Than Human: Why Robots Will—and Must—Take Our Jobs," *WIRED*, https://www.wired.com/2012/12/ff-robots-will-take-our-jobs/, accessed January 31, 2018, Kevin Kelly thinks that robots "will help us discover new jobs for ourselves, new tasks that expand who we are."

111. Nicholas Carr, *The Glass Cage: How Our Computers Are Changing Us* (New York: W. W. Norton, 2015). See especially page 60.

112. See http://www.addforums.com/forums/archive/index.php/t-18702.html, accessed February 24, 2016.

113. Greg Bayles, Salt Lake City, Utah, July 5, 2016, interviewed by Luke Fernandez and Susan Matt.

114. Carolyn Y. Johnson, "The Joy of Boredom," *Boston.com,* March 9, 2008, http://archive.boston.com/bostonglobe/ideas/articles/2008/03/09/the_joy_of_boredom/?page=full.

115. "An App a Day Keeps the Boredom Away T-Shirt," https://shop.spreadshirt.com/mrappy/an+app+a+day+keeps+the+boredom+away-A7681581, accessed January 31, 2018; "Tumblr Sponsored LITERALLY FREE Download Tumblr Never Be Bored Again GIF. . . . ," http://me.me/i/tumblr-sponsored-literally-free-download-tumblr-never-be-bored-again, accessed January 31, 2018.

116. H., Grinnell, Iowa, May 23, 2016, interviewed by Luke Fernandez and Susan Matt.

117. Skyler Coombs, Ogden, Utah, October 26, 2015, interviewed by Luke Fernandez and Susan Matt.

118. Hailee Money, Ogden, Utah, October 30, 2015, interviewed by Luke Fernandez and Susan Matt.

119. Alta Martin, Ogden, Utah, October 28, 2015, interviewed by Luke Fernandez and Susan Matt.

120. Britney Fitzgerald, "Facebook Study Explains Why We Still Spend So Many Hours Stalking Each Other," *Huffington Post,* July 5, 2012, https://www.huffingtonpost.com/2012/07/04/facebook-study-shows-we-u_n_1644061.html, accessed June 11, 2018.

121. "Why Do People Use Facebook? Mostly to Relieve Boredom, but Every Once in a While to Express Political and Social Views," *Atlantic,* April 5, 2013,

https://www.theatlantic.com/technology/archive/2013/04/why-do-people-use-facebook/274721/, accessed June 11, 2018. In a 2017 post, a blogger at Hubspot explored what she termed "the Candy Crush effect." She maintained that most people who felt boredom today "all have at least one common outlook on boredom: It can be remedied by the internet." Amanda Zantal-Wiener, "The Candy Crush Effect: How Apps for Boredom Monetized Mobile Addiction," https://blog.hubspot.com/marketing/boredom-apps-mobile-addiction, accessed January 31, 2018.

122. Q., Ogden, Utah, October 19, 2015, interviewed by Luke Fernandez and Susan Matt.

123. Money interview.

124. Martin interview.

125. Coombs interview.

126. Martin interview.

127. Anna Renkosiak, Bettendorf, Iowa, August 12, 2015, interviewed by Susan Matt.

128. H. interview.

129. Debbie and William Speigle, Ogden, Utah, June 3, 2015, interviewed by Luke Fernandez and Susan Matt.

130. Barry Gomberg, Ogden, Utah, May 25, 2015, interviewed by Luke Fernandez and Susan Matt; Emily Meredith, Oakland, California, January 30, 2016, interviewed by Luke Fernandez and Susan Matt.

131. H. interview.

132. Sherry Turkle, "The Flight from Conversation," *New York Times,* April 21, 2012, http://www.nytimes.com/2012/04/22/opinion/sunday/the-flight-from-conversation.html.

133. William Powers, *Hamlet's Blackberry: Building a Good Life in the Digital Age* (New York: HarperCollins, 2010).

134. Perhaps the most famous of these is writer Paul Miller of *The Verge* who spent an entire year offline. See Paul Miller, "I'm Still Here: Back Online after a Year without the Internet," *Verge,* May 1, 2013, http://www.theverge.com/2013/5/1/4279674/im-still-here-back-online-after-a-year-without-the-internet. Other advocates of disconnecting (aka unplugging) are reviewed in Evgeny Morozov, "Only Disconnect," *New Yorker,* October 28, 2013, http://www.newyorker.com/magazine/2013/10/28/only-disconnect-2.

135. "Tim Cook at Apple WWDC 2018 Keynote," *Singju Post,* June 6, 2018, https://singjupost.com/tim-cook-at-apple-wwdc-2018-keynote-full-transcript/

?singlepage=1, https://www.apple.com/apple-events/june-2018/, accessed June 6, 2018.

136. "Center for Humane Technology," http://humanetech.com/, accessed June 13, 2018.

137. *The Case for Boredom,* http://www.wnyc.org/story/bored-brilliant-project-part-1/, accessed January 31, 2018; "Bored . . . and Brilliant? A Challenge to Disconnect from Your Phone," http://www.wbur.org/npr/376717870/bored-and-brilliant-a-challenge-to-disconnect-from-your-phone, accessed January 31, 2018.

138. Ian Bogost, *Play Anything: The Pleasure of Limits, the Uses of Boredom, and the Secret of Games* (New York: Basic Books, 2016), 6.

139. Spacks, *Boredom,* xi, 15.

140. Younger generations can also indict older generations. Although we did not interview anybody under eighteen, younger teenagers, as Patricia Spacks has noted, sometimes "believe in an adult conspiracy dooming them to tedium." And teenagers can also accuse adults of being "bores." See Spacks, *Boredom,* 259–60, xi. As danah boyd recounts in *It's Complicated: The Social Life of Networked Teens* (New Haven, CT: Yale University Press, 2015), teens are often depicted as bored and vacuous, but a far more sympathetic description of teen behavior can be made if we look beyond this surface caricature. Teens might seem addicted to their phones and unhealthily tethered to them. But what is really going on is that teens are "addicted to each other." They use their phones to establish social connections and to take hold of whatever remaining sovereignty they can salvage from lives that their parents have overly structured.

141. Diana Lopez, Ogden, Utah, May 22, 2018, interviewed by Luke Fernandez and Susan Matt.

142. L., Ogden, Utah, October 19, 2015, interviewed by Luke Fernandez and Susan Matt.

143. Speigles interview.

144. Cooper Ferrario, Ogden, Utah, June 3, 2015, interviewed by Luke Fernandez and Susan Matt.

145. K., Ogden, Utah, October 27, 2015, interviewed by Luke Fernandez and Susan Matt.

146. Money interview.

147. Meredith interview.

148. Pezze and Salzani, *Essays on Boredom and Modernity,* 21, 5–26. Goodstein also explores the relationship between boredom and modernity; see Goodstein, *Experience without Qualities.*

4. PAY ATTENTION

1. Nicholas Carr, "Is Google Making Us Stupid?" *Atlantic,* August 2008, http://www.theatlantic.com/magazine/archive/2008/07/is-google-making-us-stupid/306868/. See also Nicholas Carr, *The Shallows: What the Internet Is Doing to Our Brains* (New York: W. W. Norton, 2011), 7, 16.

2. Kevin McSpadden, "You Now Have a Shorter Attention Span Than a Goldfish," *Time,* May 14, 2015, http://time.com/3858309/attention-spans-goldfish/, accessed April 19, 2016.

3. Steven Johnson, *Everything Bad Is Good for You: How Today's Popular Culture Is Actually Making Us Smarter* (New York: Riverhead, 2006); Clay Shirky, *Cognitive Surplus: How Technology Makes Consumers into Collaborators* (New York: Penguin, 2011); Clive Thompson, *Smarter Than You Think: How Technology Is Changing Our Minds for the Better* (New York: Penguin, 2014).

4. See Carr on the Flynn effect, *The Shallows,* 144–145. Anne Anastasi, "Intelligence as a Quality of Behavior," in *What Is Intelligence? Contemporary Viewpoints on Its Nature and Definition,* ed. Robert J. Sternberg and Douglas K. Detterman (Norwood, NJ: Ablex, 1988), 19–21; J. W. Berry, "A Cross-Cultural View of Intelligence," in Sternberg and Detterman, *What Is Intelligence?,* 35–38.

5. Psychologists have described attention as a component of intelligence. See Nicholas R. Burns, Ted Nettelbeck, and James McPherson, "Attention and Intelligence: A Factor Analytic Study," *Journal of Individual Differences* 30 (2009): 44–57; and Karl Schweizer, Helfried Moosbrugger, and Frank Goldhammer, "The Structure of the Relationship between Attention and Intelligence," *Intelligence* 33, no. 6 (November–December 2005): 589–611. Attention and working memory are linked as "critical aspects" of "cognitive capacities." See Daryl Fougnie, "The Relationship between Attention and Working Memory," in *New Research on Short-Term Memory,* ed. Noah B. Johansen (New York: Nova Science, 2008), 1–45.

6. L. A. Dexter, "The Sociology and Politics of Stupidity in Society," in *Social Problems in American Society,* ed. James M. Henslin and Larry T. Reynolds (Boston: Holbrook, 1973), 35.

7. C. F. Goodey, *A History of Intelligence and Intellectual Disability: The Shaping of Psychology in Early Modern Europe* (Farnham, UK: Ashgate, 2011), 63–64, 69, 83.

8. Darrin M. McMahon, *Divine Fury: A History of Genius* (New York: Basic Books, 2013), xi–xxii.

9. The idea of genius took on great significance during the Enlightenment. Super intelligence had once been considered only the province of God, but

during the eighteenth century, philosophers began to assert that mere mortals might possess it, as they attributed more power to man and less to God. Ibid., xviii, 69, 73–75, 81, 112, 232–33, 239. For more on the range of early modern viewpoints, see Gilbert Gonzalez, "The Historical Development of the Concept of intelligence," *Review of Radical Political Economics* 11 (1979): 45. See also Robert J. Steinberg, *Metaphors of Mind: Conceptions of the Nature of Intelligence* (Cambridge: Cambridge University Press, 1990), 27–30; and George Ross MacDonald, *Starting with Hobbes* (London: Continuum, 2009).

10. Noah Sobe, "Concentration and Civilization: Producing the Attentive Child in the Age of Enlightenment," *Pedagocica Historica* 46, no. 1–2 (April 2010): 156–57, 150; Michael Hagner, "Toward a History of Attention in Culture and Science," *MLN* 118, no. 3 (2003): 670–87, doi:10.1353/mln.2003.0054.

11. Isaac Watts, *The Improvement of the Mind* (London: W. Wilson, 1821), 151–56.

12. Samuel Read Hall, *Lectures on School-Keeping* (Boston: Richardson, Lord and Holbrook, 1829), 112–14.

13. Henry Ward Beecher, *Seven Lectures to Young Men on Various Important Subjects Delivered before the Young Men of Indianapolis, Indiana, during the Winter of 1843–44* (Indianapolis: Thomas B. Cutler, 1844), 9, 11, 2, 4, 7.

14. Daniel Drake, *Pioneer Life in Kentucky, 1785–1800,* ed. Emmet Field Horine (New York: Henry Schuman, 1948), 144–46.

15. Warren Burton, *The District School as It Was, Scenery-Showing, and Other Writings* (Boston: R. E. Marvin, 1852), 94, 56.

16. "The Yankee Schoolmaster," quoted in Alfred L. Shoemaker, *Christmas in Pennsylvania: A Folk-Cultural Study* (Mechanicsburg, PA: Stackpole, 1959), 9.

17. Daniel T. Rodgers, *The Work Ethic in Industrial America, 1850–1920* (Chicago: University of Chicago Press, 1979), 18.

18. Thomas L. Haskell, *The Emergence of Professional Social Science: The American Social Science Association and the Nineteenth-Century Crisis of Authority* (Baltimore, MD: Johns Hopkins University Press, 2000), 40.

19. Robert H. Wiebe, *The Search for Order, 1877–1920* (Princeton, NJ: Hill and Wang, 1966), xiii.

20. William B. Sprague, "Sermon on the Completion of the Atlantic Telegraph, a Sermon Addressed to the Second Presbyterian Congregation, Albany, on Sunday Morning, September 5, 1858, on the Completion of THE ATLANTIC TELEGRAPH, by William B. Sprague, D.D., Their Pastor" (Albany, NY: Charles Van Benthuysen, 1858), 22–24, Warshaw Collection of Business Americana,

Collection 60, Telegraph Box 6, Folder 7, National Museum of American History, Smithsonian Institution.

21. William Allen White, *The Autobiography of William Allen White* (New York: Macmillan, 1946), 43.

22. *Washington Times* (Washington, DC), March 12, 1919.

23. Luther H. Gulick, *Mind and Work* (New York: Doubleday, Page and Company, 1908),.57.

24. Charles Richet, "Mental Strain," *Popular Science Monthly* 37 (1890): 488.

25. Jonathan Crary, *Suspensions of Perception: Attention, Spectacle, and Modern Culture* (Cambridge, MA: MIT Press, 1999), 13–14.

26. Die Aufmerksamkiet und die Funktion der Sinnesorgane," Ebbinghaus, Zeitschr., IX, 5 and 6, January 1896, pp. 342, 343, quoted in Alice Julia Hamlin, "Attention and Distraction, Thesis Submitted to the Faculty of Cornell University for the Degree of Ph.D.," reprinted from the *American Journal of Psychology* 8, no. 1.

27. Brian F. O'Donnell and Ronald A. Cohen, "Attention: A Component of Information Processing," in *The Neuropsychology of Attention,* ed. Ronald A. Cohen (New York: Springer, 1993), 11–15.

28. W. B. Pillsbury, *Attention* (London: Swan Sonnenschein, 1908); Edward Bradford Titchener, *Lectures on the Elementary Psychology of Feeling and Attention* (New York: Macmillan, 1908); Alice Julia Hamlin, "Attention and Distraction" (Ph.D. diss., Cornell University, 1896); John J. B. Morgan, "The Overcoming of Distraction and Other Resistances," in *Archives of Psychology,* ed. R. S. Woodworth, no. 35 (February 1916), Columbia University Contributions to Philosophy and Psychology 24, no. 4 (New York: Science Press, 1916).

29. Joseph Baldwin, *Elementary Psychology and Education* (New York: Appleton, 1888), 10–11.

30. Orison Swett Marden, *How They Succeeded: Life Stories of Successful Men Told by Themselves* (Boston: Lothrop, 1901), 233.

31. Ibid., 31, 119, 120, 121.

32. Michael O'Malley, "That Busyness That Is Not Business: Nervousness and Character at the Turn of the Last Century," *Social Research* 72, no. 2 (Summer 2005): 377–78; Ron Chernow, *Titan: The Life of John D. Rockefeller, Sr.,* 2nd ed. (New York: Vintage, 2005), 174.

33. T. Sharper Knowlson, *Business Psychology: A System of Mental Training for Commercial Life* (Libertyville, IL: Sheldon University Press, 1912), 37, 65, 69–70. Emphasis in the original.

34. Charles W. Gerstenberg, *Personal Power in Business* (New York: Prentice Hall, 1922), 78.

35. Roger W. Babson, *Making Good in Business* (New York: Fleming H. Revell, 1921), 14, 61.

36. Knowlson, *Business Psychology,* 37, 38.

37. Yale College and Robert J. O'Hara, "The Yale Report of 1828, Part I: Liberal Education and Collegiate Life," 1828, http://collegiateway.org/reading/yale-report-1828/.

38. Ibid. See also Luke O. Fernandez, "Preparing Students for Citizenship: The Pedagogical Vision of Yale's Noah Porter, Harvard's Charles Eliot and Princeton's Woodrow Wilson" (Ph.D. diss., Cornell University, 1997), http://ecommons.cornell.edu/handle/1813/7983, p. 42.

39. Charles W. Eliot, "Letter to Theodore Tebbets," January 29, 1854, Harvard University. Quoted in Burton J. Bledstein, *The Culture of Professionalism: The Middle Class and the Development of Higher Education in America* (New York: W. W. Norton, 1976).

40. "A Brief Timeline of Our First Two Centuries," Harvard Law School, http://hls.harvard.edu/about/history/, accessed May 9, 2016; "The Early Years," Harvard Medical School, https://hms.harvard.edu/about-hms/history-hms/early-years, accessed May 9, 2016.

41. Haskell, *Emergence of Professional Social Science;* Laurence R. Veysey, *The Emergence of the American University* (Chicago: University of Chicago Press, 1970); Hugh Hawkins, *Pioneer: A History of the Johns Hopkins University* (Baltimore, MD: Johns Hopkins University Press, 2002).

42. Irving Babbitt, *Atlantic Monthly* 89 (1902): 776. Also in Irving Babbitt and Russell Kirk, *Literature and the American College: Essays in Defense of the Humanities,* 3rd ed. (Washington, DC: National Humanities Institute, 1986). See also Fernandez, "Preparing Students for Citizenship," 164–65.

43. John Bascom, *Things Learned by Living* (New York: G. P. Putnam's Sons, 1913), 140. Quoted in Veysey, *Emergence of the American University,* 198.

44. Ruth Nelson to her parents, February 28, 1894, in *First-Person Cornell: Students' Diaries, Letters, Email, and Blogs,* ed. Carol Kammen (Ithaca, NY: Cornell University Press, 2006), 53.

45. Adelaide Tabor Young, December 17, 1896, in ibid., 64.

46. Harold Gulvin, March 6, 1927, in ibid., 124–25.

47. Evan Randolph, March 1, 1900, in "Journal," *American Scholar* 70, no. 2 (Spring 2001): 129–30.

48. John Hunter Detmold, 1943, in Kammen, *First-Person Cornell,* 149.

49. Nile Kinnick, January 16, 1939, in *A Hero Perished: The Diary and Selected Letters of Nile Kinnick,* ed. Paul Baender (Iowa City: University of Iowa Press, 1991), 24–28.

50. Daniel Hack Tuke, *Insanity in Ancient and Modern Life with Chapters on Its Prevention* (London: Macmillan, 1878), 108–10.

51. Ibid., 163.

52. Quoted in John Duffy, "Mental Strain and 'Overpressure' in the Schools: A Nineteenth-Century Viewpoint," *Journal of the History of Medicine and Allied Sciences* 23, no. 1 (1968): 75.

53. C. F. Folsom, *The Relation of Our Public Schools to the Disorders of the Nervous System* (Boston: Ginn and Company, 1886); David G. Schuster, *Neurasthenic Nation: America's Search for Health, Happiness, and Comfort, 1869–1920* (New Brunswick, NJ: Rutgers University Press, 2011), 116.

54. Charles K. Mills, "Mental Over-Work and Premature Disease among Public and Professional Men," Toner Lectures, Lecture IX, delivered March 19, 1884 (Washington, DC: Smithsonian Institution, 1885), 1.

55. George Beard, *American Nervousness: Its Causes and Consequences, a Supplement to Nervous Exhaustion (Neurasthenia)* (New York: G. P. Putnam's Sons, 1881), 105, 114

56. Frederick MacCabe, *On Mental Strain and Overwork* (Lewes, UK: G. P. Bacon, 1875), 11.

57. Beard, *American Nervousness,* 116–17.

58. William Hammond, *Cerebral Hyperaemia the Result of Mental Strain or Emotional Disturbance* (New York: G. P. Putnam's Sons, 1878), 10, 15.

59. Ibid., 20, 21, 22.

60. Ibid., 24.

61. H. C. Wood, *Brain-Work and Overwork* (Philadelphia: Presley Blakiston, 1880), 50–51.

62. Schuster, in *Neurasthenic Nation,* discusses the idea of limited energy within the body as a whole. For this reason, neurasthenics should not masturbate or have sex because ejaculation represented a net loss in energy.

63. Charles A. Dana, "Neurasthenia," *Medical Era* 9, no. 7 (July 1891): 212; George Miller Beard, *A Practical Treatise on Nervous Exhaustion (Neurasthenia): Its Symptoms, Nature, Sequences, Treatment* (New York: L. E. B. Treat, 1888), 137. See also Schuster, *Neurasthenic Nation,* 63.

64. Charles W. Burr, "Neurasthenia. The Traumatic Neuroses and Psychoses," in William Osler, *A System of Medicine by Eminent Authorities in Great Britain, the*

United States and the Continent, Vol VII., Diseases of the Nervous System (London: Henry Froude, 1910), 725; Maurizio Macaroni and Julien Bogousslavsky, "The Borderland with Neurasthenia," in *Hysteria: The Rise of an Enigma,* ed. J. Bogousslavsky (Basel: Karger, 2014), 149–56.

65. Hammond, *Cerebral Hyperaemia,* 31, 71–72.

66. Clarke described the case histories of many young women who ventured off to college. Edward H. Clarke, *Sex in Education or, A Fair Chance for the Girls* (Boston: J. R. Osgood, 1873), 83–84, 39.

67. Silas Weir Mitchell, *Wear and Tear: Or Hints for the Overworked,* 5th ed. (Philadelphia: J.B. Lippincott Company, 1891), 31–41; G. Stanley Hall, *Adolescence: Its Psychology, and Its Relation to Physiology, Anthropology, Sociology, Sex, Crime, Religion and Education, Vol. 1* (New York: D. Appleton, 1904), 508–11.

68. Diana Martin, "The Rest Cure Revisited," *American Journal of Psychiatry* 164, no. 5 (May 2007).

69. Hammond, *Cerebral Hyperaemia,* 98, 106, 107.

70. Wood, *Brain-Work and Overwork,* 90, 87.

71. Quoted in Rodgers, *Work Ethic in Industrial America,* 67.

72. US Senate, Committee on Education and Labor, *Report upon the Relations between Labor and Capital,* 4 vols. (Washington, DC, 1885), 2:549, quoted in Rodgers, *Work Ethic in Industrial America,* 67–68.

73. Beard, *American Nervousness,* 101–3.

74. Rodgers, *Work Ethic in Industrial America,* 73.

75. As Eliot said, "When the judicious determination of a public policy depends on careful collection of facts, keen discrimination, sound reasoning, and sure foresight, our republic must soon follow, as all other civilized governments already do, the advice of highly trained men, who have made themselves, by long study and observation, experts in the matter in hand. . . . The more complicated and difficult the public business becomes, the more pressing the need of expert management; and soon any other management will be simply ruinous. Now, the experts needed are going to be trained in the American universities which . . . maintain at large centers of population well-equipped schools for all the learned and scientific professions." Fernandez, "Preparing Students for Citizenship," 183.

76. B. O. Flower, "President Eliot and Union Labor," *Arena* (1903): 192–200; Fernandez, "Preparing Students for Citizenship," 188–89; Rodgers, *Work Ethic in Industrial America,* 234–40.

77. Rodgers, *Work Ethic in Industrial America,* 67–68.

78. Frederick Winslow Taylor, *The Principles of Scientific Management* (1911; New York: W. W. Norton, 1967), 9, 59.

79. Frederick W. Taylor, "Workmen and Their Management," delivered April 20, 1909, pp. 3, 8, in Lectures and Addresses, 1908, in Box "Archives CD 791.50-DC 897, folder 823.1," Harvard Business School Archives, Baker Library, Cambridge, MA; See also Taylor, *Principles of Scientific Management,* 125–26.

80. Taylor, *Principles of Scientific Management,* 125–26, 7.

81. William James, *The Energies of Men* (New York: Moffat, Yard and Company, 1914), 14–15, 10–11.

82. Barry L. Beyerstein, "Whence Cometh the Myth That We Only Use 10% of Our Brains," in *Mind Myths: Exploring Popular Assumptions about the Mind and Brain*, ed. Sergio Della Salla (Chichester, UK: John Wiley and Sons, 1999), 12–14. See also Christian Jarrett, *Great Myths of the Brain* (Chichester, UK: John Wiley and Sons, 2015), 51–54.

83. "The Telephone: Its Use, Misuse and True Place in Our System of Communication," *Omaha (NE) Daily Bee,* April 7, 1881.

84. Annie Maude Dee Porter, Diary, February 16, 1936, Special Collections, Stewart Library, Weber State University, Ogden, UT; https://dc.weber.edu/node/13.

85. Annie Maude Dee Porter, Diary, October 29, 1940, in ibid.

86. Harvey Cantril and Gordon W. Allport, *The Psychology of Radio* (New York: Harper and Brothers, 1935), 25–26.

87. "Radio inside Campus Gates," *New York Times,* December 7, 1930.

88. Cantril and Allport, *Psychology of Radio,* 104, 105.

89. "'Both Sides' of the Radio Argument," *Literary Digest,* January 11, 1930, 26.

90. Azriel Eisenberg, *Children and Radio Programs: A Study of More Than Three Thousand Children in the New York Metropolitan Area* (New York: Columbia University Press, 1936), 6, 135.

91. Ibid., 135–36.

92. "Books and TV," *Publishers Weekly* (June 17, 1950): 2638.

93. "Effect of Television on Reading Is Estimated in Recent Surveys," *Publishers Weekly* 159 (April 21, 1951): 1708. See also "Television: Marvel or Monster?" Newsweek: The Club and Educational Bureaus, March 1949, in Folder 4, Box 7, Series 15, Allen B. Dumont Collection, Archives Center, National Museum of American History, Smithsonian Institution; "Television versus Reading," *Wilson Library Bulletin,* December 1951, 327.

94. William Schlamm, "Critical Look at Television," *American Mercury*, October 1959, 70.

95. "Television and Education" data from Dumont, Dumont Sales and Research Department, Box 53, Folder 8, Series 12, Archives Center, National Museum of American History, Smithsonian Institution.

96. "A Summary of Yearly Studies of Televiewing 1949–1963," by Paul A. Witty, Paul Kinsella, and Anne Coomer, reprinted from the Elementary English Organ of the National Council of Teachers of English, October 1963, p. 594, in Dumont Collection, Box 53, Folder 8, Series 12, in ibid.

97. Matt Richtel, *A Deadly Wandering: A Mystery, a Landmark Investigation, and the Astonishing Science of Attention in the Digital Age* (New York: William Morrow, 2015), 98–104; John Duncan, *How Intelligence Happens* (New Haven, CT: Yale University Press, 2012), 15–16; Neville Moray, "Donald E. Broadbent: 1926–1993," *American Journal of Psychology* 108, no. 1 (1995): 117–21.

98. Moray, "Donald E. Broadbent"; D. E. Broadbent, *Perception and Communication* (New York: Oxford University Press, 1987); Anne M. Treisman, "Contextual Cues in Selective Listening," *Quarterly Journal of Experimental Psychology* 12, no. 4 (October 1, 1960): 242–48, doi:10.1080/17470216008416732; "Michael Posner on the Anatomy of Attentional Networks—a Historical Perspective," n.d., https://vimeo.com/5280169; Noel Sheehy and Alexandra Forsythe, *Fifty Key Thinkers in Psychology* (New York: Routledge, 2003), 47–51.

99. Quoted in Richtel, *Deadly Wandering*, 100.

100. Matthew Smith, *Hyperactive: The Controversial History of ADHD* (London: Reaktion, 2014), 54.

101. Ibid., 62, 68.

102. American Psychiatric Association, *Diagnostic and Statistical Manual of Mental Disorders (Third Edition)*, 2nd ed. (Washington, DC: American Psychiatric Association, 1980); American Psychiatric Association, *Diagnostic and Statistical Manual of Mental Disorders: DSM-III-R*, 3rd rev. ed. (Washington, DC: American Psychiatric Association, 1987).

103. Smith, *Hyperactive*, 21.

104. Stephen P. Hinshaw and Katherine Ellison, *ADHD: What Everyone Needs to Know* (New York: Oxford University Press, 2015), 15–16, 18; Centers for Disease Control and Prevention, "Attention-Deficit / Hyperactivity Disorder (ADHD), Data and Statistics," https://www.cdc.gov/ncbddd/adhd/data.html, accessed October 7, 2017.

105. Hinshaw and Ellison, *ADHD*, 95.

106. Richard DeGrandpre, *Ritalin Nation: Rapid-Fire Culture and the Transformation of Human Consciousness* (New York: W. W. Norton, 2000), 15, 32.

107. The quote comes from Milan Kundera, *Slowness: A Novel* (New York: Harper Perennial, 1997), quoted in DeGrandpre, *Ritalin Nation,* 16, 32, 33.

108. Malcolm Gladwell, "Running from Ritalin," *New Yorker,* February 15, 1999, http://www.newyorker.com/magazine/1999/02/15/running-from-ritalin, accessed October 4, 2015.

109. Meredith Melnick, "Faking It: Why Nearly 1 in 4 Adults Who Seek Treatment Don't Have ADHD," *Time,* April 28, 2011, http://healthland.time.com/2011/04/28/faking-it-why-nearly-1-in-4-adults-who-seek-treatment-dont-have-adhd/; "Poll Results: Look Who's Doping," *Nature* 452 (2008), http://www.nature.com/news/2008/080409/full/452674a.html, accessed October 4, 2015; Henry Greely et al., "Towards Responsible Use of Cognitive Enhancing Drugs by the Healthy," *Nature* (2008): 702–5; A. D. DeSantis, E. M. Webb, and S. M. Noar, "Illicit Use of Prescription ADHD Medications on a College Campus: A Multi-methodological Approach," *Journal of American College Health* 57, no. 3 (November–December 2008): 315–24, https://www.ncbi.nlm.nih.gov/pubmed/18980888, accessed October 7, 2017.

110. Alan Schwarz, "Drowned in a Stream of Prescriptions," *New York Times,* February 2, 2013.

111. Christian Wutz and Chris Fowles, Salt Lake City, Utah, March 25, 2016, interviewed by Luke Fernandez and Susan Matt.

112. TEDx Talks, "TEDxEast—Matt Crawford—Manual Competence," https://www.youtube.com/watch?v=xdGky1JZovg, accessed April 10, 2016. While Crawford appreciates why Ritalin is dispensed, he ultimately regards it as a drug that runs counter to many people's natural inclinations: "There's wide use of psychiatric drugs to medicate boys. In particular against their natural bent toward action. 'The better to keep things on track' as the school nurse says. And this serves the interests of educators in preserving their own sanity. And I know that because I taught high school for a year and I would have loved to have had a Ritalin fogger in my classroom for the sake of maintaining order. So I get it. But what I get, I think, is that it's a rare person, male or female, whose naturally inclined to sit still for sixteen years at school and then indefinitely at work" (minute 13:54).

113. Greely et al., "Towards Responsible Use of Cognitive Enhancing Drugs," 702–5.

114. Thomas Hills and Ralph Hertwig, "Why Aren't We Smarter Already: Evolutionary Trade-Offs and Cognitive Enhancements," *Current Directions in Psychological Science* 20 (2011): 373–77.

115. Hinshaw and Ellison, *ADHD,* 104–6, ; Peter R. Breggin and Dick Scruggs, *Talking Back to Ritalin: What Doctors Aren't Telling You about Stimulants and ADHD* (Cambridge, MA: Da Capo, 2001); Lawrence H. Diller, "The Ritalin Wars Continue," *Western Journal of Medicine* 173, no. 6 (December 2000): 366–67; Judith Warner, "Ritalin Wars," *New York Times,* November 15, 2007, http://opinionator.blogs.nytimes.com/2007/11/15/ritalin-wars/.

116. Patricia Marx, "Mentally Fit," *New Yorker,* July 29, 2013.

117. http://www.lumosity.com/about, accessed May 1, 2016.

118. http://www.brainiversity.com/About.html, accessed August 4, 2016.

119. Shirky argues that information overload is a perennial aspect of the human condition—not specific to modernity. Therefore, one should not try to stem its flow so much as develop better filters to manage it. Shirky is eager to shift the emphasis away from curtailing the information stream to finding technologies and human aptitudes that can manage it better.

120. As one pundit puts it, there are "25 billion reasons you need to pay attention." See Robert Brands, "All Aboard the Sensor Revolution: How the Internet of Things Is Driving Innovation," *Huffington Post,* http://www.huffingtonpost.com/robert-f-brands/all-aboard-the-sensor-rev_b_8263454.html. See also National Science Foundation, "The Sensor Revolution: Overview," https://www.nsf.gov/news/special_reports/sensor/overview.jsp, accessed May 15, 2016.

121. David Hochman, "Reinvent Yourself: The *Playboy* Interview with Ray Kurzweil," *Playboy,* April 19, 2016, https://www.playboy.com/articles/playboy-interview-ray-kurzweil.

122. Ibid.; Nicholas Carr, "Gigantic, a Big Big Brain," *ROUGH TYPE,* April 23, 2016, http://www.roughtype.com/?p=6912.

123. Carr, "Gigantic, a Big Big Brain."

124. Adam Gazzaley and Larry D. Rosen, *The Distracted Mind: Ancient Brains in a High-Tech World* (Cambridge, MA: MIT Press, 2016); Nicholas Carr, "Question Marks of the Mysterians," *ROUGH TYPE,* September 8, 2017, http://www.roughtype.com/?p=8093.

125. "New Survey Finds Americans Care about Brain Health, but Misperceptions Abound," Michael J. Fox Foundation, September 25, 2013, https://www.michaeljfox.org/foundation/publication-detail.html?id=484&category=7, accessed August 15, 2016.

126. Interview with Duhigg by Stephen Dubner on Dubner's Freakonomics podcast. See Arwa Gunja, "How to Be More Productive," *Freakonomics,* http://freakonomics.com/podcast/how-to-be-more-productive/, accessed April 22, 2016.

127. Clive Thompson, "Meet the Life Hackers," *New York Times,* October 16, 2005, http://www.nytimes.com/2005/10/16/magazine/meet-the-life-hackers.html.

128. Ibid.; Mary Czerwinski, Edward Cutrell, and Eric Horvitz, "Instant Messaging and Interruption: Influence of Task Type on Performance," http://research.microsoft.com/apps/pubs/default.aspx?id=101782, accessed April 20, 2016; Mary Czerwinski et al., "Toward Characterizing the Productivity Benefits of Very Large Displays," http://research.microsoft.com/apps/pubs/default.aspx?id=64317, accessed April 20, 2016.

129. Chad Brooks, "10 Distractions That Kill Workplace Productivity," *Business News Daily,* June 12, 2015, http://www.businessnewsdaily.com/8098-distractions-kiling-productivity.html, accessed April 28, 2016.

130. Cheryl Conner, "Employees Really Do Waste Time at Work, Part II," *Forbes,* November 15, 2012, http://www.forbes.com/sites/cherylsnappconner/2012/11/15/employees-really-do-waste-time-at-work-part-ii/#2984cd627ad4, accessed April 28, 2016.

131. "Driven by Distractions: Why Employees Focus Is Waning at Work and What You Can Do about It," 2014 Pulse Paper, http://community.virginpulse.com/web_drivendistractions, accessed April 24, 2018.

132. "More Than 7 in 10 Americans Think Technology Has Become Too Distracting and Is Creating a Lazy Society," http://www.theharrispoll.com/health-and-life/Technology-Too-Distracting-Lazy-Society.html, accessed May 6, 2016.

133. Cathy N. Davidson, *Now You See It: How Technology and Brain Science Will Transform Schools and Business for the 21st Century* (New York: Penguin, 2011), 6, 181–87.

134. N. Katherine Hayles, "Hyper and Deep Attention: The Generational Divide in Cognitive Modes," *Profession* 2007, no. 1 (November 26, 2007): 188, https://doi.org/10.1632/prof.2007.2007.1.187.

135. Ibid.

136. Ibid., 194.

137. Crary, *Suspensions of Perception,* 13, 14.

138. Jonathan Crary, *24/7: Late Capitalism and the Ends of Sleep* (London: Verso, 2013), 15.

139. Warren and Harriet Trezevant, Oakland, California, January 31, 2016, interviewed by Luke Fernandez and Susan Matt.

140. Gregory Bayles, Salt Lake City, Utah, July 5, 2016, interviewed by Luke Fernandez and Susan Matt.

141. Barry Gomberg, Ogden, Utah, May 25, 2015, interviewed by Luke Fernandez and Susan Matt.

142. Will Xu, Grinnell, Iowa, May 23, 2016, interviewed by Luke Fernandez and Susan Matt.

143. Cooper Ferrario, Ogden, Utah, June 3, 2015, interviewed by Luke Fernandez and Susan Matt.

144. Anna Renkosiak, Bettendorf, Iowa, August 24, 2015, interviewed by Susan Matt.

145. Stephanie Christiansen, Ogden, Utah, May 20, 2015, interviewed by Luke Fernandez and Susan Matt.

146. Jack and Nancy Pleger, Phoenix, Arizona, December 15, 2014, interviewed by Luke Fernandez and Susan Matt.

147. K., Ogden, Utah, October 27, 2015, interviewed by Luke Fernandez and Susan Matt.

148. Alta Martin, Ogden, Utah, October 28, 2015, interviewed by Luke Fernandez and Susan Matt.

149. Allen Riedy, Honolulu, Hawaii, July 10, 2015, interviewed by Luke Fernandez and Susan Matt.

150. Katherine J. Schall, Ogden, Utah, October 18, 2014, interviewed by Luke Fernandez and Susan Matt.

151. Ana Chavez, Ogden, Utah, October 19, 2015, interviewed by Luke Fernandez and Susan Matt.

152. Wutz and Fowles interview.

153. N., Honolulu, Hawaii, July 9, 2015, interviewed by Luke Fernandez and Susan Matt.

154. Victor, Rocky Mountains, January 8, 2017, interviewed by Luke Fernandez and Susan Matt.

155. Xavier Stilson, Ogden, Utah, October 20, 2015, interviewed by Luke Fernandez and Susan Matt.

156. Christiansen interview.

157. Camree Larkin, Ogden, Utah, October 30, 2015, interviewed by Luke Fernandez and Susan Matt.

158. Eli Gay, Berkeley, California, January 31, 2016, interviewed by Luke Fernandez and Susan Matt.

159. See http://www.bostonmagazine.com/news/article/2014/07/29/fomo-history/. Also see Alice E. Marwick, *Status Update: Celebrity, Publicity, and Branding in the Social Media Age* (New Haven, CT: Yale University Press, 2015), 226–27.

160. Gay interview.

161. Ibid.

162. Recalling the "recession of causation" that Haskell noted in the nineteenth century, Crawford notes a similar sensation at play today: "Both as workers and consumers, we feel we move in channels that have been projected from afar by vast impersonal forces. We worry that we are becoming stupider, and begin to wonder if getting an adequate grasp on the world, intellectually, depends on getting a handle on it in some literal and active sense." See Matthew B. Crawford, *Shop Class as Soulcraft: An Inquiry into the Value of Work* (New York: Penguin, 2010), 7. Worries about complexity and how to manage it are also captured in an SAP advertisement: "While . . . advancements have improved our lives and provided us with greater opportunities for innovation than ever before, they have also accelerated the rise of . . . unprecedented and crippling complexity. . . . Complexity is becoming the most intractable issue of our time, an epidemic of wide-ranging proportions, affecting our lives, our work and even our health. . . . How do we pull ourselves out from this expanding quagmire of complexity? . . . In a world besieged by complexity, 'simple' wins. . . . [Y]ou'll hear more from SAP on the many ways we can help the world run simple by reducing and managing the complexity that is the real enemy of innovation. Simply put, it's what the world needs now." See "SAP Run Simple: Manifesto," http://global.sap.com/campaigns/digitalhub-runsimple/manifesto/index.html, accessed April 9, 2016. Finally, danah boyd says that "living in a networked world is complicated" and that complexity is not just a challenge for adults but also an emblematic experience for modern networked teens. See danah boyd, *It's Complicated: The Social Lives of Networked Teens* (New Haven, CT: Yale University Press, 2015).

163. Nicholas Carr is perhaps the best-known critic of the new attention economy and its power to distract Americans. But see also Matthew B. Crawford, *The World beyond Your Head: On Becoming an Individual in an Age of Distraction* (New York: Farrar, Straus and Giroux, 2016), 3–27; and Malcolm McCullough, *Ambient Commons: Attention in the Age of Embodied Information* (Cambridge, MA: MIT Press, 2015).

164. Adam Alter, *Irresistible: The Rise of Addictive Technology and the Business of Keeping Us Hooked* (New York: Penguin, 2017).

165. Tim Wu, *The Attention Merchants: The Epic Scramble to Get Inside Our Heads* (New York: Knopf, 2016), 6.

166. Wutz and Fowles interview.

167. Q., Ogden, Utah, October 19, 2015, interviewed by Luke Fernandez and Susan Matt.

168. Emily Meredith, Oakland, California, January 30, 2016, interviewed by Luke Fernandez and Susan Matt.

169. Carr, *Shallows,* 156–57.

170. Wutz and Fowles interview.

171. One person's distractions, in addition to being another's opportunity for financial gain, could, paradoxically, also be another's opportunity for focus. Emily Meredith saw it this way when she observed that when her kids were allowed to be distracted by screens, she herself had more time to focus: "The really tricky part is that I benefit from their time online because then I have time to myself so . . . when they spend time on games I can get something done. . . . And so I feel really ambivalent about it. . . . I don't want them on screens basically . . . but I have to admit that I have a benefit from them being on the screens." Meredith interview.

5. AWE

1. Jill Shargaa, "Please, Please, People. Let's Put the 'Awe' Back in 'Awesome,'" https://www.ted.com/talks/jill_shargaa_please_please_people_let_s_put_the_awe_back_in_awesome?language=en, accessed June 20, 2016.

2. Gwen Dewar, "A Chronic Lack of Awe," *Psychology Today* (blog), August 8, 2012, https://www.psychologytoday.com/blog/making-humans/201208/chronic-lack-awe; Paul K. Piff et al., "Awe, the Small Self, and Prosocial Behavior," *Journal of Personality and Social Psychology* 108, no. 6 (2015): 883–99, https://doi.org/10.1037/pspi0000018.

3. Paul Pearsall, *AWE: The Delights and Dangers of Our Eleventh Emotion* (Deerfield Beach, FL: Health Communications, 2007), 145–46; Paul Piff and Dacher Keltner, "Why Do We Experience Awe?" *New York Times,* May 22, 2015, http://www.nytimes.com/2015/05/24/opinion/sunday/why-do-we-experience-awe.html?nytmobile=0, accessed June 11, 2016.

4. "Awe," Oxford English Dictionary Online, http://www.oed.com/view/Entry/13911?rskey=U7Frtj&result=1&isAdvanced=false#eid. The John Templeton Foundation, which has funded a number of grants on "understanding awe and wonder," defines "awe" in this way: "Awe and wonder express our longing and our uncertainty, our fascination and our terror. They point to the transcendent and to the limits of being human." "Understanding Awe and Wonder," https://www.templeton.org/what-we-fund/funding-priorities/understanding-awe-and-wonder, accessed June 30, 2016. This definition is also quoted in Greg Bayles, "Awe Evolving: Transforming Notions of Awe in the Digital Age," *Digital America,* April 24, 2014, p. 2, http://www

.digitalamerica.org/awe-evolving-transforming-notions-of-awe-in-the-digital -age-greg-bayles/.

5. "Wonder," Oxford English Dictionary Online, http://www.oed.com.hal .weber.edu:2200/, accessed October 1, 2017.

6. "Sublime," Oxford English Dictionary Online, http://www.oed.com.hal .weber.edu:2200/, accessed October 1, 2017.

7. "Astonishment," Oxford English Dictionary Online, http://www.oed.com .hal.weber.edu:2200/view/Entry/12178?redirectedFrom=astonishment&; http://www.dictionary.com/browse/astonish?s=t, accessed October 2, 2017.

8. For a discussion of the emotional reactions to new technologies in nineteenth-century America, and particularly the feeling of the sublime that such inventions often evoked, see Brenton Malin, *Feeling Mediated: A History of Media Technology and Emotion in America* (New York: New York University Press, 2014); David E. Nye, *American Technological Sublime* (Cambridge, MA: MIT Press, 1996); John Kasson, *Civilizing the Machine: Technology and Republican Values in America, 1776–1900* (New York: Hill and Wang, 1999); and Leo Marx, *The Machine in the Garden: Technology and the Pastoral Ideal in America* (New York: Oxford University Press, 1964).

9. William B. Sprague, "Sermon on the Completion of the Atlantic Telegraph—a Sermon Addressed to the Second Presbyterian Congregation, Albany, on Sunday Morning, September 5, 1858, on the Completion of THE ATLANTIC TELEGRAPH by William B. Sprague D.D., Their Pastor" (Albany, NY: Charles Van Benthuysen, 1858), 22, 7–8, 19, Warshaw Collection of Business Americana, Collection 60, Telegraph Box 6, Folder 7, National Museum of American History, Smithsonian Institution.

10. Allen Riedy, Honolulu, Hawaii, July 10, 2015, interviewed by Luke Fernandez and Susan Matt.

11. Max Weber, "Science as a Vocation," in *From Max Weber: Essays in Sociology*, ed. H. H. Gerth and C. Wright Mills (New York: Oxford University Press, 1946), 129–56. The literature on disenchantment is vast, and not all agree as to when or if Europeans and Americans became disenchanted. Some suggest it was a phenomenon that began in the seventeenth and eighteenth centuries; others point to the late nineteenth or twentieth centuries, and still other scholars maintain that disenchantment never occurred. A good introduction is Michael Saler, "Modernity and Enchantment: A Historiographic Review," *American Historical Review* 111, no 3 (June 2006): 692–716. Taking up Weber's claim that enchantment has not disappeared so much as "retreated," in Landy and Saler's *The Re-Enchantment of the World,* the authors suggest that disenchantment has not actually happened. Rather, it's moved to new venues. See

Joshua Landy and Michael Saler, eds., *The Re-Enchantment of the World: Secular Magic in a Rational Age* (Stanford, CA: Stanford University Press, 2009). See also George Ritzer, *Enchanting a Disenchanted World: Continuity and Change in the Cathedrals of Consumption,* 3rd ed. (Thousand Oaks, CA: Sage, 2010); and Bruno Latour, *We Have Never Been Modern,* trans. Catherine Porter (Cambridge, MA: Harvard University Press, 1993).

12. For a discussion of the significance of the classical tradition to American selfhood and to American education, see John Shields, *The American Aeneas: Classical Origins of the American Self* (Knoxville: University of Tennessee Press, 2001); and Caroline Winterer, *The Culture of Classicism: Ancient Greece and Rome in American Intellectual Life, 1780–1910* (Baltimore, MD: Johns Hopkins University Press, 2002).

13. I. Bernard Cohen, *Benjamin Franklin's Science* (Cambridge, MA: Harvard University Press, 1990), 157; "Two Boston Puritans on God, Earthquakes, Electricity, and Faith, 1755-1756," National Humanities Center, 2009, nationalhumanitiescenter.org/pdf; Eleanor M. Tilton, "Lightning Rods and the Earthquake of 1755," *New England Quarterly* 13, no. 1 (March 1940): 85-97; Philip Dray, *Stealing God's Thunder: Benjamin Franklin's Lightning Rod and the Invention of America* (New York: Random House, 2005), xvii, 105-11.

14. Early railroad passengers wondered at the strange power the machines possessed. Harriet Beecher Stowe wrote of a train trip in 1842, describing the locomotive as "spitting fire and smoke like some great fiend monster." She added, "I do think these steam concerns border a little too much on the super-natural to be agreeable." Letter from Harriet Beecher Stowe to Georgiana May Sykes, August 29, 1842, in *Life and Letters of Harriet Beecher Stowe*, ed. Annie Fields (Boston: Houghton, Mifflin, 1897), 107-8. Capturing similar sentiments was Nathaniel Hawthorne's short story "The Celestial Railroad," which parodied John Bunyan's *Pilgrim's Progress.* Hawthorne's pilgrims, unlike Bunyan's, journeyed by rail instead of by foot, and instead of ending in heaven, realized too late that the railroad was taking them to hell.

15. Leo Marx, "Does Improved Technology Mean Progress?" *Technology Review,* January 1987, 33-41.

16. As Leo Marx, the eminent historian of technology, has observed, these events led to a certain degree of "technological pessimism" in the American public. Leo Marx, "The Idea of 'Technology' and Postmodern Pessimism," in *Technology, Pessimism, and Postmodernism,* ed. Yaron Ezrahi, Everett Mendelsohn, and Howard Segal, Sociology of the Sciences 17 (Dordecht, Netherlands: Kluwer Academic Publishers, 1994), 11-28, https://doi.org/10.1007/978-94-011-0876-8_2. Also see Merritt Roe Smith and Leo Marx, eds., *Does Technology*

Drive History? The Dilemma of Technological Determinism (Cambridge, MA: MIT Press, 1994).

17. Theologian Philip Hefner suggests that, at root, "technology is about being finite and mortal. We create technology in order to compensate for our finitude. Because technology can outlive us and be stronger than we are, more accurate, and faster, the very existence of our technology reminds us of our finitude and mortality." Because of its "engagement with finitude and death, technology becomes almost explicitly religious." Philip Hefner, "Technology and Human Becoming," *Zygon* 37 (September 2002): 658–59.

18. David F. Noble, *The Religion of Technology: The Divinity of Man and the Spirit of Invention* (New York: Penguin, 1999), 94.

19. Frances O. J. Smith, "Report," *Electro-Magnetic Telegraph* (1838) 2; and Frances O. J. Smith, "The Post Office Department: Considered with Reference to Its Condition, Policy, Prospects, and Remedies," *Hunt's Merchant Magazine* 11–12 (December 1844–February 1845): 151, all quoted in Richard J. John, *Network Nation: Inventing American Telecommunications* (Cambridge, MA: Belknap Press of Harvard University Press, 2010), 35–36, 42.

20. John, *Network Nation,* 34–43.

21. For more context on this, see Paul Gilmore, *Aesthetic Materialism: Electricity and American Romanticism* (Stanford, CA: Stanford University Press, 2009); and Nye, *American Technological Sublime.*

22. For references to the "lightning line," see, for instance, "Telegraphic Communication with Washington," *New York Herald,* November 20, 1845; "The Telegraph," *Columbian Fountain* (Washington, D.C.), July 29, 1846; "Lightning Line Complete!" *Northern Galaxy* (Middlebury, VT), September 22, 1846; and "Atlantic Lake and Mississippi Telegraph," *Boon's Lick Times* (Fayette, MO), December 19, 1846, among many others.

23. Amos Kendall to William Stickney, June 1, 1855, in *Autobiography of Amos Kendall,* ed. William Stickney (Boston: Lee and Shepard, 1872), 551–52; Scott M. Cutlip, *Public Relations History: From the 17th to the 20th Century: The Antecedents* (New York: Routledge, 2009), 113.

24. Alvin F. Harlow, *Old Wires and New Waves* (1936; New York: Arno and the *New York Times,* 1971), 157.

25. Joseph A. Copp, *The Atlantic Telegraph: As Illustrating the Providence and Benevolent Designs of God. A Discourse Preached in the Broadway Church, Chelsea, August 8, 1858* (Boston: T. R. Marvin and Son, 1858), 11.

26. Sprague, "Sermon."

27. Rufus W. Clark, *The Atlantic Telegraph: A Discourse Preached in the South Congregational Church, Brooklyn, by Rev. Rufus W. Clark* (New York: Sheldon, Blakeman, 1858), 5–7.

28. Henry M. Field, *History of the Atlantic Telegraph* (New York: Charles Scribner, 1869), 419.

29. Caroline Cowles Richards, Diary, August 1858, in Richards, *Village Life in America, 1858–1872 Including the Period of the American Civil War as Told in the Diary of a School-Girl* (New York: Henry Holt, 1913), 101–2.

30. Letter from Elizabeth Emma Stuart to William Chapman Baker, March 4, 1858, in *Stuart Letters of Robert and Elizabeth Sullivan Stuart and Their Children 1819–1864: With an Undated Letter Prior to July 21, 1813*, vol. 2 (New York: Privately published, 1961), 855; "My Telegram," Telegraph Box 5, Folder "Western Union-poem," Warshaw Collection of Business Americana, National Museum of American History, Smithsonian Institution.

31. Letter from Sister Saint Francis Xavier to her mother, August 16, 1855; Letter from Sister Saint Francis Xavier to Charles le Fer de la Motte, March 1851, both in *The Life and Letters of Sister St. Francis Xavier (Irma LeFer de la Motte) of the Sisters of Providence of Saint Mary-of-the-Woods, Indiana,* rev. ed. (St Louis, MO: B. Herder, 1917), 387, 307–8.

32. Catherine M. Trowbridge, "The Invisible Wires," *Telegrapher,* December 15, 1866.

33. Jeffrey Sconce, *Haunted Media: Electronic Presence from Telegraphy to Television* (Durham, NC: Duke University Press, 2000), 20–28; John Durham Peters, *Speaking into the Air: A History of the Idea of Communication* (Chicago: University of Chicago Press, 2000), 94–96.

34. Julia LeGrand Waitz, Diary, January 1863, in *The Journal of Julia LeGrand, New Orleans 1862–1863,* ed. Kate Mason Rowland and Agnes E. Croxall (Richmond, VA: Everett Waddey, 1911), 98–99.

35. Quoted in Daniel J. Czitrom, *Media and the American Mind: From Morse to McLuhan* (Chapel Hill: University of North Carolina Press, 1982), 12.

36. Quoted in Harlow, *Old Wires and New Waves,* 77.

37. Frederick Douglass, "Lecture on Pictures" (1861), in *Picturing Frederick Douglass: An Illustrated Biography of the Nineteenth-Century's Most Photographed American,* ed. John Stauffer, Zoe Trodd, and Celeste-Marie Bernier (New York: W. W. Norton, 2015), 141.

38. Anne Lynch Botta to James Anthony Froude, March 28, 1874, in *Memoirs of Mrs. Anne C. L. Botta: Written by Her Friends: With Selections from Her Correspondence*

and from Her Writing in Prose and Poetry, ed. Vincenzo Botta (New York: J. Selwin Tait and Sons, 1893), 301–2.

39. Delia Locke, Diary, September 18, 1858, in Delia Locke Diary, 1858–1861, Locke-Hammond Family Papers, MSS 110, Holt-Atherton Special Collections, University of the Pacific, http://www.pacific.edu/University-Libraries/Find/Holt-Atherton-Special-Collections/Delia-Locke-Diaries.html, accessed April 15, 2013.

40. Delia Locke, May 23, 1864, in Delia Locke Diary, 1862–1869, ibid.

41. Delia Locke, April 25, 1871, in Delia Locke Diary, 1870–1874, ibid.

42. Delia Locke, October 10, 1871, in ibid.

43. Delia Locke, December 23, 1871, in ibid.

44. Delia Locke, November 28, 1874, in ibid.

45. Delia Locke, December 26, 1875, in Delia Locke Diary, 1875–79, ibid.

46. Delia Locke, February 9, 1878; September 18, 1879; and April 22, 1879, in ibid.

47. June 9, 1862, in Mary Boykin Chesnut, *A Diary from Dixie, as Written by Mary Boykin Chesnut, Wife of James Chesnut, Jr., United States Senator from South Carolina, 1859–1861, and Afterward an Aide to Jefferson Davis and a Brigadier-General in the Confederate Camp,* ed. Isabella D. Martin and Myrta Lockett Avary (New York: D. Appleton and Company, 1906), 177.

48. *Adair (KY) County News,* August 15, 1906.

49. David Henkin notes this was the case with the postal service as well. David Henkin, *The Postal Age: The Emergence of Modern Communications in Nineteenth-Century America* (Chicago: University of Chicago Press, 2006).

50. Nye, *American Technological Sublime,* 283.

51. "Our Modern Prometheus," *Journal of the Telegraph* 2, no. 10 (April 15, 1869): 112.

52. Alexander Bain, *The Emotions and the Will* (London: Longmans, Green, 1865), 47, 48.

53. David Inglis, "Remarkable Exaggeration of the Sense of Awe," *New York Medical Journal* 67 (1898): 464.

54. James H. Leuba, "Fear, Awe and the Sublime in Religion: A Chapter in the Study of Instincts, Impulses, and Motives in Religious Life," *American Journal of Religious Psychology and Education* 2, no. 1 (March 1906): 2, 3, 11–12, 15, 18, 23.

55. William McDougall, *An Introduction to Social Psychology* (Boston: John Luce, 1917), 312.

56. Weber, "Science as a Vocation," 139.

57. See, for instance, John White Chadwick, *The Revelation of God and Other Sermons* (Boston: George H. Ellis, 1889), 78; and:H. J. Wilmot-Buxton, "Everyday Miracles," in *Plain Preaching for a Year* Third Series, Vol. II., ed. Edmund Fowle (London: W. Skeffington and Sons, 1882), 128–29.

58. "Sound Sent by Wire," *New York Tribune,* March 31, 1877, 4.

59. "Growth of the Telephone Service," *New York Times,* December 23, 1901.

60. "The Magic Flight of Thought," AT&T ad, 1914, AT&T Book 109, Box 21, Folder 2, N. W. Ayer Advertising Records, Archives Center, National Museum of American History, Smithsonian. See also American Telephone and Telegraph Company, "The Story of a Great Achievement: Telephone Communication from Coast to Coast," Bancroft Library, University of California, Berkeley.

61. Thomas Edison, "The Phonograph and Its Future," *North American Review* 126, no. 262 (May–June 1878): 533–34.

62. Quoted in Susan J. Douglas, *Listening In: Radio and the American Imagination, from Amos 'n' Andy and Edward R. Murrow to Wolfman Jack and Howard Stern* (New York: Random House, 1999), 46.

63. "Miscellaneous Readings: Echoes from Dead Voices. Wonderful Possibilities," *Farmer's Cabinet,* March 5, 1878.

64. Evan Eisenberg, *The Recording Angel: Music, Records and Culture from Aristotle to Zappa,* 2nd ed. (New Haven, CT: Yale University Press, 2005), 46.

65. Douglas, *Listening In,* 43–52.

66. "Suggest Radio Sounds Will Go On Forever—British Experts Picture Men Years Hence Hearing the Voices of Those Long Dead," *New York Times,* December 18, 1927.

67. Anne O'Hare McCormick, "Radio: A Great Unknown Force," *New York Times,* March 27, 1932.

68. "Telephone Insanity," *New York Times,* August 19, 1890.

69. *Blue-Grass Blade* (Lexington, KY), January 29, 1905.

70. *Spanish Fork (UT) Press,* November 2, 1905.

71. "Hello Central, Give Me Heaven," Duke University Library Digital Collections, https://library.duke.edu/digitalcollections/hasm_n0311/, accessed February 2, 2018.

72. "Child's Query," *Valentine (NE) Democrat,* July 6, 1911. See also *Edgefield (SC) Advertiser,* October 10, 1894.

73. *Somebody's Stenog* (1923), George H. Clark Radioana Collection, Series 169, Box 399, Folder 2, National Museum of American History, Smithsonian Institution.

74. Jeffrey Sconce notes another difference between hopes for the telegraph and telephone versus hopes for the radio: "Although no less awestruck than the speculative narratives and supernatural theories that greeted telegraphy, fantastic accounts of wireless technology were decidedly more anxious, pessimistic, and melancholy. In the first three decades of the new century, a variety of paranormal theories, technologies, and fictions challenged the otherwise wholly enthusiastic celebration of the emerging medium by suggesting an eerie and even sinister undercurrent to the new electronic worlds forged by wireless." Sconce, *Haunted Media,* 61–62.

75. W. J. McEvoy, "Radio," George H. Clark Radioana Collection, Series 189, Box 402, Folder 3, National Museum of American History, Smithsonian Institution.

76. If the most sublime experiences have not declined, they have at least, as Weber put it, "retreated from public life": "The fate of our times is characterized by rationalization and intellectualization and, above all, by the 'disenchantment of the world.' Precisely the ultimate and most sublime values have retreated from public life either into the transcendental realm of mystic life or into the brotherliness of direct and personal human relations." See Weber, "Science as a Vocation," 129–56.

77. Joel Dinerstein, "Technology and Its Discontents: On the Verge of the Posthuman," *American Quarterly* 58, no. 3 (September 2006): 573.

78. Nye, *American Technological Sublime*; Vincent Mosco, *The Digital Sublime: Myth, Power, and Cyberspace* (Cambridge, MA: MIT Press, 2005).

79. Thomas L. Friedman, *The World Is Flat: A Brief History of the Twenty-First Century* (New York: Farrar, Straus and Giroux, 2005).

80. Various theorists argue that awe is less capable of being triggered when it is mediated or witnessed as a reproduction rather than when it is experienced directly. For a good (albeit not comprehensive) survey of these theorists, see Bayles, "Awe Evolving," 6–7. Nye makes similar claims in his study of what he calls the "consumer sublime" in *American Technological Sublime.* Nye argues that the technological sublime "'ought to' have gone into terminal decline after Hiroshima." But instead Americans continue to visit the Kennedy Space Center, they go in droves to the Smithsonian Air and Space Museum, and they visit highly technologized spaces like Disneyland and Las Vegas. However, as Nye notes, these spectacles are not really as awe inspiring (or more aptly sublime) as the kinds of encounters with technology that earlier Americans experienced. They "reveal not the existence of God, not the power of nature, not the majesty of human reason, but the titillation of representation itself" (291).

81. Jane McGonigal, *Reality Is Broken: Why Games Make Us Better and How They Can Change the World* (New York: Penguin, 2011), 95–115.

82. Ibid., 101.

83. Ibid., 104.

84. Bayles, "Awe Evolving," 2.

85. Gregory Bayles, Salt Lake City, Utah, July 5, 2016, interviewed by Luke Fernandez and Susan Matt.

86. Eric Whitacre, "A Virtual Choir 2,000 Voices Strong," 2011, https://www.ted.com/talks/eric_whitacre_a_virtual_choir_2_000_voices_strong?language=en; Alexander Varty, "Eric Whitacre Thinks Big with Choral Music," *Georgia Straight Vancouver's News & Entertainment Weekly*, October 24, 2012, http://www.straight.com/arts/eric-whitacre-thinks-big-choral-music.

87. Although it is not inflected with the same communitarian sentiments as Whitacre's music, a viral video of Funtwo (an electric guitarist) playing a rock version of Pachelbel's Canon inspires a similar sense of awe in Virginia Heffernan: "Over and over the guitarist's left hand articulated strings with barely perceptible movements, sounding and muting notes almost simultaneously, and playing complete arpeggios through a single stroke with his right hand. Funtwo's accuracy and velocity seemed record-breaking, but his mouth and jawline—to the extent that they were visible—looked impassive, with none of the exaggerated grimaces of heavy metal guitar heroes. The contrast between the soaring bravado of the undertaking and the reticence of the guitarist gave the 5-minute, 20-second video a gorgeous solemnity. Like a celebrity sex tape or a Virgin Mary sighting, the video drew hordes of seekers with diverse interests and attitudes. . . . Now, with nearly 7.35 million views—and a spot in the site's 10 most-viewed videos of all time—Funtwo's performance would be platinum many times over. From the perch it's occupied for months on YouTube's 'most discussed' list, it generates a seemingly endless stream of praise (riveting, sick, better than Hendrix), exegesis, criticism, footnotes, skepticism, anger and awe." Virginia Heffernan, "Web Guitar Wizard Revealed at Last," *New York Times*, August 27, 2006, http://www.nytimes.com/2006/08/27/arts/television/27heff.html. Also see Virginia Heffernan, *Magic and Loss: The Internet as Art* (New York: Simon and Schuster, 2016).

88. Comments on "A Virtual Choir 2,000 Voices Strong," 2011, https://www.youtube.com/watch?v=2NENlXsW4pM.

89. Warren and Harriet Trezevant, Oakland, California, January 31, 2016, interviewed by Luke Fernandez and Susan Matt.

90. Bayles, July 5, 2016, interview.

91. "What Is the Best Comment in Source Code You Have Ever Encountered?" *Stackoverflow.com,* September 18, 2011, https://stackoverflow.com/questions/184618/what-is-the-best-comment-in-source-code-you-have-ever-encountered/482129#482129.

92. Gregory Jackson, "Mystery / Mastery, Knowledge as Power, Long Distance Calls, and the Voss Admonition," *LinkedIn Pulse*, April 15, 2016, https://www.linkedin.com/pulse/mysterymastery-knowledge-power-long-distance-calls-voss-greg-jackson.

93. Adam Kaslikowski, Salt Lake City, Utah, April 13, 2016, interviewed by Luke Fernandez and Susan Matt.

94. Bayles July 5, 2016, interview.

95. Christian Wutz and Chris Fowles, Salt Lake City, Utah, March 25, 2016, interviewed by Luke Fernandez and Susan Matt. The technology journalist Dan Lyons, who wrote *Disrupted: My Misadventure in the Start-Up Bubble* and who also was a writer for the sitcom *Silicon Valley,* offers similar recollections of his experience working for a startup named Hubspot in 2013. As Lyons recalls, workers often invoked the expression "one plus one equals three" as a way of describing the magical qualities that they aspired to embed in their software. See Terry Gross, "Laid-Off Tech Journalist Joins a Start-Up, Finds It's Part Frat, Part Cult," *NPR.org,* April 5, 2016, http://www.npr.org/2016/04/05/473097951/laid-off-tech-journalist-joins-a-start-up-finds-its-part-frat-part-cult; and Dan Lyons, *Disrupted: My Misadventure in the Start-Up Bubble* (New York: Hachette, 2016).

96. Ritzer, *Enchanting a Disenchanted World*, 7–8, 92.

97. Ray Kurzweil, *The Singularity Is Near: When Humans Transcend Biology* (New York: Penguin, 2006), 258, 325; Brad Reed, "A Guide to All the Insane Predictions Made by Google's New Engineering Director," *BGR* (blog), December 18, 2012, http://bgr.com/2012/12/18/ray-kurzweil-predictions-overview-257980/.

98. Shots of Awe, "Welcome to Shots of Awe," https://www.youtube.com/watch?v=VmJVcRoROKI, accessed June 25, 2016.

99. Shots of Awe, "Welcome to Shots of Awe"; Shots of Awe, "We Are the Gods Now—Jason Silva at Sydney Opera House," https://www.youtube.com/watch?v=cF2VrefjIjk, accessed June 25, 2016.

100. Shots of Awe, "We Are the Gods Now."

101. This is in keeping with Peter Stearns's observations in *American Cool,* where he suggests that twentieth-century Americans learned to adopt a more restrained and less effusive emotional style. See Peter N. Stearns, *American*

Cool: Constructing a Twentieth-Century Emotional Style (New York: New York University Press, 1994).

102. Bayles July 5, 2016, interview.

103. Skyler Coombs, Ogden, Utah, October 26, 2015, interviewed by Luke Fernandez and Susan Matt.

104. Xavier Stilson, Ogden, Utah, October 20, 2015, interviewed by Luke Fernandez and Susan Matt.

105. Anna Renkosiak, Bettendorf, Iowa, August 12, 2015, interviewed by Susan Matt.

106. Cooper Ferrario, Ogden, Utah, June 3, 2015, interviewed by Luke Fernandez and Susan Matt.

107. Barry Gomberg, Ogden, Utah, May 25, 2015, interviewed by Luke Fernandez and Susan Matt.

108. Ana Chavez, Ogden, Utah, October 19, 2015, interviewed by Luke Fernandez and Susan Matt.

109. Gregory Noel, Ogden, Utah, May 17, 2018, interviewed by Luke Fernandez and Susan Matt.

110. Emily Meredith, Oakland, California, January 30, 2016, interviewed by Luke Fernandez and Susan Matt.

111. Katherine J. Schall, Ogden, Utah, October 18, 2014, interviewed by Luke Fernandez and Susan Matt.

112. Camree Larkin, Ogden, Utah, October 30, 2015, interviewed by Luke Fernandez and Susan Matt.

113. Meredith interview.

114. Eli Gay, Berkeley, California, January 31, 2016, interviewed by Luke Fernandez and Susan Matt.

115. Eddie Baxter, Ogden, Utah, May 24, 2018, interviewed by Luke Fernandez and Susan Matt.

116. Ibid.

117. EBSCO host Psych Info results list, accessed June 29, 2016.

118. Michelle Shiota, Dacher Keltner, and John Oliver, "Positive Emotion Dispositions Differentially Associated with Big Five Personality and Attachment Style," *Journal of Positive Psychology* 1, no. 2 (April 2006): 61–71. See also http://fetzer.org/sites/default/files/images/stories/pdf/selfmeasures/Self_Measures_for_Personal_Growth_and_Positive_Emotions_DPES_AWE_SUBSCALE.pdf.

119. Piff and Keltner, "Why Do We Experience Awe?"

120. "The Art and Science of Awe," *Greater Good*, http://greatergood.berkeley.edu/news_events/event/the_art_and_science_of_awe, accessed June 20, 2016.

121. Andy Tix, "The Loss of Awe," *Psychology Today*, January 27, 2016, https://www.psychologytoday.com/blog/the-pursuit-peace/201601/the-loss-awe; Andy Tix, "The Replacement of Awe," *Psychology Today*, February 3, 2016, https://www.psychologytoday.com/blog/the-pursuit-peace/201602/the-replacement-awe; Andy Tix, "Reflections on Mystery and Awe," https://mysteryandawe.wordpress.com/, accessed June 20, 2016.

122. Curiously, the possibility that technology might also be a trigger for awe is relatively absent from their work.

123. Piff et al., "Awe, the Small Self, and Prosocial Behavior."

124. Piff and Keltner, "Why Do We Experience Awe?" Also see Tix, "Loss of Awe"; Tix, "Replacement of Awe"; and Tix, "Reflections on Mystery and Awe." One of the most recent and telling examples of the way that awe is mediated through technology is illustrated in a September 2017 *New York Times* article reviewing the introduction of the iPhone X. As James Poniewozik tells it, an Apple senior vice president "showed how an app could overlay an image of the constellations on a live camera picture of the sky. This got him excited. 'This isn't some generic sky!' he said. 'This is the sky around you!'" Poniewozik's reactions, in contrast, are mixed: "I don't know. It looked like a generic sky to me: clouds, blue, sunlight. But it *did* look better on the phone—or rather, on an image of a phone, projected on a screen in Cupertino, Calif., streamed to my computer. It was luminous. The constellation labels, appearing and fading, gave it a look of magical whimsy. I looked up through the skylight over my desk. Yep. That iPhone sky looked way better than the garbage regular sky that I could see through my garbage human eyes. This enhancement of reality is what each video-streamed Apple event sells, more than any particular iPhone or set-top box. If advertising once told us that 'Things go better with Coke,' this event—a jewel box for Apple's products and the people who use them—says that 'Things look better with Apple.'" In a review of Apple, it's a matter of course that Poniewozik places his main emphasis on the way that the Apple technology served to augment nature. But what he also mentions is a "garbage sky" and, by allusion, the way that a hollower experience is brought about by the lights of modern civilization. In effect, nature's awe is at once diminished and augmented by modern technology. That, undoubtedly, is a mixed and bittersweet message. But in the conclusion of his review, Poniewozik reflects that "another year from now, I'll set another reminder to watch another Apple event, believing somewhere deep down that with one more upgrade, I might be perfected." If the awe that nature once evoked could convey the perfectness of

the universe, and the fallen state of man, the awe that Apple's technology evokes conveys the perfectibility of man. See James Poniewozik, "At the Apple Keynote, Selling Us a Better Vision of Ourselves," *New York Times,* September 12, 2017, https://www.nytimes.com/2017/09/12/arts/television /apple-event-iphone.html.

125. In fact, some historians suggest that it was because nineteenth-century Americans believed themselves to be in control of their own fates that they often rejected the myths and tragedies of the ancient Greeks, in which protagonists' lives and actions were constrained by capricious and jealous gods. Such limits on human agency and individual will seemed out of step with American mores. Lawrence Levine, *Highbrow, Lowbrow: The Emergence of Cultural Hierarchy in America* (Cambridge, MA: Harvard University Press, 1990).

126. Henry Petroski, *To Engineer Is Human: The Role of Failure in Successful Design* (New York: Vintage, 1992), 216–17.

127. "Committed to Raising Hubris Awareness," http://www.daedalustrust .com/, accessed August 4, 2017.

128. There are also perils in the recovery of awe. When a technological object is made magical, that magic can obfuscate how the object was made and how its manufacture might perpetuate onerous working conditions. (For example, the magic in Apple's products might prevent consumers from considering the labor practices of the company's Foxconn contractor in China.) And when magical technology serves to reduce friction (an effect that Warren Trezvenant and other designers have associated with magic), that lack of friction can encourage complacency in the face of social practices whose injustice might otherwise be more transparent and more likely to be resisted. To these concerns in *To Save Everything, Click Here*, writer Evgeny Morozov suggests that technology must incorporate more elements of "adversarial design" that intentionally build in social friction (and presumably less magic) as a way of combating this complacency. And in "Too Much Magic, Too Little Social Friction: Why Objects Shouldn't Be Enchanted," philosopher Evan Selinger observes, "The more we're inclined to see technology as wizardry, the less disposed we are to demystifying the illusions that obscure why some people get to enjoy hocus pocus at other people's expense." These concerns deserve consideration. But given the prosocial potential in awe, they should not trump its pursuit altogether. See Evgeny Morozov, *To Save Everything, Click Here: The Folly of Technological Solutionism* (New York: PublicAffairs, 2014); Evan Selinger, "Too Much Magic, Too Little Social Friction: Why Objects Shouldn't Be Enchanted," *Los Angeles Review of Books,* https://lareviewofbooks.org/article /much-magic-little-social-friction-objects-shouldnt-enchanted/, accessed July 13, 2016; and Luke Fernandez, "I.T. in the University: Solutionism,

Adversarial Design, and the Politics of Usability," *I.T. in the University* (blog), July 23, 2013, http://itintheuniversity.blogspot.com/2013/07/solutionism-adversarial-design-and.html.

6. ANGER RISING

1. Greg Bayles, Salt Lake City, Utah, July 5, 2016, interviewed by Luke Fernandez and Susan Matt.

2. Greg Bayles, Salt Lake City, Utah, November 1, 2017, interviewed by Luke Fernandez and Susan Matt.

3. Ibid.

4. Franklin Foer, *World without Mind: The Existential Threat of Big Tech* (New York: Penguin, 2017).

5. Douglas Rushkoff, *Throwing Rocks at the Google Bus: How Growth Became the Enemy of Prosperity* (New York: Penguin, 2017), 2.

6. "Activists Block Tech Bus Commute, Say E-scooters Treated Better than Homeless," *San Francisco Examiner,* May 31, 2018, http://www.sfexaminer.com/activists-block-tech-bus-commute-say-e-scooters-treated-better-homeless/, accessed June 9, 2018.

7. According to a 2018 survey, in the wake of reports about Facebook and Cambridge Analytica, roughly 10 percent of Americans deleted their Facebook accounts. See Joe Osborne, "Nearly One in 10 Americans Have Deleted their Facebook Accounts, Survey Says," April 13, 2018, https://www.techradar.com/news/nearly-one-in-10-us-facebook-users-have-deleted-their-accounts-survey-says, accessed May 20, 2018. For instance, Viviana Felix, the diversity affairs officer for the city of Ogden, Utah, told us, "A lot of communities of color that are more engaged with the political system in the United States were upset. They were upset. I had a couple friends say 'I'm deleting my Facebook account; if you want to get ahold of me, ask me for my number." Viviana Felix, Ogden, Utah, May 16, 2018, interviewed by Luke Fernandez and Susan Matt.

8. Cass R. Sunstein, *Republic.com* (Princeton, NJ: Princeton University Press, 2002); danah boyd, "The Not-So-Hidden Politics of Class Online," *Microsoft Research,* June 30, 2009, https://www.microsoft.com/en-us/research/publication/the-not-so-hidden-politics-of-class-online. See also Eli Pariser, *The Filter Bubble: How the New Personalized Web Is Changing What We Read and How We Think* (New York: Penguin, 2012).

9. "How Digital Media Fuels Moral Outrage and What to Do about It," *Christian Science Monitor,* September 22, 2017, https://www.csmonitor.com/Technology/2017/0922/How-digital-media-fuels-moral-outrage-and-what-to

-do-about-it, accessed January 22, 2018; "Angry Social Media Posts Are Never a Good Idea. How to Keep Them in Check," *Conversation,* January 10, 2017, https://theconversation.com/angry-social-media-posts-are-never-a-good-idea -how-to-keep-them-in-check-71016; "The Internet Isn't Making Us Dumb. It's Making Us Angry," *Washington Post,* September 16, 2013, https://www .washingtonpost.com/news/the-switch/wp/2013/09/16/the-internet-isnt -making-us-dumb-its-making-us-angry/?utm_term=.a11d41cc7800.

10. D. M., Ogden, Utah, November 19, 2017, interviewed by Luke Fernandez and Susan Matt.

11. Jeff Hellrung, Ogden, Utah, November 8, 2017, interviewed by Luke Fernandez and Susan Matt.

12. See, for instance, David Konstan, *The Emotions of the Ancient Greeks: Studies in Aristotle and Classical Literature* (Toronto: University of Toronto Press, 2006), chap. 2; William V. Harris, *Restraining Rage: The Ideology of Anger Control in Classical Antiquity* (Cambridge, MA: Harvard University Press, 2004); Carol Zisowitz Stearns and Peter N. Stearns, *Anger: The Struggle for Emotional Control in America's History* (Chicago: University of Chicago Press, 1986); Barbara Rosenwein, ed., *Anger's Past: The Social Uses of an Emotion in the Middle Ages* (Ithaca, NY: Cornell University Press, 1998); and Nicole Eustace, *Passion Is the Gale: Emotion, Power, and the Coming of the American Revolution* (Chapel Hill: University of North Carolina Press for the Omohundro Institute, 2008).

13. "Anger," Online Etymology Dictionary, https://www.etymonline.com /word/anger, accessed January 27, 2018.

14. "Outrage," Online Etymology Dictionary, https://www.etymonline.com /word/outrage, accessed January 27, 2018.

15. Stearns and Stearns, *Anger,* 19.

16. Ibid.; Michael E. Woods, *Emotional and Sectional Conflict in the Antebellum United States* (New York: Cambridge University Press, 2014).

17. Stearns and Stearns, *Anger,* 19; John Robinson, "On Anger," in *The Works of John Robinson, Pastor of the Pilgrim Fathers, with a Memoir and Annotations*, vol. 1, ed. Robert Ashton (London: John Snow, 1851), http://oll.libertyfund.org/titles /1737.

18. Eustace, *Passion Is the Gale,* 184. Barbara Rosenwein and Lester Little point out that during the medieval era, God's wrath was perceived as controlled, purposeful, and moral; it stood in contrast to the uncontrollable, viceful anger of humans. See Barbara H. Rosenwein, "Introduction" and "Controlling Paradigms," in Rosenwein, *Anger's Past,* 1–8, 233–34; Lester K. Little, "Anger in Monastic Curses," in Rosenwein, *Anger's Past,* 9–35.

19. Cotton Mather, *Febrifugium. An essay for the cure of ungoverned anger: in a sermon preached, at the proposal and on the occasion, of a man under a sentence of death, for a murder committed by him in his anger. At Boston, 23. d. III. m. 1717* (Boston: Printed by J. Allen, for Benjamin Gray, at the corner shop, on the north-side of the town-house, 1717), 5, https://quod.lib.umich.edu/e/evans/N01597.0001.001/1:3?rgn=div1;view=fulltext.

20. Rosenwein and contributors to her volume trace a similar divide during the medieval era—aristocrats' anger was generally more legitimate than the rage of peasants. See Rosenwein, "Introduction," 5; Richard E. Barton, "'Zealous Anger' and the Renegotiation of Aristocratic Relationships in Eleventh- and Twelfth-Century France, in Rosenwein, *Anger's Past,* 153–70; and Paul Freedman, "Peasant Anger in the Late Middle Ages," in Rosenwein, *Anger's Past,* 171–88.

21. Eustace, *Passion Is the Gale,* 34, 40, 158, 142, 156–60, 172–197. Eustace describes Franklin's views in more detail. For Franklin, see Benjamin Franklin, "Plain Truth; Or Serious Considerations on the Present State of the City of Philadelphia and Province of Pennsylvania. By a Tradesman of Philadelphia," in *The Works of Benjamin Franklin,* vol. 3, ed. Jared Sparks (Chicago: Townsend Mac Coun, 1882), 13–14.

22. John Hancock, "The Boston Massacre," in *American Eloquence: A Collection of Speeches and Addresses by the Most Eminent Orators of America,* vol. 1, ed. Frank Moore (New York: D. Appleton and Son, 1872), 227; for discussions of nineteenth-century accounts of Hancock's anger, see Woods, *Emotional and Sectional Conflict,* 120, 126–27.

23. Eustace offers eighteenth-century examples of differing anger rules for differing social classes; however, she argues that these differences were eroded in the mid-eighteenth century. Based on our research, they still seem apparent in the nineteenth century. Eustace, *Passion Is the Gale,* 156–65, 196–99.

24. James W. C. Pennington, *The Fugitive Blacksmith; or, Events in the History of James W. C. Pennington, Pastor of a Presbyterian Church, New York, Formerly a Slave in the State of Maryland,* 2nd ed. (1849; New York: Arno, 1968), 6, 7.

25. Barbara Welter, "The Cult of True Womanhood: 1820–1860," *American Quarterly* 18, no. 2, pt. 1 (Summer 1966): 151–74; Stearns and Stearns, *Anger,* 36–68.

26. Alice B. Neal, "The First Quarrel," *Godey's Ladies Book* 47 (1853): 139; Stearns and Stearns, *Anger,* 40–41.

27. Stearns and Stearns, *Anger,* 49–51; Peter N. Stearns, *American Cool: Constructing a Twentieth-Century Emotional Style* (New York: New York University Press, 1994), 23–25; Woods, *Emotional and Sectional Conflict,* 86.

28. "Angry Letters," *Bellows (VT) Fall Times,* June 7, 1867; see also the *Alexandria (VA) Gazette,* August 1, 1859; *Nashville (TN) Patriot,* August 5, 1859; and *Weekly Mississippian,* September 28, 1859.

29. Elisabeth Robinson Scovil, "The Art of Letter-Writing," *Ladies' Home Journal* 5 (1893): 2.

30. John Allen Wyeth, *With Sabre and Scalpel: The Autobiography of a Soldier and Surgeon* (New York: Harper Brothers, 1914), 8–9.

31. Hamlin Garland, *A Son of the Middle Border* (New York: Macmillan, 1923), 70, 177–78, 185–86.

32. Michael E. McGerr, *The Decline of Popular Politics: The American North, 1865–1928* (New York: Oxford University Press, 1988), 14.

33. Ibid., 41.

34. David Grimsted, *American Mobbing, 1828–1861: Toward Civil War* (New York: Oxford University Press, 1998); Kimberly K. Smith, *The Dominion of Voice: Riot, Reason, and Romance in Antebellum Politics* (Lawrence: University Press of Kansas, 1999); see also Jill LePore, "How We Used to Vote," *New Yorker,* October 13, 2008.

35. David Paul Nord, "Newspapers and American Nationhood, 1776–1826," *Proceedings of the American Antiquarian Society,* 1991, 397, 392–93.

36. "The Crimes of Federalism," *Kalida (OH) Venture,* March 5, 1846.

37. "The Reign of Free Whisky Begun," *Evansville (IN) Daily Journal,* December 24, 1855.

38. "The Rising Tide," *Press and Tribune* (Chicago, IL), May 31, 1860.

39. "Business Revival," *Indiana State Sentinel,* February 28, 1877.

40. The historians Stuart Blumin and Glenn Altschuler suggest that the papers did this seasonally, and when elections were not looming, the papers adopted a less polarizing and more inclusive tone. See Glenn C. Altschuler and Stuart M. Blumin, "'Where Is the Real America?' Politics and Popular Consciousness in the Antebellum Era," *American Quarterly* 49, no. 2 (June 1997).

41. Woods, *Emotional and Sectional Conflict,* 128, 128n.

42. Ibid., 130.

43. "Ap Shenkin," *Maumee (OH) City Express,* December 28, 1839; *Richmond (VA) Enquirer,* September 13, 1842; "Indignation Meeting," *Indiana State Sentinel,* July 3, 1845; "Race Issue Rends Quinquennial Here," *Evening Star* (Washington, DC), May 6, 1925.

44. *Richmond (VA) Enquirer,* February 10, 1852.

45. Richard Garrity, "A Page in Oklahoma History: An Eyewitness Account of the Constitutional Convention," *Chronicles of Oklahoma* 60, no. 3 (1982): 346.

46. David Grimsted, "Ante-Bellum Labor: Violence, Strike, and Communal Arbitration," *Journal of Social History* 19, no. 1 (Autumn 1985): 5–28.

47. "Probably Murder," *Wheeling (WV) Daily Register,* January 24, 1873.

48. "Lad Was Struck with Shovel and Is Near Death," *Bennington (VT) Evening Banner,* May 26, 1919.

49. "Anger as a Moral Force," *Mt. Sterling (KY) Advocate,* October 5, 1897.

50. G. Stanley Hall, "A Study of Anger," *American Journal of Psychology* 10, no. 4 (July 1899): 586, 587.

51. Ibid., 579, 588.

52. G. Stanley Hall, "How Rage, Anger, and Hatred Help Us to Success," *Richmond (VA) Times Dispatch,* August 15, 1915.

53. Hall, "A Study of Anger," 573–74.

54. Ibid., 575, 577.

55. Ida B. Wells, "Lynch Law," in *The Reason Why the Colored American Is Not in the World's Columbian Exposition, Selected Works of Ida B. Wells-Barnett* (New York: Oxford University Press, 1991), 75–77; Patricia A. Schechter, "'All the Intensity of My Nature': Ida B. Wells, Anger, and Politics," *Radical History Review* 70 (1998): 68.

56. "Impudent Negro," *Paducah (KY) Sun,* November 2, 1901.

57. "Hot Coffee in Negro's Face," *Laclede (MO) Blade,* July 29, 1905.

58. "Negroes Anger Wilson," *Seattle (WA) Star,* November 13, 1914.

59. Herman J. Stich, "Two Minutes of Optimism, Get Angry Aright," *Evening Public Ledger* (Philadelphia, PA), May 9, 1921.

60. Frank Crane, "Anger," *Evening Star* (Washington, DC), December 18, 1923.

61. "Keep Well," *Liberal (KS) Democrat,* August 16, 1917.

62. "A Dangerous Enemy, Poisonous Mental Reptile," *Washington Times,* October 26, 1919.

63. "The Curse and the Cure of Anger," *Abilene (KS) Weekly Reflector,* February 3, 1916 (from a sermon delivered by Dr. Dutton at the M. E. Church on January 30).

64. Angel Kwolek-Folland, *Engendering Business: Men and Women in the Corporate Office, 1870–1930* (Baltimore, MD: Johns Hopkins University Press, 1998), 190. For discussions of the rise of organizational society and its emotional demands, see Susan J. Matt, *Keeping Up with the Joneses: Envy in American Consumer Society, 1890–1930* (Philadelphia: University of Pennsylvania Press,

2003), chap. 2; Arlie Russell Hochschild, *The Managed Heart: Commercialization of Human Feeling* (1983; Berkeley: University of California Press, 2012); Daniel Rodgers, *The Work Ethic in Industrial America, 1850–1920* (Chicago: University of Chicago Press, 1974, 1978), 25; and Stearns, *American Cool,* 123.

65. Stearns and Stearns, *Anger,* 116.

66. H. Meltzer, "Human Relations and Morale in Industry, Round Table, 1944," *American Journal of Orthopsychiatry* 15, no. 2 (April 1945): 329, 331.

67. Stearns, *American Cool,* 237–38.

68. Stearns and Stearns, *Anger,* 126.

69. Studs Terkel, *Working: People Talk about What They Do All Day and How They Feel about What They Do* (1972; New York: New Press, 2004), xxxi–xxxiii.

70. Hochschild, *Managed Heart,* 25, 28–29.

71. Bruce Felton, "Earning It: When Rage Is All the Rage," *New York Times,* March 15, 1998.

72. Pam Belluck, "Ideas and Trends; 'Rages' Maybe Aren't What They're Cracked Up to Be," *New York Times,* August 1, 1999; Julie Dunn, "Personal Business: Diary; Early Alert on Workplace Rage," *New York Times,* December 17, 2000.

73. Stearns, *American Cool,* 264–84.

74. Paul Schubert, *The Electric Word: The Rise of Radio* (1928; New York: Arno, 1971), 303, 304.

75. Hadley Cantril and Gordon W. Allport, *The Psychology of Radio* (1935; New York: Arno and the *New York Times,* 1971), 21.

76. Lawrence W. Levine and Cornelia R. Levine, *The Fireside Conversations: America Responds to FDR during the Great Depression* (Berkeley: University of California Press, 2010), 4–5.

77. "The Radio Mind Seeks Relief," *New York Times,* July 17, 1932.

78. Letter to President Roosevelt, March 10, 1937, in Levine and Levine, *Fireside Conversations.*

79. Donald Warren, *Radio Priest: Charles Coughlin, The Father of Hate Radio* (New York: Free Press, 1996), 2; "Text of Father Coughlin's Address to Townsendites," *New York Times,* July 17, 1936, p. 6, http://archives.lib.cua.edu/res/docs/education/politics/pdfs/24-coughlin-townsend-speech.pdf; Susan J. Douglas, *Listening In: Radio and the American Imagination, from Amos 'n' Andy and Edward R. Murrow to Wolfman Jack and Howard Stern* (New York: Random House, 1990), 128.

80. "'Somebody Must be Blamed': Father Coughlin Speaks to the Nation," http://historymatters.gmu.edu/d/5111/, accessed February 4, 2018.

81. David Goodman, "Beyond Hate Speech: Charles Coughlin, Free Speech and Listeners' Rights," *Patterns of Prejudice* 49, no. 3 (2015): 218–19, 220, 222–23.

82. Douglas, *Listening In,* 287.

83. "Number of News / Talk / Information Radio Stations Fluctuates between 1,900 and 2,000," Pew Research Center, http://www.journalism.org/chart/number-of-newstalkinformation-radio-stations-fluctuates-between-1900-and-2000/, accessed December 3, 2017.

84. Douglas, *Listening In,* 289.

85. Murray B. Levin, *Talk Radio and the American Dream* (Lexington, MA: Lexington, 1987), xiii–xiv.

86. Douglas, *Listening In,* 289.

87. Jeffrey M. Berry and Sarah Sobieraj, "Understanding the Rise of Talk Radio," *PS: Political Science and Politics* 44, no. 4 (2011): 762–67.

88. James Reston, "Washington: The Rising Spirit of Protest," *New York Times,* March 19, 1965.

89. Julian Zelizer, "How America Got Polarized," *CNN,* August 3, 2015, http://www.cnn.com/2015/08/03/opinions/zelizer-buckley-vidal-debates-polarized/index.html, accessed December 3, 2017.

90. S. I. Hayakawa, "Social Change through TV," *Saturday Evening Post,* March 1, 1974, 46. See also "Hidden TV Messages Create Social Discontent," *Intellect* 104 (February 1976): 350–51.

91. James G. Webster, "Beneath the Veneer of Fragmentation: Television Audience Polarization in a Multichannel World," *Journal of Communication* (June 2005): 366–68.

92. Elihu Katz, "And Deliver Us from Segmentation," *Annals of the American Academy of Political and Social Science* 546 (July 1996): 25.

93. Webster, "Beneath the Veneer of Fragmentation," 370.

94. Markus Prior, "Media and Political Polarization," *Annual Review of Political Science* 16 (2013): 101–27.

95. Robert D. Putnam, *Bowling Alone: The Collapse and Revival of American Community* (New York: Simon and Schuster, 2000).

96. "Road Rage," Oxford English Dictionary Online, http://www.oed.com/view/Entry/246415?redirectedFrom=%22road+rage%22#eid, accessed June 4, 2018. Sociologist Barry Glassner makes a case that the incidence of road rage events has been greatly exaggerated. See Barry Glassner, *Fear: Why Americans are Afraid of the Wrong Things: Crime, Drugs, Minorities, Teen Moms, Killer Kids, Mutant*

Microbes, Plane Crashes, Road Rage, & So Much More (New York: Basic Books, 1999), 3–9.

97. Maria L. Garase, *Road Rage* (New York: LFB Scholarly, 2006), 10–11, 1–4.

98. "Mad, Bad and on the Road," *Economist,* July 24, 1997; Garase, *Road Rage,* 2.

99. Garase, *Road Rage,* 19.

100. Andy Greenberg, "It's Been 20 Years since This Man Declared Cyberspace Independence," *WIRED,* February 8, 2016, https://www.wired.com/2016/02/its-been-20-years-since-this-man-declared-cyberspace-independence/.

101. John Perry Barlow, "A Declaration of the Independence of Cyberspace," February 8, 1996, https://www.eff.org/cyberspace-independence, accessed June 12, 2018.

102. Greenberg, "It's Been 20 Years."

103. Barlow, "Declaration of the Independence of Cyberspace."

104. Ibid.

105. Ibid.

106. "Civility in America VII: The State of Civility," https://www.webershandwick.com/news/article/civility-in-america-vii-the-state-of-civility, accessed December 19, 2017; Sue Scheff, "Shame Nation: The Rise of Incivility in America," *Huffington Post* (blog), June 13, 2017, https://www.huffingtonpost.com/entry/shame-nation-the-rise-of-incivility-in-america_us_591357d3e4b0e070cad70b2f, accessed June 12, 2018.

107. Carroll Doherty, "Key Takeaways on Americans' Growing Partisan Divide over Political Values," *Pew Research Center* (blog), October 5, 2017, http://www.pewresearch.org/fact-tank/2017/10/05/takeaways-on-americans-growing-partisan-divide-over-political-values/; Jocelyn Kiley, "In Polarized Era, Fewer Americans Hold a Mix of Conservative and Liberal Views," *Pew Research Center* (blog), October 23, 2017, http://www.pewresearch.org/fact-tank/2017/10/23/in-polarized-era-fewer-americans-hold-a-mix-of-conservative-and-liberal-views/.

108. Alan Abramowitz and Steven Webster, "All Politics Is National," annual meeting of the Midwest Political Science Association, Chicago, Illinois, April 16, 2015, http://stevenwwebster.com/research/all_politics_is_national.pdf; Amber Phillips, "Why US Elections Are All about Voting against Something," *Washington Post,* June 5, 2015, https://www.washingtonpost.com/news/the-fix/wp/2015/06/05/why-u-s-elections-are-all-about-voting-against-something-and-not-for-it/.

109. Lee Rainie et al., "The Future of Free Speech, Trolls, Anonymity and Fake News Online," *Pew Research Center: Internet, Science & Tech* (blog), March 29,

2017, http://www.pewinternet.org/2017/03/29/the-future-of-free-speech-trolls-anonymity-and-fake-news-online/, accessed June 12, 2018.

110. "For Online Ranters, Anger Begets Anger," *All Tech Considered,* July 18, 2013, https://www.npr.org/sections/alltechconsidered/2013/07/18/202967700/for-online-ranters-anger-begets-anger, accessed January 19, 2018.

111. Susanna Schrobsdorff, "The Rage Flu: Why All This Anger Is Contagious and Making Us Sick," *Time Magazine,* June 29, 2017, http://time.com/4838673/anger-and-partisanship-as-a-virus/, accessed January 15, 2018.

112. Douglas, *Listening In,* 317.

113. Anna Renkosiak, Bettendorf, Iowa, November 11, 2017, interviewed by Susan Matt.

114. Leslie Jamison, "I Used to Insist I Didn't Get Angry; Not Anymore," *New York Times Magazine,* January 17, 2018, https://www.nytimes.com/2018/01/17/magazine/i-used-to-insist-i-didnt-get-angry-not-anymore.html?hp&action=click&pgtype=Homepage&clickSource=story-heading&module=second-column-region®ion=top-news&WT.nav=top-news&_r=0, accessed January 17, 2018.

115. "The Silence Breakers," *Time Magazine,* December 18, 2017, http://time.com/time-person-of-the-year-2017-silence-breakers/?xid=homepage, accessed January 18, 2018.

116. Yolanda Pierce, "Righteous Anger, Black Lives Matter, and the Legacy of King," *Berkley Forum,* January 16, 2018, https://berkleycenter.georgetown.edu/responses/righteous-anger-black-lives-matter-and-the-legacy-of-king, accessed January 17, 2018.

117. Katherine J. Schall, Ogden, Utah, October 30, 2017, interviewed by Luke Fernandez and Susan Matt.

118. Jennifer, Iowa, December 28, 2017, interviewed by Susan Matt.

119. Jim Dabakis, Salt Lake City, Utah, November 17, 2017, interviewed by Luke Fernandez and Susan Matt.

120. Ibid.

121. Todd Weiler, Layton, Utah, December 1, 2017, interviewed by Luke Fernandez and Susan Matt.

122. Zeynep Tufekci, *Twitter and Tear Gas: The Power and Fragility of Networked Protest* (New Haven, CT: Yale University Press, 2017), xxvii.

123. The phrase "the internet is angry" comes from a satirical video on YouTube: https://www.youtube.com/watch?v=rRT3whzCvl0, accessed January 28, 2018.

124. Weiler interview.

125. Dabakis interview.

126. Renkosiak interview, November 11, 2017.

127. Jennifer, interview.

128. Schall interview, October 30, 2017..

129. Sunstein, *Republic.com;* boyd, "Not-So-Hidden Politics of Class Online"; Pariser, *Filter Bubble.*

130. Colin Parker, Ogden, Utah, November 28, 2017, interviewed by Luke Fernandez and Susan Matt.

131. Bayles interview, November 1, 2017.

132. Parker interview.

133. Schall interview, October 30, 2017.

134. Phil from Iowa, December 23, 2017, interviewed by Susan Matt.

135. Erica, January 4, 2018, interviewed by Luke Fernandez and Susan Matt.

136. "Narcissistic Personality Disorder," *Psychology Today,* April 17, 2017, https://www.psychologytoday.com/conditions/narcissistic-personality-disorder, accessed January 19, 2018.

137. Renkosiak interview, November 11, 2017.

138. Parker interview.

139. Lee Sproull and Sara Kiesler, "Reducing Social Context Cues: Electronic Mail in Organizational Communications," *Management Science* 32, no. 11 (November 1986): 1492–1512.

140. Barry Gomberg, Ogden, Utah, October 27, 2017, interviewed by Luke Fernandez and Susan Matt.

141. Cooper Ferrario, Ogden, Utah, June 3, 2015, interviewed by Luke Fernandez and Susan Matt.

142. H., Grinnell, Iowa, May 23, 2016, interviewed by Luke Fernandez and Susan Matt.

143. JulieAnn Carter-Winward and Kent Winward, Ogden, Utah, October 29, 2017, interviewed by Luke Fernandez and Susan Matt.

144. Phil, interview.

145. Chuck and Susan Theiling, Bettendorf, Iowa, December 28, 2017, interviewed by Susan Matt.

146. D. M., Ogden, Utah, November 19, 2017, interviewed by Luke Fernandez and Susan Matt.

147. Eddie Baxter, Ogden, Utah, May 24, 2018, interviewed by Luke Fernandez and Susan Matt.

148. Will Xu, Grinnell, Iowa, May 23, 2016, interviewed by Luke Fernandez and Susan Matt.

149. Eulogio Alejandre, Ogden, Utah, May 15, 2018, interviewed by Luke Fernandez and Susan Matt.

150. Gomberg interview, October 27, 2017.

151. Hellrung interview, November 8, 2017.

152. Warren Olney, "After Alabama, What's Next for the GOP?" *To the Point*, December 14, 2017, http://www.kcrw.com/news-culture/shows/to-the-point, accessed January 23, 2018.

153. Theilings interview.

154. Bayles interview, November 1, 2017.

155. Phil interview.

156. Parker interview.

157. Barbara Rosenwein, "Worrying about Emotions in History," *The American Historical Review* 107, no. 3 (June 2002): 821–45.

158. Lorena Blas, "Weinstein Scandal: Uma Thurman Finally Speaks, Posting, 'You Don't Deserve a Bullet,'" *USA TODAY*, accessed December 28, 2017, https://www.usatoday.com/story/life/people/2017/11/23/harvey-weinstein-scandal-uma-thurman-finally-speaks-posting-you-dont-deserve-bullet/891675001/.

159. Instagram post by Uma Thurman, https://www.instagram.com/p/Bb2h0hBlV3T/, accessed January 23, 2018; Blas, "Weinstein Scandal."

160. "Matt Damon Thinks We Need to 'Correct' the 'Culture of Outrage' over Sexual Harassment," *SPIN*, https://www.spin.com/2017/12/matt-damon-sexual-harassment-comments/, accessed December 28, 2017.

161. E., Ogden, Utah, December 2, 2017, interviewed by Luke Fernandez and Susan Matt.

162. Erica interview.

163. Schall interview, October 30, 2017.

164. Dabakis interview.

165. Theilings interview.

166. Gregory Noel, Ogden, Utah, May 17, 2018, interviewed by Luke Fernandez and Susan Matt.

167. Jennifer interview.

168. Bayles interview, November 1, 2017.

169. Leon F. Seltzer, "What Your Anger May Be Hiding," *Psychology Today,* July 11, 2008, https://www.psychologytoday.com/blog/evolution-the-self/200807/what-your-anger-may-be-hiding.

170. Ryan C. Martin, Kelsey Ryan Coyier, Leah M. Van Sistine, and Kelly L. Schroeder, "Anger on the Internet: The Perceived Value of Rant-Sites," *Cyberpsychology, Behavior, and Social Networking* 16 (2013): 119–22; "Why Ranting Online Doesn't Help to Manage Anger," *PBS Newshour,* February 28, 2014, https://www.pbs.org/newshour/science/many-rants-online, accessed January 20, 2018.

171. Stearns and Stearns, *Anger.*

CONCLUSION

1. Kurzweil, in Doug Wolens, *The Singularity* (2012), documentary, http://www.imdb.com/title/tt2073120/.

2. McKibben, in ibid.

3. Langdon Winner, *The Whale and the Reactor: A Search for Limits in an Age of High Technology* (Chicago: University of Chicago Press, 1989), xi.

Acknowledgments

Books are rarely if ever the products of single individuals, and this one is no exception. Many organizations and people aided us over the course of the seven years it took to research and write this book, and we are grateful to them.

First, thanks to Jeff Dean, Executive Editor at Harvard University Press. His early enthusiasm for the project, as well as his thought-provoking and probing comments and questions along the way, made this a better book. Thanks as well to Stephanie Vyce, Esther Blanco Benmaman, and Michael Higgins, all at Harvard, and production editor Angela Piliouras. Their care in turning the manuscript into a finished book is much appreciated.

The book's beginnings were aided by a start-up grant from the National Endowment for the Humanities for a project titled "Concentration in the Humanities." The grant funded a course that we co-taught with our colleague Scott Rogers, the development of a "concentration browser," and the creation of a "distraction lab." That project, which we conducted in the 2011–2012 academic year, prompted us to think about attention, distraction, and human connection, and how those experiences have changed across different technological eras. During that same academic year, William Powers came to Weber State University to talk about his recently published book, *Hamlet's BlackBerry*. His thoughts on "digital maximalism" and "Walden Zones" were an inspiration to us, and we have appreciated his ongoing support since the project's inception.

Various conferences, and the people who hosted them, also refined the ideas in this book.

Nathan Jurgenson and the other folks who host the yearly Theorizing the Web conference helped us hone our chapter on loneliness and avoid the fallacy Jurgenson calls digital dualism. We appreciated the feedback we received when we presented our work at the conference in 2014.

The annual Utah Digital Humanities Symposium gave us a forum for exploring our ideas about boredom and anger in presentations in 2016 and 2018. We are grateful for the comments we received, and we are also indebted to the people and the institutions who have hosted the symposium over the last couple of years, in particular to Jeremy Browne, who founded it in 2016.

Portions of this book were also presented at meetings of the Organization of American Historians, the American Historical Association, the Society for US Intellectual History, and the Council on Contemporary Families.

We were invited to present our work at the University of Tübingen. We are grateful to Georg Schild, Daniel Menning, and the University of Tübingen History Department for providing the resources that made this possible. We also taught a course at Tübingen on emotions and technology in American history. Our students' thoughtful comments on the subject offered us comparative perspectives that enriched our vision.

Archivists across the country helped us in our research. Thanks to archivists at the National Museum of American History and the Smithsonian, who pointed us to rich collections during research visits there, to George Miles at the Beinecke Rare Books and Manuscripts Library at Yale University, to staff at the Bancroft Library at UC Berkeley, to librarians at Baker Library at the Harvard Business School, and to archivists at Boston College.

At our own institution, Weber State University, we incurred many debts. Joan Hubbard, the former university librarian, and Wendy Holliday, the current university librarian, provided working space as well as access to vast numbers of books and archival materials. Many drafts of the chapters were written in the library's carrels. Kathy Payne, Misty Edwards, Debbie Stephenson, Sandy Andrews, Sara Pomeroy, Heather Hootman, Chris Vilches, and the rest of the Stewart Library staff were generous with their time and resources. Sarah Singh and Jamie Weeks at Weber State University offered us access to their well-curated collections.

Thanks also to Joe Salmond for help with some of the images.

We are especially grateful for Weber State University's commitment to faculty research and the people on campus who foster this culture, including Mike Vaughan, Madonne Miner, Dave Ferro, and Francis Harrold. In 2015, with the help of our colleague Eric Amsel, we organized a mini-symposium at Weber State headlined by Matt Richtel of the *New York Times,* as well as the cognitive neuroscientist David Strayer, founder of the Center for the Prevention of Distracted Driving at the University of Utah. Richtel had just published *A Deadly Wandering,* a book that focuses on a driving-and-texting tragedy that happened forty miles north of us here in Utah. That work, and his many articles on texting and driving for the *New York Times,* provided us with both local and national perspectives on the American crisis of attention.

We had the opportunity to coteach a course on the themes in this book to Weber State honors students in 2017. Thanks to Dan Bedford for facilitating that and to the excellent honors students for their participation.

We are grateful to our dear friends and colleagues on campus and off, including David Ferro, Marjukka Ollilainen, Stephen Francis, Brady Brower, Vikki Deakin, Sara Dant, Jeff Richey, Branden Little, Richard Sadler, Gene Sessions, John Sillito, Nathan Rives, Kathryn MacKay, Eric Swedin, Michael Wutz, Mark Stevenson, Jean Norman, Hugo Valle, Julia Panko, Brian Rague, Matt Romaniello, Jenny Eckenbrecht, Jenna Daniels, Angela Christensen, Brian Jacobs, Jennifer Serventi, Adam Wolfson, and Christine Y. Todd, for the many scintillating conversations and insights that they have provided on themes related to this book.

We also appreciate the help (and often the hospitality) of those we relied on to broker interviews across various states. They include Bill Weinberg, Rebecca Egger, Carol McNamara, Ken Forsberg, Michael Montesano, Greg Lewis, Adrienne Andrews, Paul Hibbeln, Michael Wutz, Marilee Rohan, and Lauren Fowler.

The people we interviewed gave generously of their time and their confidences.

Friends and colleagues from academic years long past also have our thanks. Isaac Kramnick, who was a member of both our dissertation committees, deserves a special note. We first met during a political theory seminar he led in the spring of 1990 at Cornell University. His erudite descriptions of classical liberalism and civic republicanism were formative for both of us—and if you read the text closely you might be able to detect these ideologies' latent presence. Our other graduate advisors continue to shape the way we see the world—deep thanks to Joan Jacobs Brumberg, R. Laurence Moore, Stuart Blumin, the late Michael Kammen, and the late Theodore Lowi.

We're also grateful to Peter N. Stearns for his friendship and advice and for reading a draft of this work in its entirety; to Alexis McCrossen, for her feedback on Chapter 2; to Barbara Rosenwein, for her thoughtful response to Chapter 6; to Gary Cross, for conversations about Chapter 3; and to the anonymous readers for Harvard University Press, whose comments strengthened the book.

Thanks also to L. M. Sacasas and his wonderful blog thefrailestthing.com.

Our siblings aided us as well. Luke's older sister, Lisa, lent her constant support, read and edited occasional passages, fed us useful citations, and wowed us with her knowledge of the publishing business back east. It was from her, back when Luke and Lisa were in high school, that Luke first learned that emotions could be intellectualized. Susan's sister, Betsy Matt Turner, helped us find interview subjects, offered interesting citations, and provided editorial feedback as well as good advice and support more generally. Andrew Fernandez helped us arrange interviews and generously provided us hospitality.

Finally, our parents, James and Renate Fernandez and Barbara J. Matt and the late Joseph Matt, played important roles in the development of this book. Their love, enthusiasm, interest, and encouragement sustained us as we worked; their examples of collaboration inspired us; and their respect for books, letters, universities, and the life of the mind had foundational and enduring effects on both of us.

Index